SpringerWienNewYork

CISM COURSES AND LECTURES

The series presents lecture notes, monographs, edited works and proceedings in the field of Mechanics, Engineering, Computer Science and Applied Mathematics.
Purpose of the series is to make known in the international scientific and technical community results obtained in some of the activities organized by CISM, the International Centre for Mechanical Sciences.

INTERNATIONAL CENTRE FOR MECHANICAL SCIENCES

COURSES AND LECTURES - No. 539

NONDETERMINISTIC MECHANICS

EDITED BY

ISAAC ELISHAKOFF
FLORIDA ATLANTIC UNIVERSITY, BOCA RATON, USA

CHRISTIAN SOIZE
UNIVERSITÉ PARIS-EST, MARNE-LA-VALLÉE, FRANCE

SpringerWienNewYork

This volume contains 100 illustrations

SPIN 86122669

All contributions have been typeset by the authors.

ISBN 978-3-7091-1305-9 SpringerWienNewYork

PREFACE

It has been recognized for several decades that uncertainty and absence of determinism play an important role in engineering sciences; to-date, three basic techniques have been developed to deal with various uncertainties, namely in variation and scatter in uncertain system parameters such as mechanical properties, geometric parameters, boundary conditions, in model uncertainties induced by modeling errors as well as actions such as the impact of earthquakes, wind loads, imperfect road profiles, or turbulence experienced by aircraft. These methodologies are (a) probabilistic or stochastic modeling; (b) fuzzy sets based analysis, and (c) anti-optimization of structures.

Spectacular advances have been recorded in stochastic mechanics based in the construction of stochastic models of uncertainties as soon as the probability theory can be used; likewise industrial applications have been developed worldwide in using fuzzy sets and logic for devising reliable machines and components, and a relatively recent new-old field has emerged, referred to as anti-optimization (also known as convex modeling of uncertainty, ellipsoidal modeling, guaranteed approach, maxmin, and worst case analysis)—that identifies uncertainty with boundedness. Interval analysis is the specific and simplest form of dealing with the best and worst scenarios under uncertainty. Imprecise probabilities combine both set and stochastic uncertainties, to obtain bounds of probabilities or expectations of variables.

However, the question which analysis is preferable for researchers and engineers is short of a consensus, as one can anticipate.

The aim of this book is to present the current state of the art of nondeterministic mechanics in its various forms. The topics range from stochastic problems to fuzzy sets; from linear to nonlinear problems; from specific methodologies to combinations of various techniques; from theoretical considerations to practical applications.

It is specially designed to illuminate the various aspects of above three techniques and deepen the discussion of their pros and cons.

The book is divided in three parts. Part 1 is devoted to stochastic analysis; it contains papers by Umberto Alibrandi and Giuseppe Ricciardi; Christian Soize; Isaac Elishakoff and Lova Andriamasy. Part 2 is devoted to nonstochastic analysis with papers by Thomas Haag and Michael Hanss; and by Alberto Bernardini and Fulvio Tonon.

Part 3 consists of one paper by Michael Oberguggenberger, dealing with combined methods.

If this volume stimulates further mutual and useful dialogue between proponents of differing methodologies, with demarcation of the area in which an approach ought to be preferred—our effort will be amply rewarded.

Isaac Elishakoff and Christian Soize, editors

CONTENTS

Part 1
Stochastic Models

Stochastic Methods in Nonlinear Structural Dynamics

Umberto Alibrandi* and Giuseppe Ricciardi*

* Department of Civil Engineering, University of Messina, Italy

Abstract The uncertainties are inherent in any structural problem. Here attention is focused only on the uncertain nature of the dynamic actions and its consequences on the structural response. In the framework of stochastic dynamics, only three methods are the most used: the Moment Equation Method (MEM), the Stochastic Linearization (SL) and the Monte Carlo Simulation (MCS). The MEM in conjuction with a closure method (CM) allows to obtain the response statistical moments, but it increases in complexity as the problem dimension increases. The SL is easily applied to a large variety of engineering problems. Providing information limited to the first two moments of the system response, unfortunately it suffers of accuracy in the case of strongly nonlinear behavior. MCS is able to give additional information on the structural response, yielding estimates for the probability density function of the nonlinear response, but it is computationally expensive. In this paper some improvements of these three methods are presented, which allow to overcome the aforementioned drawbacks.

1 Introduction

In many cases of engineering interest, the structural systems are subject to dynamic excitations characterized by complex random fluctuations, that can not be adequately represented by deterministic models. Stochastic methods are able to represent adequately these type of excitations as stochastic processes, giving a characterization of the structural response in terms of statistical moments or in terms of the probability density function (pdf). This is a very complicated task to be accomplished, and in literature much research has been devoted to this topic in the last several decades, giving rise to methods based on: (i) the numerical solution of Fokker-Planck equation, such as path integration (Naess and Johnsen, 1993; Wehner and Wolfer, 1983), cell mapping (Hsu, 1987) and finite element method (Spencer and Bergman, 1993), (ii) the Moment Equation Method (MEM), (iii) the

Stochastic Linearization (SL), (iv) the Monte Carlo Simulation (MCS). The latter ones (MEM, SL and MCS) have gained wide popularity by their applicability to general nonlinear systems with many degrees of freedom (MDOF) and they will be described in detail in the following subsections.

1.1 Moment Equation Method

In the MEM approach the response statistical characterization is given by the response moments or by other quantities related to the former such as cumulants or quasi-moments (Stratonovich, 1967; Ibrahim, 1985). This method consists in writing differential equations for the response statistical moments of any order, taking advantage from the Fokker-Planck equation or from the Itï¿œ differential rule.

When dealing with nonlinear systems a serious problem arises in the MEM approach, which stems from the need of knowing the response pdf to close the hierarchy of the resulting equations. In order to overcome this difficulty, the so called closure methods (CM) are used. The key idea is to approximate the response pdf in a series form, truncating it at a certain term. The coefficients of the above mentioned series can be written as functions of the response moments or of the response cumulants or of the response quasi-moments. Thus, neglecting the terms beyond a given order is equivalent to make moments or cumulants or quasi-moments zero, which makes the moment equations solvable. In this way the moments of higher order are expressed in terms of moments of lower order by means of nonlinear relationships.

Cumulant neglect closure method has been independently developed by Wu and Lin (1984) and Ibrahim et al. (1985), which is certainly the most popular among the closure methods. Central moment and cumulant closure methods have been proposed by Bellman and Richardson (1968) and Sancho (1970a), (1970b) in a mathematical context. Bover (1978) and Sperling (1979) independently presented the quasi-moments closure method. Crandall (1980) presented the Hermite moment closure method. All these closure methods lead to a set of differential moment equations which are nonlinear. This nonlinearity is a major drawback in the case of systems with many degrees of freedom. This shortcoming has been overcome by Grigoriu (1999), which closed the moment equations taking advantage from MCS in order to estimate lower order moments, obtaining so a set of linear equations whose unknowns are the higher-order moments.

1.2 Stochastic Linearization

The SL method is the most versatile method for the analysis of general nonlinear structures under random excitations. In almost 60 years since its virtually simultaneous presentations by Booton (1953), Kazakov (1956) and Caughey (1959), SL has been widely applied in the study of various nonlinear structures. For example, the monographs by Roberts and Spanos (1991) and Socha (2008) and many review papers (Socha and Soong, 1991; Elishakoff, 2000; Falsone and Ricciardi, 2003) written in these last years confirm its success.

The basic idea of SL is to replace the original nonlinear system by an equivalent linear one, whose determination is performed by minimizing the difference between the two systems in some statistical sense. The SL exhibits different forms based on the adopted pdf for the evaluation of the coefficients that appear in the linearized system (Kozin, 1988).

The Gaussian Stochastic Linearization (GSL) is based on the hypothesis of Gaussianity of the response process and it allows approximating the second order moments of the response. In this way, the probabilistic characterization of the equivalent linear system response reduces to the evaluation of its Gaussian properties. Against relatively little numerical efforts, unfortunately the GSL method gives accurate results for weakly nonlinear systems only. This drawback is due to the inadequacy of the Gaussian assumption to represent the non-Gaussian characteristic of the response for systems that exhibit strong nonlinear behaviour.

Kozin (1988) introduced the concept of "true" linearization. He showed that, if the averages appearing in the expressions of the equivalent coefficients are evaluated by the exact probability density function of the response, the "true" stochastic linearization leads to the exact results in terms of response covariances. Starting from these observations, alternative approaches have been developed based on a more realistic representation of the non Gaussian probability density function of the response process. Beaman and Hedrick (1981) improved the accuracy of the GSL method by using the classical Gram-Charlier series expansion of the unknown probability density function of the response, which includes up to fourth order terms. The coefficients of the series expansion are approximately evaluated by solving the non-linear system of the moment equations up to fourth order. Pradlwarter (1991) proposed a numerical method based on non-Gaussian stochastic linearization and on the Fokker-Planck equation, requiring nonlinear transformations in order to consider the non-Gaussian properties of the stochastic response. In the non-Gaussian linearization method proposed by Chang (1992), the non-Gaussian density is built as the weighted sum of undetermined Gaussian densities. The undetermined Gaussian parameters are then derived through solving a set of non-linear algebraic moment relations. Lee (1995) improved the stochastic linearization results for the Duffing oscillator by performing a non-Gaussian closure scheme, based on the abridged Edgeworth series expansion of the probability density of the response. Hurtado and Barbat (1996) proposed an improved non-Gaussian stochastic linearization for the Bouc-Wen-Baber hysteretic model by using mixed discrete-continuous Gaussian distributions. Recently, Crandall (2004) proposed the use of non-Gaussian distributions for the stochastic linearization of the power law and the double-well Duffing oscillators.

1.3 Monte Carlo Simulation

Whereas only biased estimates for the first two statistical moments of the response are obtainable by the SL method, MCS yields unbiased estimates for the probability density function of the nonlinear response. The first applications in structural dynamics are due to Shinozuka (1972) and Shinozuka and Wen (1972) and later extended by other authors (Grigoriu, 1993; Grigoriu, 1998; Spanos and Zeldin, 1998; Schueller and Spanos, 2000).

MCS is very robust and relatively easy to apply. A sample size of few hundred independent realizations is generally sufficient to obtain a suitable estimate for the first lower order moments and to provide information on the shape of the distribution. MCS becomes an increasingly attractive method as the problem dimension increases. Hence MCS is often the only feasible solution for real engineering problems with large dimensions. Besides being more efficient than analytical-based approaches, it has the advantages that the tools of deterministic analysis can be fully exploited. Against its efficiency, the computational costs required are often very high. Clearly, the computational efforts increase with the dimension and the complexity of the nonlinear structural models. Hence, efficient techniques have been proposed in order to reduce the computational burden, such as variance reduction procedures (Roberts, 1986), importance sampling (Tanaka, 1998; Au and Beck, 2001), controlled MCS (Harnpornchai et al., 1999).

1.4 Outline

The MEM is difficult to apply in the case of complex structures, increasing dramatically the number of nonlinear equations to be solved. The SL is less accurate in the case of structures with high nonlinearity. The MCS is in general time consuming, when higher order statistics or the probability densities are of interest. In this paper, the aforementioned methods are called in their classical form first. Then, some improved versions are proposed for overcoming these problems.

2 Stochastic Structural dynamics

2.1 Motion equations of the structural system

Linear structure. A general form of the equations of motion of an n d.o.f. linear system is as follows:

$$\mathbf{M}\ddot{\mathbf{u}}(t) + \mathbf{C}\dot{\mathbf{u}}(t) + \mathbf{K}\mathbf{u}(t) = \mathbf{G}_{\mathrm{e}}\mathbf{F}_e(t) \tag{1}$$

where $\mathbf{u}(t)$ is the n_s-vector of generalized displacement, $\mathbf{M}$, $\mathbf{C}$ and $\mathbf{K}$ are the mass, damping and stiffness matrices, of order $n_s \times n_s$, $\mathbf{G}_{\mathrm{e}}$ is an influence matrix, of order $n_s \times n_e$, and $\mathbf{F}_{\mathrm{e}}(t)$ is the n_e-vector of external generalized forces.

Nonlinear structure. The response $\mathbf{u}(t)$ of a general dynamical system with nonlinear behavior subjected to an external force vector $\mathbf{F}_{\mathrm{e}}(t)$ satisfies the following nonlinear differential equations:

$$\mathbf{M}\ddot{\mathbf{u}}(t) + \mathbf{h}[\mathbf{u}(t), \dot{\mathbf{u}}(t)] = \mathbf{G}_{\mathrm{e}}\mathbf{F}_e(t) \tag{2}$$

where $\mathbf{h}[\mathbf{u}(\mathbf{t}), \dot{\mathbf{u}}(t)]$ is the n_s-vector of non-linear restoring forces, which depends on the displacements and velocities.

External and parametric excitation. A more general form of a system of differential equations of motion can be given as:

$$\mathbf{M}\ddot{\mathbf{u}}(t) + \mathbf{h}[\mathbf{u}(t), \dot{\mathbf{u}}(t)] = \mathbf{G}_e\mathbf{F}_e(t) + \mathbf{G}_p[\mathbf{u}(t), \dot{\mathbf{u}}(t)]\mathbf{F}_p(t) \tag{3}$$

where $\mathbf{F}_{\mathrm{p}}(t)$ is the n_p-vector of parametric excitations, modulated by the matrix $\mathbf{G}_p[\mathbf{u}(\mathbf{t}), \dot{\mathbf{u}}(\mathbf{t})]$, of order $n_s \times n_p$, which depends on the system response.

The state vector approach. By introducing the vector of state variables $\mathbf{Z}(t) = [\; \mathbf{u}^T(t) \quad \dot{\mathbf{u}}^T(t) \;]^T$, of order $2n_s$, eq.(3) can be written in the following first order form:

$$\dot{\mathbf{Z}}(t) = \mathbf{a}[\mathbf{Z}(t)] + \mathbf{V}[\mathbf{Z}(t)]\mathbf{F}(t) \tag{4}$$

where $\mathbf{F}(t) = [\; \mathbf{F}_e^T(t) \quad \mathbf{F}_p^T(t) \;]^T$ is the vector of external and parametric excitations, of order $n_f = n_e + n_p$, while $\mathbf{a}[\mathbf{Z}(t)]$ and $\mathbf{V}[\mathbf{Z}(t)]$ are a $2n_s$-vector and a $2n_s \times n_f$ matrix, respectively, given as:

$$\mathbf{a}[\mathbf{Z}(t)] = \begin{bmatrix} \mathbf{0} \\ -\mathbf{M}^{-1}\mathbf{h}[\mathbf{u}(t), \dot{\mathbf{u}}(t)] \end{bmatrix}, \tag{5}$$

$$\mathbf{V}[\mathbf{Z}(t)] = \begin{bmatrix} \mathbf{0} & \mathbf{0} \\ \mathbf{M}^{-1}\mathbf{G}_e & \mathbf{M}^{-1}\mathbf{G}_p[\mathbf{u(t)}, \dot{\mathbf{u}}\mathbf{(t)}] \end{bmatrix} \tag{6}$$

that are functions of the response vector of state variables $\mathbf{Z}(t)$.

2.2 Filtered excitations

The vector of random excitations is frequently assumed as a stationary Gaussian stochastic process, characterized by a Power Spectral Density (PSD) matrix $\mathbf{S_{FF}}(\omega)$ (or by its inverse Fourier transform, the correlation matrix $\mathbf{R_{FF}}(\tau)$). It can be considered as the response of a system of linear filter differential equations excited by m Gaussian white noises:

$$\dot{\mathbf{F}}(t) = \mathbf{A}_f\mathbf{F}(t) + \mathbf{V}_f\mathbf{W}(t) \tag{7}$$

The matrices $\mathbf{A}_f$ and $\mathbf{V}_f$ are constant. The random excitations $\mathbf{W}(t) = [\ W_1(t) \quad W_2(t) \quad \ldots \quad W_m(t)\ ]^T$ are zero-mean Gaussian white noise processes, fully characterized by the second order cross-correlation functions $E[W_k(t_1)W_l(t_2)] = Q_{kl}\delta(t_1 - t_2)$, $E[\cdot]$being stochastic averaging, $\delta(\cdot)$ the Dirac's delta function, $Q_{kl} = 2\pi S_{kl}$ the constant strengths of the white noise processes, S_{kl} the cross-power spectral density of $W_k(t)$ and $W_l(t)$.

In general, the vector of random excitations can be considered as the response of a system of nonlinear filter differential equations excited by m white noises:

$$\dot{\mathbf{F}}(t) = \mathbf{a}_f[\mathbf{F}(t)] + \mathbf{V}_f[\mathbf{F}(t)]\mathbf{W}(t) \tag{8}$$

where the vector $\mathbf{a}_f[\mathbf{F}(t)]$ and the matrix $\mathbf{V}_f[\mathbf{F}(t)]$ depend on the excitation vector $\mathbf{F}(t)$.

An example: the earthquake excitation. As an example, the earthquake excitation on a structure is given as:

$$\mathbf{G}_e\mathbf{F}_e(t) = -\mathbf{M}\boldsymbol{\tau}\ddot{u}_g(t) \tag{9}$$

where $\ddot{u}_g(t)$ is the ground acceleration and τ is an influence vector.

In the model of the earthquake excitation given by Kanai-Tajimi (KT) (Kanai, 1957; Tajimi, 1960), the ground acceleration $\ddot{u}_g(t)$ is characterized by the following PSD:

$$\mathrm{S}_{KT}(\omega;\omega_g, \varsigma_g, S_w) = \frac{\mathrm{S}_w(\omega_g^4 + 4\varsigma_g^2\omega_g^2\omega^2)}{(\omega_g^2 - \omega^2)^2 + 4\varsigma_g^2\omega_g^2\omega^2} \tag{10}$$

where ω_g is the soil natural frequency, that determines the dominant range of input frequencies, while ς_g is the ground damping ratio, that indicates the sharpness of the power spectral density shape. The KT seismic model can be considered as a white noise excitation at bedrock level, with constant power spectrum S_w, filtered through the overlaying soil deposits. The seismic excitation $\ddot{u}_g(t)$ is a stationary stochastic filtered white noise process obtained as solution of the following set of linear filter equations:

$$\left\{ \begin{array}{c} \ddot{u}_g(t) = \omega_g^2 q_g(t) + 2\varsigma_g\omega_g\dot{q}_g(t) \\ \ddot{q}_g(t) + 2\varsigma_g\omega_g\dot{q}_g(t) + \omega_g^2 q_g(t) = w(t) \end{array} \right. \tag{11}$$

The well known drawback of the KT model is that its PSD function approaches S_w when the frequency tends to zero, which is not consistent with the power spectrum of the real earthquake records.

The Clough-Penzien (CP) model overcomes this drawback (Clough and Penzien, 1975), by introducing a further filter with two additional parameters ω_p and ς_p ; the CP-PSD is given by:

$$\mathrm{S}_{CP}(\omega;\omega_p,\varsigma_p,\omega_g,\varsigma_g,\mathrm{S}_w) = \tilde{\mathrm{S}}_{CP}(\omega;\omega_p,\varsigma_p)\cdot \mathrm{S}_{KT}(\omega;\omega_g,\varsigma_g,\mathrm{S}_w) \tag{12}$$

$$\tilde{\mathrm{S}}_{CP}(\omega;\omega_p,\varsigma_p)= \frac{\omega^4}{(\omega_p^2-\omega^2)^2+4\varsigma_p^2\omega_p^2\omega^2} \tag{13}$$

The linear filter equations of the CP model are:

$$\left\{ \begin{array}{c} \ddot{u}_g(t) = \omega_g^2 q_g(t) + 2\varsigma_g\omega_g\dot{q}_g(t) - \omega_p^2 q_p(t) - 2\varsigma_p\omega_p\dot{q}_p(t) \\ \ddot{q}_p(t) + 2\varsigma_p\omega_p\dot{q}_p(t) + \omega_p^2 q_p(t) = \omega_g^2 q_g(t) + 2\varsigma_g\omega_g\dot{q}_g(t) \\ \ddot{q}_g(t) + 2\varsigma_g\omega_g\dot{q}_g(t) + \omega_g^2 q_g(t) = w(t) \end{array} \right. \tag{14}$$

2.3 Dynamic equation of the whole system

The differential equations of the structure and of the filter, eqs.(4) and (8), can be cast in the following general form:

$$\dot{\mathbf{X}}(t) = \mathbf{f}[\mathbf{X}(t)] + \mathbf{G}[\mathbf{X}(t)]\mathbf{W}(t) \tag{15}$$

where $\mathbf{X}(t) = [\ \mathbf{Z}^T(t) \quad \mathbf{F}^T(t)\ ]^T$ is the vector of order $n = 2n_s + n_f$ collecting the displacements and velocities of the structure and the components of the filtered external and parametric excitations, while $\mathbf{f}[\mathbf{X}(t)]$ and $\mathbf{G}[\mathbf{X}(t)]$ are an n-vector and an $n \times m$ matrix, respectively, given as:

$$\mathbf{f}[\mathbf{X}(t)] = \begin{bmatrix} \mathbf{a}[\mathbf{Z}(t)] + \mathbf{V}[\mathbf{Z}(t)]\mathbf{F}(t) \\ \mathbf{a}_f[\mathbf{F}(t)] \end{bmatrix} \tag{16}$$

$$\mathbf{G}[\mathbf{X}(t)] = \begin{bmatrix} \mathbf{0} \\ \mathbf{V}_f[\mathbf{F}(t)] \end{bmatrix} \tag{17}$$

that, in general, are functions of the vector $\mathbf{X}(t)$. Note that in the case of linear filter equations, $\mathbf{G}[\mathbf{X}(t)] \equiv \mathbf{G} = \text{cost}$, then the whole system is subjected to white noises external excitations only.

2.4 Itï¿œ stochastic differential equations

Let us consider a non-linear system whose dynamic behavior, described by the time evolution of the n-vector of state variables $\mathbf{X} = \mathbf{X}(t)$, is governed by the system (15) of n first-order differential equations (in scalar form):

$$\frac{d}{dt}X_i(t) = f_i[\mathbf{X}(t)] + \sum_{k=1}^{m} G_{ik}[\mathbf{X}(t)]W_k(t) \tag{18}$$

where $X_i(t)$ (with $i = 1, 2, \ldots, n$) are the components of the response state vector $\mathbf{X}(t)$, while $W_k(t)$ (with $k = 1, 2, \ldots, m$) are m random excitations. Functions $f_i[\mathbf{X}(t)]$ and $G_{ik}[\mathbf{X}(t)]$ of the state vector $\mathbf{X}(t)$ are generally non-linear. If $G_{ik}[\mathbf{X}(t)]$ are constant, i.e. $G_{ik}[\mathbf{X}(t)]{=}G_{ik}$, then the system is said to be excited by additive excitations only; on the contrary, if the functions $G_{ik}[\mathbf{X}(t)]$ depend on the response process $\mathbf{X}(t)$ the system is said to be excited by multiplicative excitations.

The white noise processes $W_k(t)$ are not mean square Riemann integrable and, consequently, eq.(18) has no traditional mathematical meaning. The latter can be considered to be formally equivalent to the following Itï¿œ-type stochastic differential equations (Lin, 1967; Ibrahim, 1985):

$$dX_i(t) = m_i[\mathbf{X}(t)]dt + \sum_{k=1}^{m} G_{ik}[\mathbf{X}(t)]d\xi_k(t) \tag{19}$$

where $\xi_k(t)$ (with $k = 1, 2, \ldots, m$) are m Wiener processes whose increments are characterized by cross-correlation functions $E[d\xi_k(t_1)d\xi_l(t_2)] = Q_{kl}\delta(t_1 - t_2)dt_1dt_2$, while $m_i[\mathbf{X}(t)]$ are the drift coefficients related to the coefficients of the equations of motion (18) by the following expression:

$$m_i[\mathbf{X}(t)] = f_i[\mathbf{X}(t)] + z_i[\mathbf{X}(t)] \tag{20}$$

In particular, the second terms on the right-hand side of eq.(20), given by

$$z_i[\mathbf{X}(t)] = \frac{1}{2}\sum_{j=1}^{n}\sum_{k=1}^{m}\sum_{l=1}^{m} Q_{kl}G_{jl}[\mathbf{X}(t)]\frac{\partial}{\partial X_j}G_{ik}[\mathbf{X}(t)] \tag{21}$$

are the well-known Wong–Zakai (or Stratonovich) correction terms (Wong and Zakai, 1965; Stratonovich, 1967). From eq.(21) it is easy to note that the Wong–Zakai correction terms $z_i[\mathbf{X}(t)]$ vanish in the case of purely additive excitations only.

By introducing the standard Wiener processes $B_j(t)$ (with $j = 1, 2, \ldots, n$), whose increments $dB_j(t)$ are characterized by the cross-correlation functions $E[dB_j(t_1)dB_l(t_2)] = \delta(t_1 - t_2)dt_1dt_2$, eq.(19) can be rewritten in the following equivalent Itï¿œ form:

$$dX_i(t) = m_i[\mathbf{X}(t)]dt + \sum_{j=1}^{n}\sigma_{ij}[\mathbf{X}(t)]dB_j(t) \tag{22}$$

where $\sigma_{ij}[\mathbf{X}(t)]$ are the diffusion coefficients related to the coefficients of the equations of motion (18) by the following expressions:

$$\sigma_{ij}^2[\mathbf{X}(t)] = \sum_{k=1}^{m}\sum_{l=1}^{m} Q_{kl}G_{ik}[\mathbf{X}(t)]G_{jl}[\mathbf{X}(t)] \tag{23}$$

3 Moment equation method (MEM)

3.1 The moment equations of the response

Exact solutions for non-linear systems subjected to additive or both additive and multiplicative random excitations are difficult to obtain. In many cases, therefore, it is necessary to adopt approximate solutions. In the case in which the response of the system can be represented by a Markov process, a frequently used approximate solution is given by the moment equation method (MEM). This approach is based on the use of the Itï¿œ differential rule (Lin, 1967), given by:

$$d\varphi[\mathbf{X}(t)] = \sum_{i=1}^{n}\left\{\left[m_i[\mathbf{X}(t)]dt + \sum_{j=1}^{n}\sigma_{ij}[\mathbf{X}(t)]dB_j(t)\right]\frac{\partial\varphi[\mathbf{X}(t)]}{\partial X_i}\right\}$$

$$+\frac{1}{2}\sum_{i=1}^{n}\sum_{j=1}^{n}\left[\sigma_{ij}^2[\mathbf{X}(t)]\frac{\partial^2\varphi[\mathbf{X}(t)]}{\partial X_i\partial X_j}dt\right] \tag{24}$$

where $\varphi[\mathbf{X}(t)]$ is a differentiable function of the response vector process $\mathbf{X}(t)$. By applying the stochastic average to both members of eq.(24) and dividing by dt, the following differential equation governing the evolution of the average of the function $\varphi(\mathbf{X}) = \varphi[\mathbf{X}(t)]$ is obtained:

$$\frac{d}{dt}E[\varphi(\mathbf{X})] = \sum_{i=1}^{n}\left[m_i(\mathbf{X})\frac{\partial\varphi(\mathbf{X})}{\partial X_i}\right] + \frac{1}{2}\sum_{i=1}^{n}\sum_{j=1}^{n}E\left[\sigma_{ij}^2(\mathbf{X})\frac{\partial^2\varphi(\mathbf{X})}{\partial X_i\partial X_j}\right] \tag{25}$$

Setting $\varphi(\mathbf{X}) = X_1^{k_1}X_2^{k_2}\dots X_n^{k_n}$, the differential equation (25) rule all the moments of order $k = k_1 + k_2 + \dots + k_n$ of the response state vector $\mathbf{X}(t)$. For linear systems, the moment equations can be easily solved. Unfortunately, for non-linear systems, these equations constitute an infinite hierarchy; in fact the moment differential equations up to a given order contain moments of higher order and, consequently, this system is not solvable.

Two simple examples: the linear oscillator and the nonlinear Duffing oscillator. Let us consider the following linear oscillator subjected to a white noise:

$$\ddot{U}(t) + \beta\dot{U}(t) + kU(t) = \sqrt{2\beta}W(t) \tag{26}$$

where β and k are constants and $W(t)$ is a Gaussian white noise with intensity $q = 2\pi S_W$, S_W being the constant power spectral density. In terms of the state variables we have:

$$\dot{\mathbf{X}}(t) = \mathbf{DX}(t) + \mathbf{G}W(t) \tag{27}$$

where

$$\mathbf{X}(t) = \begin{bmatrix} X_1(t) \\ X_2(t) \end{bmatrix} = \begin{bmatrix} U(t) \\ \dot{U}(t) \end{bmatrix}, \quad \mathbf{D} = \begin{bmatrix} 0 & 1 \\ -k & -\beta \end{bmatrix}, \quad \mathbf{G} = \begin{bmatrix} 0 \\ \sqrt{2\beta} \end{bmatrix} \tag{28}$$

The stationary response is Gaussian and the complete probabilistic characterization is given by the second order moment of displacement and velocity. By applying eq.(25), the following second order moment equations result:

$$\frac{d}{dt}E[X_1^2] = 2E[X_1X_2] \tag{29}$$

$$\frac{d}{dt}E[X_1X_2] = -kE[X_1^2] + E[X_2^2] - \beta E[X_1X_2] \tag{30}$$

$$\frac{d}{dt}E[X_2^2] = -2\beta E[X_2^2] - 2kE[X_1X_2] + 2\beta q \tag{31}$$

Eqs.(29)-(31) constitute a closed set of equations whose unknowns are the second order moments $E[X_1^2]$, $E[X_1X_2]$ and $E[X_2^2]$. It can be solved by a numerical scheme in order to obtain the time evolution of the response moments. The stationary solution can be easily determined by solving the set of algebraic equations resulting from eqs. (29)-(31) where the left-hand sides are set to zero. The stationary second order moment are:

$$E[X_1^2] = \frac{q}{k}, \quad E[X_1X_2] = 0, \quad E[X_2^2] = q \tag{32}$$

Let us consider now the following nonlinear Duffing oscillator:

$$\ddot{U}(t) + \beta\dot{U}(t) + kU(t) + \epsilon kU^3(t) = \sqrt{2\beta}W(t) \tag{33}$$

where $\epsilon > 0$ is a nonlinear parameter. In this case the response is not Gaussian and, in principle, all the moments of every order are needed in order to characterize the response. Operating in a similar way, by using the Itï¿œ differential rule the moment equations of every order can be written. For example, the second order moment differential equations are:

$$\frac{d}{dt}E[X_1^2] = 2E[X_1X_2] \tag{34}$$

$$\frac{d}{dt}E[X_1X_2] = -kE[X_1^2] - \epsilon kE[X_1^4] + E[X_2^2] - \beta E[X_1X_2] \tag{35}$$

$$\frac{d}{dt}E[X_2^2] = -2\beta E[X_2^2] - 2kE[X_1X_2] - 2\epsilon kE[X_1^3X_2] + 2\beta q \tag{36}$$

These equations cannot be solved, because moments of fourth order appear. This circumstance is also evident in the moment equations of every order r, where the moments of order $r+2$ appear. Then the moment differential equations constitute an infinite hierarchy of equations, whose solution requires the use of a closure scheme in order to express the higher order moments in terms of the lower order ones.

3.2 Closure schemes

The moment equation method applied to nonlinear systems requires the use of a closure scheme in order to become solvable the hierarchy of the equations, by expressing in an approximate way the higher order moments in terms of lower order ones. This goal can be obtained by introducing a suitable approximation of the pdf of the response.

The Gaussian Closure (GC). The simplest closure method is given by the Gaussian approximation and the resulting scheme is called Gaussian Closure (GC). The Gaussian probability density function is given as:

$$p_{\mathbf{X}}^{G}(\mathbf{x}) = \frac{1}{\sqrt{(2\pi)^n Det(\mathbf{\Sigma_X})}} exp\left[-\frac{1}{2}(\mathbf{x}-\mu_{\mathbf{X}})^T \mathbf{\Sigma}_{\mathbf{X}}^{-1}(\mathbf{x}-\mu_{\mathbf{X}})\right] \tag{37}$$

where $\mu_{\mathbf{X}} = E[\mathbf{X}]$ is the mean vector and $\mathbf{\Sigma_X} = E[(\mathbf{X}-\mu_{\mathbf{X}})(\mathbf{X}-\mu_{\mathbf{X}})^T]$ is the covariance matrix of the response. In this way, the moment of order grather than two can be expressed in term of the moment up to second order by means of nonlinear relationships. The moment equation method in conjuction with the GC consider the second order moment equations only.

An example: the Gaussian Closure applied to the nonlinear Duffing oscillator. As an illustration, the Duffing oscillator is considered. In the second order moment equations (34)-(36) the fourth order moments $E[X_1^4]$ and $E[X_1^3 X_2]$ appear. From the Gaussian hyphothesis, they can be expressed in an approximate way in terms of the second order moments as follows:

$$E[X_1^4] \approx E_G[X_1^4] = 3(E[X_1^2])^2 \tag{38}$$

$$E[X_1^3 X_2] \approx E_G[X_1^3 X_2] = 3E[X_1^2]E[X_1X_2] \tag{39}$$

In the stationary state, the algebraic second order moment equations reduce to the following set of nonlinear equations:

$$E[X_1X_2] = 0 \tag{40}$$

$$-kE[X_1^2] - 3\epsilon k(E[X_1^2])^2 + E[X_2^2] - \beta E[X_1X_2] = 0 \tag{41}$$

$$-2\beta E[X_2^2] - 2kE[X_1X_2] - 6\epsilon kE[X_1^2]E[X_1X_2] + 2\beta q = 0 \tag{42}$$

whose solution is:

$$E[X_1^2] = \frac{1}{6\epsilon}\left(\sqrt{1+12\epsilon\frac{q}{k}} - 1\right), \quad E[X_1X_2] = 0, \quad E[X_2^2] = q \tag{43}$$

that constitutes an approximation of the second order moments of the response.

Non Gaussian Closure (NGC) methods. For strongly nonlinear systems, non Gaussian Closure (NGC) methods are necessary to obtain an adequate approximation of the response. They are based on different representation of the non Gaussian characteristics of the response. Among all, the most used schemes are the cumulant closure (CC) method and the quasi-moment closure (QMC) method. The first closure method is based on the truncation of the series expansion of the log-characteristic function of the response process, whose coefficients are the cumulants; the second is based on the truncation of the A-type Gram-Charlier series expansion of the non-Gaussian probability density function of the response, whose coefficients are the quasi-moments (or, alternatively, the Hermite moments) of the response process. For both methods, the resulting equations are nonlinear.

3.3 The modified Quasi-Moment Closure (QMC) method

Recently (Muscolino et al., 2003), a modified version of the quasi-moment closure method has been proposed. The method takes advantage of the great accuracy of the Monte Carlo Simulation (MCS) in evaluating the first two moments of the response process by considering just few samples. The quasi-moment neglect closure is used to close the infinite hierarchy of the moment differential equations of the response process. Moreover, in order to determine the higher order statistical moments of the response, the second-order probabilistic information given by MCS in conjunction with the quasi-moment neglect closure leads to a set of linear differential equations. The method has been developed for a more general case of external non Gaussian Poisson white noise excitations (DiPaola and Falsone, 1993). The case of the Gaussian white noise input can be considered as a particularization of the non Gaussian case. As an illustration, the scalar case is proposed, the extension to MDOF systems being available in Muscolino et al. (2003), taking advantage from the kronecker algebra.

Poisson white noise excitation. Let us consider the differential equation describing a one-dimensional system subjected to a purely external Poisson white noise process, given by:

$$\frac{d}{dt}X(t) = f[X(t)] + GI(t) \tag{44}$$

where $I(t)$ is a scalar Poisson white noise process (Grigoriu, 1987; Ricciardi, 1994), that is a sequence of impulses with random amplitude and arriving at random times, given by

$$I(t) = \sum_{k=1}^{N(t)} Y_k \delta(t - t_k) \tag{45}$$

the amplitudes $\{Y_k\}$ being a family of random variables, mutually independent and independent of the time instants t_k , with prescribed distribution $p_Y(y)$, $\delta(t)$ the Dirac's delta function, t_k the random time arrivals. In eq.(45), $N(t)$ is a counting Poisson process with parameter λ. The probabilistic characterization of the Poisson white noise process $I(t)$ can be given in terms of its correlation functions (Stratonovich, 1963):

$$C_r[I(t_1), I(t_2), \ldots, I(t_r)] = \lambda E[Y^r]\delta(t_1 - t_2)\delta(t_1 - t_3)\ldots\delta(t_1 - t_r) \tag{46}$$

It follows that $I(t)$ is a non-normal delta-correlated process, with intensity coefficients related to the probabilistic characterization of the random variables $\{Y_k\}$ and of the counting Poisson process $N(t)$. The Poisson white noise process $I(t)$ is not mean square Riemann integrable and, consequently, eq.(44) has no traditional mathematical meaning. The latter can be considered to be formally equivalent to the following generalized Itï¿œ-type stochastic differential equation:

$$dX(t) = f[X(t)]dt + Gd\gamma(t) \tag{47}$$

where $d\gamma(t)$ is an increment of the compound Poisson process $\gamma(t)$, defined by

$$\gamma(t) = \sum_{k=1}^{N(t)} Y_k U(t - t_k) \tag{48}$$

$U(t)$ being the unit step function.

Moment equations. The Itï¿œ differential rule generalized to the case of non Gaussian delta correlated input process leads to the following moment differential equations (DiPaola and Falsone, 1993):

$$\dot{m}_r[X(t)] = rE[X^{r-1}f(X)] + \sum_{k=1}^{r} L_{r,k} m_{r-k}[X(t)]G^k \lambda E[Y^k] \tag{49}$$

where $\dot{m}_r[X(t)] = E[X^r(t)]$ is the moment of order r and $L_{r,k} = r!/[k!(r-k)!]$. Note that in the case of zero-mean Gaussian white noise input, the sum in eq.(49) needs to be performed only for $k = 2$. In this case $q = \lambda E[Y^2]$ is usually defined as the strength of the white process.

For simplicity's sake, let us assume that the deterministic nonlinear function $f[X(t)]$ is a polynomial of p-th order:

$$f[X(t)] = \sum_{j=1}^{p} a_j X^j \tag{50}$$

By means of this equation, we can rewrite eq.(49) as follows:

$$\dot{m}_r[X(t)] = r \sum_{j=1}^{p} a_j m_{r+j-1}[X(t)] + \sum_{k=1}^{r} L_{r,k} m_{r-k}[X(t)] G^k \lambda E[Y^k] \tag{51}$$

As a consequence of the nonlinearity of the system, in the first summation of the right member of eq.(51) moments until the $(r+p-1)$-th order appear. It follows that the differential equation governing the evolution of statistics of order r will now include statistics of higher order than r. As an example, if cubic non-linearity is considered, the differential eq.(51) contains moments of $(r+2)$-th order. The numerical solution of eq.(51) can be obtained only if a Gaussian or NGC is performed (Muscolino, 1993). The most used closure techniques, such as the cumulant and the quasi-moment neglect closure, lead to a set of nonlinear equations. This fact represents the major drawback in the numerical applications, since the uniqueness of the solution is not guaranteed, as emphasized by Wojtkiewicz et al. (1996).

Non Gaussian Closure. In Muscolino et al. (2003), a new closure technique is proposed, based on the modification of the well-known quasi-moment neglect closure and taking advantage on the great accuracy of the MCS in the evaluation of the second-order statistics of the response. The new closure technique leads to a set of linear equations for the evaluation of higher order statistical moments. Indeed, the relationships between quasi-moments and statistical moments can be written as follows (Muscolino, 1993) (for $s > 2$):

$$b_s[X] = m_s[X] + (-1)^s \sum_{\substack{i=0,2\ldots(s=even)\\ i=3,5\ldots(s=odd)}}^{s-2} B_{s,i}\sigma_X^{s-i} m_i[X]$$

$$+(-1)^s \sum_{\substack{i=0,2\ldots(s=even)\\ i=3,5\ldots(s=odd)}}^{s} B_{s,i}\sigma_X^{s-i}\left\{\sum_{r=1}^{i}(-1)^r\frac{i!}{r!(i-r)!}m_{i-r}[X]\mu_X^r\right\} \tag{52}$$

in which $b_s[X]$ are the quasi-moments of the response process and

$$B_{s,s} = (-1)^s, \qquad B_{s,i} = (1+i)B_{s-1,i+1} - B_{s-1,i-1}. \tag{53}$$

If the mean μ_X and the standard deviation σ_X of the response process are approximated by MCS, that is $\mu_X = m_1 \cong \widetilde{\mu}_X$ and $\sigma_X = \sqrt{m_2 - m_1^2} \cong \widetilde{\sigma}_X$, and if a quasi-moments closure of order R is adopted, the statistical moments of order higher than R can be evaluated from eq.(52) assuming the quasi-moments of order higher than R equal to zero ($b_s[X] = 0,\ s > R$) and obtaining linear relationships. Then, we obtain a set of R–2 linear differential equations, whose unknowns are the statistical moments of the response process of order greater than two, up to order R. The unique solution of the set of linear differential equations can then be obtained by traditional numerical procedures. Moreover, since the number of realization required to approximate accurately the first two statistical moments of the response is relatively small, the method is very competitive to find additional probabilistic information in terms of moments of higher order than two.

An example: the nonlinear half-oscillator. As an example, let us consider as an application the following nonlinear half-oscillator (Muscolino et al., 2003):

$$\frac{d}{dt}X(t) = -\nu X(t) - \epsilon X^3(t) + I(t) \tag{54}$$

where ν and ϵ are positive constant and $I(t)$ is a Poisson white process, with the impulse occurrence defined by the parameter λ. The impulse amplitude is assumed as a standardized Gaussian random variable Y . Then, the moments of odd order are zero and the moments of even order are $E[Y^{2k}] = (2k-1)!!$. Taking into account that $f[X(t)] = -\nu X(t) - \epsilon X^3(t)$ is an odd function, the moments of odd order of the response process are zero and only the moments of even order must be evaluated. The stationary problem was solved for the following parameter set: $\nu = \epsilon = 1$, $\lambda = 30$. The approximate variance $\widetilde{\sigma}_X^2 = 2.289$ of the response process has been evaluated by MCS with 100 samples only and exploiting the ergodicity of the response process. The obtained value is a very good approximation

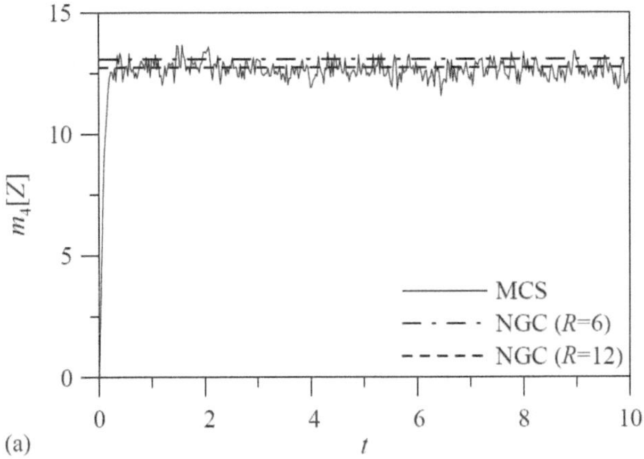

(a)

Figure 1. Comparison of the proposed method with the non-stationary MCS results for different order of NGC: fourth-order moment of the response process.

of the assumed exact value determined by MCS with 10^9 samples. The algebraic stationary moment equations of order $r = 4, 6, \ldots, R$ have been considered and the closure method has been performed, by setting to zero the quasi-moment of order $(R+2)$ and approximating the moment of $(R+2)$-th order. The resulting linear algebraic moment equations have been solved and the stationary values of the approximate moments have been compared with non-stationary MCS results with $10,000$ samples.

In Figs. 1 and 2 the fourth-order and the sixth-order approximate stationary moments are plotted for different values of R, revealing the good approximation of results compared with non-stationary solutions by simulation (with zero initial condition). In Table 1, the stationary moments of the response process up to $10-th$ order evaluated by the proposed method for different closure orders are compared with MCS results with 10^9 samples, assumed as exact solutions. This comparison reveals the good performances of the method and the accuracy increases as the closure order increases.

4 The Stochastic Linearization (SL)

The stochastic linearization (SL) is the most used approximate method for the analysis of nonlinear structural systems under random excitations (Roberts and Spanos, 1991; Socha and Soong, 1991; Elishakoff, 2000; Fal-

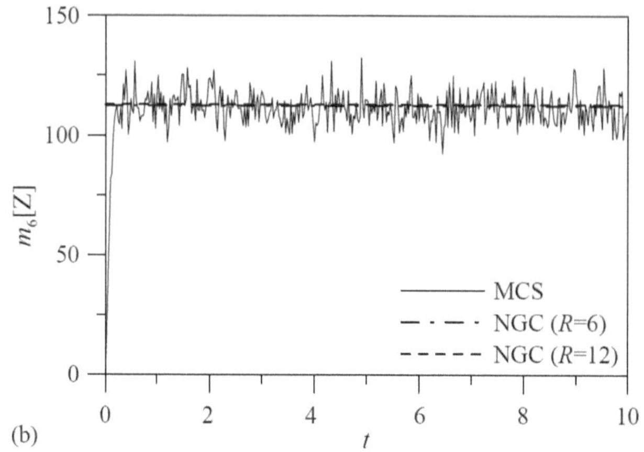

Figure 2. Comparison of the proposed method with the non-stationary MCS results for different order of NGC: sixth-order moment of the response process.

sone and Ricciardi, 2003). This success is due to the wide applicability of the SL method in solving real stochastic structural dynamic problems, often characterized by a large number of degrees of freedom and by complex mathematical models to represent adequately the non-linear structural behavior.

Booton Booton (1953), Kazakov (1956) and Caughey (1959) have independently introduced the method around to the half of the past century. The basic idea is to replace the original non-linear system by an equivalent linear one, whose determination is performed by minimizing the difference between the two systems in some statistical sense. The SL method exhibits different forms based on the adopted probability density function for the evaluation of the coefficients that appear in the linearized system (Kozin, 1988).

4.1 The Gaussian Stochastic Linearization (GSL)

In this section the basic concepts of the GSL method for the stochastic structural dynamics are given. Let us consider a nonlinear structural system excited by external excitations whose dynamic behavior is governed by the following system of first-order differential equations:

$$\dot{\mathbf{X}}(t) = \mathbf{f}[\mathbf{X}(t)] + \mathbf{G}\mathbf{W}(t) \tag{55}$$

Table 1. Stationary moments of the response evaluated by the proposed method for different order of NGC compared with assumed exact values by MCS (10^9 samples)

	$m_4[X]$	$m_6[X]$	$m_8[X]$	$m_{10}[X]$
$R=4$ $e(\%)$	13.74 (8.96)	176.1 (58.50)	-	-
$R=6$ $e(\%)$	13.08 (3.73)	112.4 (1.17)	715 (48.34)	-
$R=8$ $e(\%)$	12.85 (1.90)	112.7 (1.44)	1441 (4.12)	30,591 (31.52)
$R=10$ $e(\%)$	12.76 (1.19)	112.7 (1.44)	1435 (3.68)	24,460 (5.16)
$R=12$ $e(\%)$	12.73 (0.95)	112.8 (1.53)	1432 (3.47)	24,440 (5.07)
Exact (MCS)	**12.61**	**111.1**	**1384**	**23,260**

where $\mathbf{W}(t) = [\ W_1(t) \quad W_2(t) \quad \ldots \quad W_m(t)\ ]^T$ are zero-mean Gaussian white noise processes, with $E[\mathbf{W}(t_1)\mathbf{W}^T(t_2)] = \mathbf{Q}\delta(t_1 - t_2)$, $\mathbf{Q}$ being the $(m \times m)$-matrix collecting the strengths of the white noise processes.

Following the basic idea of stochastic linearization, it is necessary to approximate the original nonlinear system by an appropriate linear one. By assuming that the nonlinearity is odd and the response is with zero mean, eq.(55) is replaced by the following linearized one:

$$\dot{\mathbf{X}}(t) = \mathbf{A}_{eq}\mathbf{X}(t) + \mathbf{G}\mathbf{W}(t) \tag{56}$$

where the matrix $\mathbf{A}_{eq}$ must be chosen in such a way that the difference between the systems given in eq.(55) and (56) is minimum in some statistical sense. By following the classical approach of SL, the difference is measured on the motion equation and the quantity to be minimised is the mean square error, that is:

$$\mathbf{e} = \mathbf{f}[\mathbf{X}(t)] - \mathbf{A}_{eq}\mathbf{X}(t), \qquad E[\mathbf{e}^T\mathbf{e}] = \min \tag{57}$$

The minimization conditions impose that:

$$\nabla_{\mathbf{A}_{eq}} E[\mathbf{e}^T\mathbf{e}] = \mathbf{0} \tag{58}$$

where $\nabla_{\mathbf{A}_{eq}}$ is a $n \times n$ differential operator whose (i, j)-component is the partial derivative made with respect to the (i, j)-component of the matrix $\mathbf{A}_{eq}$. By replacing the first of eqs.(57) into eq. (58), after some algebra we obtained:

$$\mathbf{A}_{eq} = E[\mathbf{f}(\mathbf{X})\mathbf{X}^T] \left\{E[\mathbf{X}\mathbf{X}^T]\right\}^{-1} \tag{59}$$

that constitutes the expression of the equivalent matrix $\mathbf{A}_{eq}$ given by the SL method. The latter exhibits different forms based on the adopted probability density function for the approximation of the averages appearing in eq.(59).

In the classical GSL method the averages appearing in eq.(59) are evaluated taking into account the Gaussianity of the response process. In this way, as the response has been assumed to be Gaussian and with zero mean, the averages are approximated as:

$$E[\,\cdot\,] \cong E[\,\cdot\,]_G = \int_{\Re_n} (\,\cdot\,) p_{\mathbf{X}}^{G}(\mathbf{x}) d\mathbf{x} \tag{60}$$

where the Gaussian probability density function is given by eq.(37) with $\mu_{\mathbf{X}} = \mathbf{0}$. Then, the expression (59) of the equivalent matrix becomes:

$$\mathbf{A}_{eq}^{G}(\boldsymbol{\Sigma}_{\mathbf{X}}^{\mathrm{G}}) = E_G[\mathbf{f}(\mathbf{X})\mathbf{X}^T][\boldsymbol{\Sigma}_{\mathbf{X}}^{\mathrm{G}}]^{-1} \tag{61}$$

In this way, the equivalent matrix $\mathbf{A}_{eq}$ is expressed in terms of the statistical moments until the second order of the response only. These moments can be evaluated as solution of the following non-linear Lyapunov-type differential equation:

$$\dot{\boldsymbol{\Sigma}}_{\mathbf{X}}^{\mathrm{G}} = \mathbf{A}_{eq}^{G}\boldsymbol{\Sigma}_{\mathbf{X}}^{\mathrm{G}} + \boldsymbol{\Sigma}_{\mathbf{X}}^{\mathrm{G}}\mathbf{A}_{eq}^{G} + \mathbf{G}\mathbf{Q}\mathbf{G}^T \tag{62}$$

It is worth noticing that eq.(62) is nonlinear, because the matrix depends on the second order moments, that, in turn, depend on the probability density function adopted. Hence a numerical procedure for solving the nonlinear equations must be applied.

4.2 The Non Gaussian Stochastic Linearization (NGSL)

Against relatively little numerical efforts, unfortunately the GSL method gives accurate results for weakly non-linear systems only and, as recently highlighted, it suffers of unacceptable errors in the case of large structures (Micaletti et al., 1998). An improved approximation can be obtained if the relaxation of the Gaussian response assumption is introduced. In this way, the averages in eq.(59) are determined by adopting a better approximation of the probability density function of the response. The procedure proposed by Ricciardi (2007) uses as a probability density function a modified A-type Gram-Charlier series expansion, where the covariance matrix is not unknown, but it is assumed equal to the covariance matrix derived by the GSL method. In order to present the aforementioned method, a standardization procedure is needed.

Standardization. The following coordinate transformation is adopted:

$$\mathbf{X}(t) = \mathbf{\Gamma}_G \mathbf{Y}(t), \qquad \mathbf{\Gamma}_G = \mathbf{\Psi}_G \mathbf{\Lambda}_G^{1/2} \tag{63}$$

where the full matrix $\mathbf{\Psi}_G$ and the diagonal matrix $\mathbf{\Lambda}_G$ are evaluated by solving the following eigenproblem related to the covariance matrix $\mathbf{\Sigma}_{\mathbf{X}}^{\mathrm{G}}$:

$$\mathbf{\Sigma}_{\mathbf{X}}^{\mathrm{G}} \mathbf{\Psi}_G = \mathbf{\Psi}_G \mathbf{\Lambda}_G \tag{64}$$

with the normality condition $\mathbf{\Psi}_G^T \mathbf{\Psi}_G = \mathbf{I}_n$. In eq.(64), the covariance matrix $\mathbf{\Sigma}_{\mathbf{X}}^{\mathrm{G}}$ is supposed known, as solution of the GSL method.

It can be easily shown that the vector process $\mathbf{Y}(t)$ has approximated covariance matrix $\mathbf{\Sigma}_{\mathbf{Y}}^{\mathrm{G}} = \mathbf{I}_n$, derived by the GSL. Moreover, by adopting the coordinate transformation eq.(63), eq.(55) can be rewritten as follows:

$$\dot{\mathbf{Y}}(t) = \mathbf{r}[\mathbf{Y}(t)] + \mathbf{U}\mathbf{W}(t) \tag{65}$$

where

$$\mathbf{r}[\mathbf{Y}(t)] = \mathbf{\Gamma}_G^{-1} \mathbf{f}[\mathbf{\Gamma}_G \mathbf{Y}(t)], \qquad \mathbf{U} = \mathbf{\Gamma}_G^{-1} \mathbf{V} \tag{66}$$

By following the classical approach of the SL method, eq.(65) is replaced by the following linearized one:

$$\dot{\mathbf{Y}}(t) = \mathbf{B}_{eq} \mathbf{Y}(t) + \mathbf{U}\mathbf{W}(t) \tag{67}$$

and by minimizing the mean square error, we find the following form of the equivalent matrix $\mathbf{B}_{eq}$:

$$\mathbf{B}_{eq} = E[\mathbf{r}(\mathbf{Y})\mathbf{Y}^T] \left\{ E[\mathbf{Y}\mathbf{Y}^T] \right\}^{-1} \tag{68}$$

If GSL method is adopted, in order to evaluate the averages appearing in eq.(68), $E[\mathbf{Y}\mathbf{Y}^T] \cong E_G[\mathbf{Y}\mathbf{Y}^T] = \mathbf{I}_n$; then, the following expression of the equivalent matrix is obtained:

$$\mathbf{B}_{eq}^G = E_G[\mathbf{r}(\mathbf{Y})\mathbf{Y}^T] \tag{69}$$

It is emphasized that, by introducing the coordinate transformation, the problem is simplified, as the vector process $\mathbf{Y}(t)$ has uncorrelated components with unit variances.

Non Gaussian probability density function. In order to improve the GSL results, the NGSL method approximate the averages appearing in eq.(18) by using the A-type Gram-Charlier series expansion of the non-Gaussian probability density function, given as

$$p(\mathbf{y}) \cong p_{NG}(\mathbf{y}) = p_G(\mathbf{y})\times$$

$$\times\left[1+\sum_{\substack{j=2,4,\ldots \\ i_1+i_2+\ldots+i_n=j}}^{M}\frac{1}{i_1!i_2!\ldots i_n!}C_{i_1,i_2,\ldots,i_n}H_{i_1}(y_1)H_{i_2}(y_2)\ldots H_{i_n}(y_n)\right] \tag{70}$$

where M is the truncation order and $p_G(\mathbf{y})$ is the joint Gaussian probability density function of the uncorrelated random processes $Y_\ell(\ell = 1, 2, \ldots, n\ )$, defined as:

$$p_G(\mathbf{y}) = \prod_{\ell=1}^{n} p_G(y_\ell) = \prod_{\ell=1}^{n}\frac{1}{\sqrt{2\pi}}\exp\left[-\frac{y_\ell^2}{2}\right] \tag{71}$$

In eq.(70), $H_k(y_\ell)$ is the one-dimensional Hermite polynomial defined as follows:

$$H_k(y_\ell) = (-1)^k \exp\left[\frac{1}{2}y_\ell^2\right]\frac{d^k}{dy_\ell^k}\left\{\exp\left[-\frac{1}{2}y_\ell^2\right]\right\} \tag{72}$$

and $C_{i_1,i_2,\ldots,i_n}$ is the modified Hermite moment, defined as:

$$C_{i_1,i_2,\ldots,i_n} = E[H_{i_1}(Y_1)H_{i_2}(Y_2)\ldots H_{i_n}(Y_n)] \tag{73}$$

Improved equivalent matrix and covariance matrix. The NGSL method takes into account the non-Gaussian character of the response process and concurs to estimate the expression of the equivalent matrix of the linearized system in a more accurate way. By adopting the non-Gaussian probability density function (70), $\mathbf{B}_{eq}$ assumes the following form:

$$\mathbf{B}_{eq}^{NG} = E_{NG}[\mathbf{r}(\mathbf{Y})\mathbf{Y}^T]\left\{E_{NG}[\mathbf{Y}\mathbf{Y}^T]\right\}^{-1} \tag{74}$$

After some algebra, the equivalent matrix assumes the following form:

$$\mathbf{B}_{eq}^{NG} = \mathbf{R}^{NG}[\mathbf{I}_n + \mathbf{S}]^{-1} \tag{75}$$

where the matrices $\mathbf{S}$ and $\mathbf{R}^{NG}$ are given as:

$$\mathbf{S} = \begin{bmatrix} C_{2,0,\dots 0} & C_{1,1,\dots 0} & C_{1,0,\dots 1} \\ C_{1,1,\dots 0} & C_{0,2,\dots 0} & C_{0,1,\dots 1} \\ \\ C_{1,0,\dots 1} & C_{0,1,\dots 1} & C_{0,0,\dots 2} \end{bmatrix} \tag{76}$$

$$\mathbf{R}^{NG} = \mathbf{R}^{G} + \sum_{\substack{j=2,4,\dots \\ i_1+i_2+\dots+i_n=j}}^{M} \frac{1}{i_1! i_2! \dots i_n!} C_{i_1,i_2,\dots,i_n} \mathbf{R}_{i_1,i_2,\dots,i_n} \tag{77}$$

The matrix $\mathbf{R}^G$ and the (j,ℓ)-th element of the matrix $\mathbf{R}_{i_1,i_2,\dots,i_n}$ can be easily evaluated by performing the following averages:

$$\mathbf{R}^{G} = E_G[\mathbf{r}(\mathbf{Y})\mathbf{Y}^T] \tag{78}$$

$$(\mathbf{R}_{i_1,i_2,\dots,i_n})_{j\ell} = \hat{E}_G[r_j(\mathbf{Y})Y_\ell H_{i_1}(Y_1)H_{i_2}(Y_2)\dots H_{i_n}(Y_n)] \tag{79}$$

where $\hat{E}_G[\,\cdot\,] = \int_{\Re_n}(\,\cdot\,)p_G(\mathbf{y})d\mathbf{y}$ that involves simple one-dimensional integrals, generally known in closed form. Note that if the non-linear term $r_j(\mathbf{Y})$ is a polynomial expression, the averages are a combination of double factorials, taking advantage of the particular structure of the $\mathbf{Y}$-space.

By using the coordinate transformation, we find the following expression of the equivalent matrix in the original space:

$$\mathbf{A}_{eq}^{NG} = \mathbf{\Gamma}_G\mathbf{R}^{NG}[\mathbf{I}_n + \mathbf{S}]^{-1}\mathbf{\Gamma}_G^{-1} \tag{80}$$

This expression represents a non-Gaussian improved version of the equivalent matrix determined by means of the GSL method. A better approximation of the covariance matrix $\mathbf{\Sigma}_{\mathbf{Y}}^{\mathrm{NG}}$ can be evaluated as follows:

$$\mathbf{\Sigma}_{\mathbf{Y}}^{\mathrm{NG}} = E[\mathbf{Y}\mathbf{Y}^T] \cong E_{NG}[\mathbf{Y}\mathbf{Y}^T] = [\mathbf{I}_n + \mathbf{S}] \tag{81}$$

and the expression of the covariance matrix $\mathbf{\Sigma}_{\mathbf{X}}^{\mathrm{NG}}$ in the original space is:

$$\mathbf{\Sigma}_{\mathbf{X}}^{\mathrm{NG}} = \mathbf{\Gamma}_G[\mathbf{I}_n + \mathbf{S}]\mathbf{\Gamma}_G^T \tag{82}$$

This equation represents the NGSL solution for the covariance matrix, which requires the knowledge of the series coefficients $C_{i_1,i_2,\ldots,i_n}$. As shown later, they can be simply approximated as solution of a system of linear equations.

Approximation of the modified A-type Gram-Charlier series coefficients. The modified Hermite moments $C_{i_1,i_2,\ldots,i_n}$ are defined as the averages of products of one-dimensional Hermite polynomials. They can be evaluated by using the Itï¿œ rule or, alternatively, they can be approximated by taking advantage by the theory of Markov process. The probability density function $p_{\mathbf{Y}}(\mathbf{y})$ of the vector process $\mathbf{Y}(t)$ is the solution of the Fokker-Planck-Kolmogorov (FPK) equation (Risken, 1989):

$$\frac{\partial}{\partial t} p_{\mathbf{Y}}(\mathbf{y}) = \mathcal{L} p_{\mathbf{Y}}(\mathbf{y}) \tag{83}$$

where

$$\mathcal{L}(\,\cdot\,) = \frac{\partial}{\partial y_i}\left[r_i(\mathbf{y})(\,\cdot\,)\right] - \frac{1}{2} Q_{\ell s} U_{i\ell} U_{js} \frac{\partial^2}{\partial y_i \partial y_j}(\,\cdot\,) \tag{84}$$

is the FPK operator (where repeated indices mean summation convention).

It can be easily shown that the differential equation governing the time evolution of the average of a function $\varphi(\mathbf{Y})$ can be written as follows

$$\dot{E}\left[\varphi(\mathbf{y})\right] = E\left\{\mathcal{L}^+\left[\varphi(\mathbf{y})\right]\right\} \tag{85}$$

where

$$\mathcal{L}^+(\,\cdot\,) = r_i(\mathbf{y})\frac{\partial}{\partial y_i}(\,\cdot\,) + \frac{1}{2} Q_{\ell s} U_{i\ell} U_{js} \frac{\partial^2}{\partial y_i \partial y_j}(\,\cdot\,) \tag{86}$$

is the adjoint Fokker-Planck-Kolmogorov (AFPK) differential operator.

An approximation of the modified A-type Gram-Charlier coefficients can be obtained by choosing $\varphi(\mathbf{Y}) = H_{j_1}(Y_1)H_{j_2}(Y_2)\ldots H_{j_n}(Y_n)$, for $j_1 + j_2 + \ldots + j_n = 2, 4, \ldots, M$, with $M \geq 4$. Then, the following differential equations of $C_{j_1,j_2,\ldots,j_n} = E[H_{j_1}(Y_1)H_{j_2}(Y_2)\ldots H_{j_n}(Y_n)]$ are obtained:

$$\dot{C}_{j_1,j_2,\ldots,j_n} = E\left\{\mathcal{L}^+\left[H_{j_1}(Y_1)H_{j_2}(Y_2)\ldots H_{j_n}(Y_n)\right]\right\} \tag{87}$$

In the stationary state the average quantities are not time dependent and consequently $\dot{C}_{j_1,j_2,\ldots,j_n} = 0$. Therefore, the last equation can be simplified as follows:

$$\sum_{\substack{j=2,4,\ldots \\ i_1+i_2+\ldots+i_n=j}}^{M} D^{j_1,j_2,\ldots,j_n}_{i_1,i_2,\ldots,i_n} C_{i_1,i_2,\ldots,i_n} + b^{j_1,j_2,\ldots,j_n} = 0 \tag{88}$$

where

$$D^{j_1,j_2,\ldots,j_n}_{i_1,i_2,\ldots,i_n} = \frac{1}{i_1!i_2!\ldots i_n!}\times$$

$$\times\left\{\hat{E}_G\left[r_i(\mathbf{Y})H_{i_1}(Y_1)H_{i_2}(Y_2)\ldots H_{i_n}(Y_n)\frac{\partial}{\partial Y_i}\left(H_{j_1}(Y_1)H_{j_2}(Y_2)\ldots H_{j_n}(Y_n)\right)\right]\right.$$

$$\left.+\hat{E}_G\left[H_{i_1}(Y_1)H_{i_2}(Y_2)\ldots H_{i_n}(Y_n)\frac{\partial^2}{\partial Y_i \partial Y_j}\left(H_{j_1}(Y_1)H_{j_2}(Y_2)\ldots H_{j_n}(Y_n)\right)\right]\right\} \tag{89}$$

and

$$b^{j_1,j_2,\ldots,j_n} = \hat{E}_G\left[r_i(\mathbf{Y})\frac{\partial}{\partial Y_i}\left(H_{j_1}(Y_1)H_{j_2}(Y_2)\ldots H_{j_n}(Y_n)\right)\right]$$

$$+\frac{1}{2}Q_{\ell s}U_{i\ell}U_{js}\hat{E}_G\left[\frac{\partial^2}{\partial Y_i \partial Y_j}\left(H_{j_1}(Y_1)H_{j_2}(Y_2)\ldots H_{j_n}(Y_n)\right)\right] \tag{90}$$

In the last equations the repeated indices mean summation convention, with $i, j = 1, 2, \ldots, n$ and $\ell, s = 1, 2, \ldots, m$. Note that eqs.(88) constitute a system of linear algebraic equations with unknowns $C_{j_1,j_2,\ldots,j_n}$, whose solution is an approximation of these coefficients.

The NGSL applied to the power-law nonlinear oscillator. Let us consider the following non-linear oscillator subjected to a Gaussian white noise input with intensity q:

$$\ddot{U}(t) + \beta\dot{U}(t) + g_\gamma\left|U(t)\right|^\gamma sgn[U(t)] = \sqrt{2\beta}W(t) \tag{91}$$

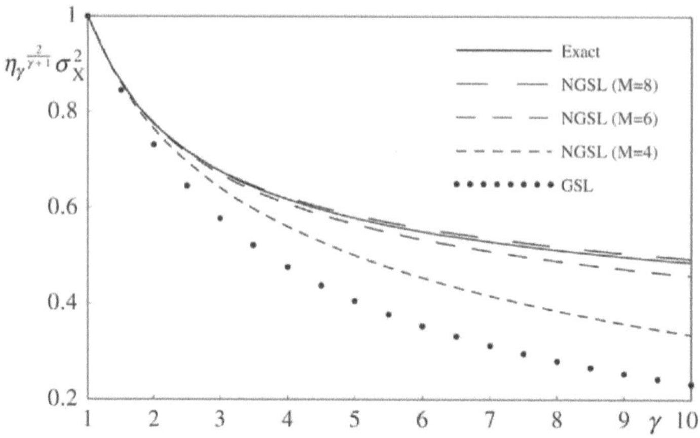

Figure 3. Displacement variance versus γ for the hardening power-law oscillator.

where g_γ is a positive constant and $\gamma > 0$ is a parameter related to the non-linear restoring force. If $0 < \gamma < 1$ or $\gamma > 1$, the oscillator exhibits softening or hardening behavior, respectively. Roberts and Spanos (1991) considered such an oscillator by the Gaussian stochastic linearization method. The exact variance of the displacement is:

$$\sigma_{U,ex}^2 = \left(\frac{\gamma+1}{\eta_\gamma}\right)^{\frac{2}{\gamma+1}} \Gamma\left(\frac{3}{\gamma+1}\right)\Gamma^{-1}\left(\frac{1}{\gamma+1}\right) \tag{92}$$

where $\eta_\gamma = g_\gamma/q$ and $\Gamma(\cdot)$ is the Gamma function. The GSL method leads to the following approximation:

$$\sigma_{U,G}^2 = \frac{1}{2}\left[\frac{\sqrt{\pi}}{\eta_\gamma\Gamma(1+(\gamma/2))}\right]^{\frac{2}{\gamma+1}} \tag{93}$$

This value has been assumed as a first approximation for the NGSL method.

In Fig. 3 and Fig. 4 the displacement variance and the relative error are plotted versus the non-linear parameter γ, in the case of hardening behavior of the oscillator ($\gamma > 1$). In these figures the NGSL results, for different closure order ($M = 4, 6, 8$), are compared with the GSL results, revealing that the NGSL method leads to better results, also for low closure order ($M = 4$). Accurate results are obtained for higher order closures, with errors

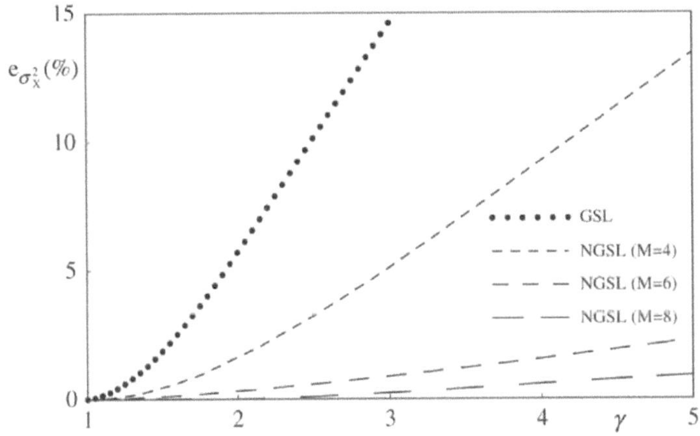

Figure 4. Relative error of displacement variance versus γ for hardening the power-law oscillator.

lesser then 1% for $1 < \gamma < 5$ and $M = 8$. From these figures, it appears that the NGSL method seems to give results very close to the exact ones as M increases. This is due to the fact that the series approximation used by the NGSL method converges to the exact probability density function of the response, given as:

$$p_{U,ex}(u) = \frac{1}{2}\left(\frac{\eta_\gamma}{\gamma+1}\right)^{\frac{1}{\gamma+1}} \Gamma^{-1}\left(\frac{\gamma+2}{\gamma+1}\right) \exp\left(-\eta_\gamma \frac{|u|^{\gamma+1}}{\gamma+1}\right) \tag{94}$$

In Fig. 5 and Fig. 6 the displacement variance and the relative error are plotted, in the case of softening behavior of the oscillator ($0 < \gamma < 1$). From these figures the lack of convergence of the series expansion used by the NGSL is evident. However, in the range $0.2 < \gamma < 1$, the NGSL method gives better results than the GSL, with errors lesser than 5% for all the closure orders considered ($M = 4, 6, 8$). In particular, for $M = 4$, the error is lesser than 5% in the entire range $0 < \gamma < 1$.

The NGSL applied to the double-well Duffing oscillator. Let us consider the following Duffing oscillator:

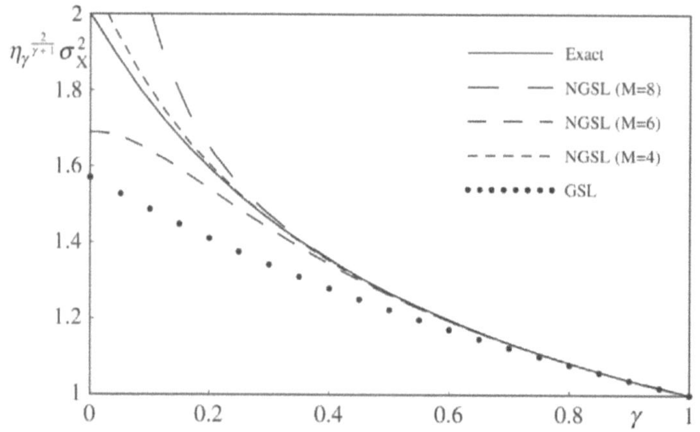

Figure 5. Displacement variance versus γ for the softening power-law oscillator.

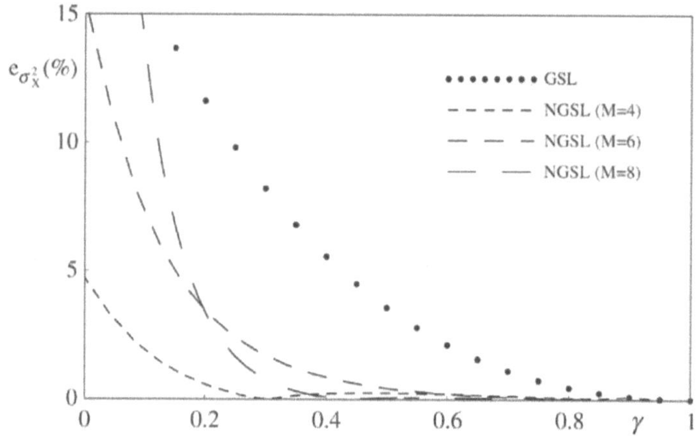

Figure 6. Relative error of displacement variance versus γ for the softening power-law oscillator.

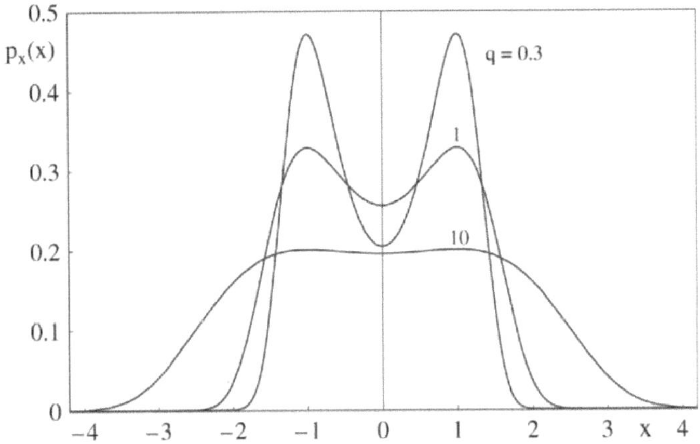

Figure 7. Probability density functions of displacement of the double-well Duffing oscillator, for different levels of the white noise excitation.

$$\ddot{U}(t) + \beta\dot{U}(t) - U(t) + U^3(t) = \sqrt{2\beta}W(t) \tag{95}$$

which exhibits two positions of stable equilibrium at $\bar{u}_{1,2} = \pm 1$. The corresponding potential energy function has two potential wells, symmetrically disposed about the origin and with minima at $\bar{u}_{1,2} = \pm 1$. The exact marginal probability density function of displacement is:

$$p_{U,ex}(u) = N \exp\left[\frac{1}{q}\left(\frac{u^2}{2} - \frac{u^4}{4}\right)\right] \tag{96}$$

being N a normalization constant. For low white noise excitation (q sufficiently small), the response of this oscillator tends to stay within one of the potential wells. In this case, the probability density becomes bi-modal, exhibiting two symmetric bumps. For high level of excitation, these bumps tend to disappear (see Fig. 7). Roberts and Spanos (1991) have studied such an oscillator by the GSL, Alaoui-Ismaili and Bernard (1997) and Ricciardi and Elishakoff (2002) by two different versions of the stochastic local linearization.

The Gaussian stochastic linearization leads to the following approximate value of the variance of displacement

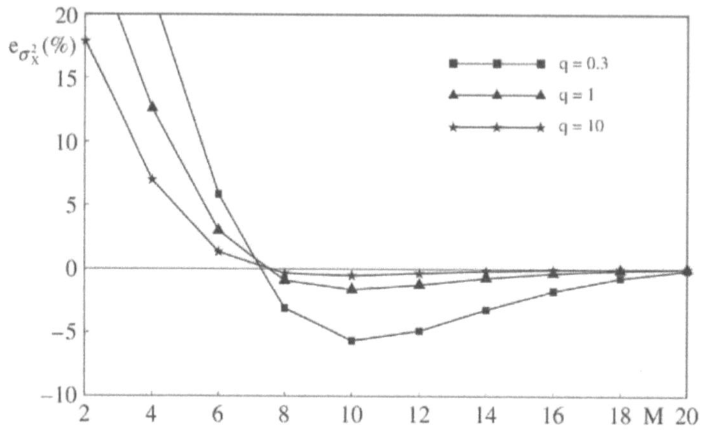

Figure 8. Relative error of displacement variance versus M for the double-well Duffing oscillator, for different levels of excitation.

$$\sigma^2_{U,G} = \frac{1 + \sqrt{1 + 12q}}{6} \tag{97}$$

In Fig. 8 the percentage error of the displacement variance by the NGSL method is plotted versus the closure order of the modified Gram-Charlier series expansion, for three different values of the intensity of the white noise: $q = 0.3$, $q = 1$ and $q = 10$. It appears that the convergence to the exact solution is faster in the case of high intensity of excitation, being almost uni-modal the probability density function of the displacement. On the contrary, in the case of low intensity of excitation, a higher closure order of the series expansion is needed in order to obtain a comparable accuracy, being slower the convergence in the case of bi-modal probability density function.

Multistory frame structure subjected to a filtered Gaussian white noise input. As a third example, a multistory frame structure subjected to a stationary excitation is considered. The earthquake is modeled as the response of a Kanai-Tajimi-like filter driven by a stationary Gaussian white noise input (Fig. 9). Non-linear effects in the columns behavior are included, modeled by restoring forces proportional to the cube of the transversal displacements through a parameter ϵ. The motion equations of this kind of system are:

$$\mathbf{M}\ddot{\mathbf{u}}(t) + \mathbf{C}\dot{\mathbf{u}}(t) + \mathbf{K}\mathbf{u}(t) + \mathbf{h}[\mathbf{u}(t)] = \mathbf{M}\tau[2\varsigma_g\omega_g\dot{q}_g(t) + \omega_g^2 q_g(t)] \quad (98)$$

$$\ddot{q}_g(t) + 2\varsigma_g\omega_g\dot{q}_g(t) + \omega_g^2 q_g(t) = -W(t) \quad (99)$$

where $\mathbf{M}$, $\mathbf{C}$ and $\mathbf{K}$ are the $p \times p$ mass, damping and stiffness matrices of the system, respectively; $\mathbf{h}[\mathbf{u}(t)]$ is the p-vector collecting the non-linear restoring forces; $\mathbf{u}(t)$ is the p-vector collecting the storey displacements; τ is the influence forcing p-vector; ς_g , ω_g and $q_g(t)$ are the damping, the circular frequency and the displacement of the ground, respectively, which is modeled like a linear filter subjected to a stationary Gaussian white noise $W(t)$. Assuming non-linear behavior, the elastic restoring forces are related to the interstory displacements $\hat{\mathbf{u}}$ by non-linear relationships. Introducing the lateral stiffness k_j of the j-th story, the story stiffness is assumed as $k_j = k_j^0(1+\epsilon_j\hat{u}_j^2)$, k_j^0 being the initial stiffness, $\hat{u}_j$ the$j$-th interstory displacement and ϵ_j the j-th non-linear parameter.

For simplicity's sake, the application has been performed for a two-story shear-type frame. The following data have been assumed: $m_1 = m_2 = 10^4 Kg$, $k_1^0 = k_2^0 = 6 \times 10^6 N/m$, the 2×2 damping matrix $\mathbf{C}$ has been evaluated for a classically damped system, setting the modal damping ratios $\varsigma_1 = \varsigma_2 = 0.05$. The parameters selected for the Tajimi-Kanai-like filter are: $\varsigma_g = 0.6$, $\omega_g = 5\pi s^{-1}$, $S_0 = 0.0406471 m^2/s^3$.

In a first step the GSL method has been developed and then the proposed NGSL method has improved the approximation of the variances of the system. The truncation orders chosen have been $M = 4, 6$, which implied the solution of a system of 147 and 609 linear equations, respectively, for the evaluation of the coefficients $C_{j_1,j_2,\ldots,j_n}$.

Two analyses have been performed, characterized by two different levels of non-linearity in the interstory shear restoring forces: a) $\epsilon_1 = \epsilon_2 = \epsilon = 5 \times 10^2$; b) $\epsilon_1 = \epsilon_2 = \epsilon = 2 \times 10^3$; the results have been compared with the "reference" ones obtained through a Monte Carlo Simulation (MCS) with 10^5 samples. In Table 2 and Table 3 the variances of the "reference" and approximated displacements and velocities are listed, respectively, for the two cases analysed. Moreover, the corresponding percentage errors and the global percentage errors relative to the covariance matrix of the structural response only are reported. Note that the proposed NGSL method improves the results given by the GSL, even if a low closure order is adopted ($M = 4, 6$). The improvement is more remarkable especially for the covariances

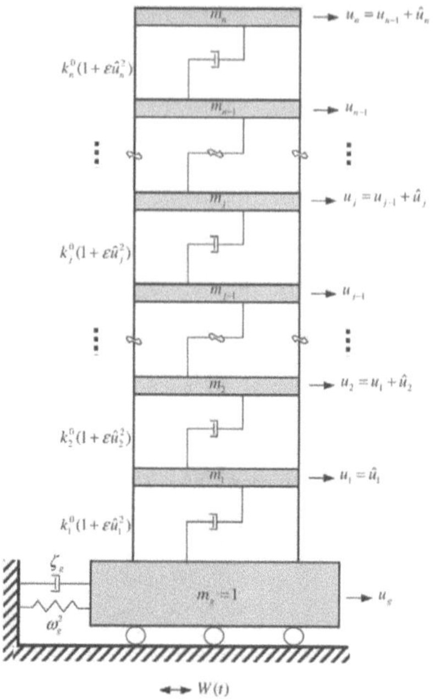

Figure 9. Schematic sketch of a multistorey frame non-linear structure.

Table 2. Variances of the response of the non-linear frame structure for $\epsilon = 5 \times 10^2$

	GSL	NGSL ($M = 4$)	NGSL ($M = 6$)	MCS (10^5samples)
$\sigma^2_{\hat{u}_2}$	2.12×10^{-4}	2.16×10^{-4}	2.16×10^{-4}	2.21×10^{-4}
$e(\%)$	(3.85)	(2.22)	(2.36)	
$\sigma^2_{\hat{u}_1}$	1.07×10^{-4}	1.07×10^{-4}	1.07×10^{-4}	1.10×10^{-4}
$e(\%)$	(2.45)	(2.72)	(2.90)	
$\sigma^2_{\dot{\hat{u}}_1}$	5.83×10^{-2}	5.90×10^{-2}	5.92×10^{-2}	6.09×10^{-2}
$e(\%)$	(4.24)	(3.01)	(2.71)	
$\sigma^2_{\dot{\hat{u}}_2}$	3.19×10^{-2}	3.40×10^{-2}	3.51×10^{-2}	3.63×10^{-2}
$e(\%)$	(12.21)	(6.35)	(3.52)	

that are affected by higher errors. From the values of the global error, it is evident the tendency of the convergence of the covariance matrix to the exact solution.

5 The Independent Component Kernel Density (IC-KD) method in stochastic dynamics

In this section a new method is proposed, based on information derived from a MCS with few samples. This information is used to make a suitable coordinate transformation that allows the projection into a space of independent random variables (Hyvarinen et al., 2001), called Independent Components (ICs). The application of the Kernel Density Maximum Entropy Method (KD-MEM) (Alibrandi and Ricciardi, 2008) to each IC allows to obtain a good approximation of the pdf of the response. The method will be described in detail in the following sections.

5.1 The moment problem (MP)

In this section we briefly sketch the classical moment problem (MP), which consists in determining under what conditions a sequence of numbers

Table 3. Variances of the response of the non-linear frame structure for $\epsilon = 2 \times 10^3$

	GSL	NGSL ($M = 4$)	NGSL ($M = 6$)	MCS (10^5samples)
$\sigma^2_{\hat{u}_2}$	1.29×10^{-4}	1.35×10^{-4}	1.36×10^{-4}	1.37×10^{-4}
$e(\%)$	(6.05)	(1.31)	(1.09)	
$\sigma^2_{\hat{u}_1}$	7.57×10^{-5}	7.73×10^{-5}	7.70×10^{-5}	7.82×10^{-5}
$e(\%)$	(3.18)	(1.24)	(1.60)	
$\sigma^2_{\dot{\hat{u}}_1}$	4.59×10^{-2}	4.69×10^{-2}	4.77×10^{-2}	4.89×10^{-2}
$e(\%)$	(6.12)	(3.95)	(2.50)	
$\sigma^2_{\dot{\hat{u}}_2}$	2.89×10^{-2}	3.14×10^{-2}	3.43×10^{-2}	3.63×10^{-2}
$e(\%)$	(20.21)	(13.41)	(5.40)	

(usually represented by a sequence of sample moments) is a sequence of moments for any distribution. Every distribution $F_X(x)$ of a random variable X has a sequence of moments given by integration

$$\mu_k = \int_\Omega x^k dF_X, \qquad k = 0, 1, 2, \ldots \tag{100}$$

Eq. (100) is stated with respect to a positive measure $F_X(x)$, as classically made in mathematical literature. In the following it will be assumed that the probability measure is regular, so that it is represented through the probability density function $p_X(x)$, whose support is Ω; consequently, in (100) $dF_X(x) = f_X(x)dx$, and moreover, in order to satisfy the normalization condition, the moment of order zero is $\mu_0 = 1$.

Of course, not every sequence of numbers is a moment sequence. For instance, it is well known that the second moment must be non-negative. The MP is to determine, given some sequence of numbers, whether it is a moment sequence, that is to say, whether the numbers given are indeed the moments of a non-negative function or distribution, with the given support.

More formally, let $\mathscr{M}$ be the "moment space", that is the set of all sequences of real numbers $\{\mu_k\}$ for which the infinite system of (100) has at least one solution $F_X(x)$. The task is to determine the conditions which must be satisfied by the assigned numbers μ_k , (with $k = 1, 2, \ldots$) in order that $\{\mu_k\} \in \mathscr{M}$. However, in practice only a finite number$M$ of moments is known. In this case we have the "truncated moment problem", that is to determine if a sequence of M real numbers μ_1, μ_2, ..., μ_n corresponds to a sequence of M moments of some pdf. The corresponding set $\mathscr{M}^M$ is called as "truncated moment space of order M". The existence conditions of at least one pdf involve $\mathscr{M}^M$, more precisely: (i) if the point $\mu \equiv (\mu_1, \mu_2, \ldots, \mu_M)$ is in the interior of $\mathscr{M}^M$, there are infinitely many distributions that admit μ_1, μ_2, ..., μ_n as moments; (ii) if $\mu \in \partial\mathscr{M}^M$ ($\partial\mathscr{M}^M$ is the boundary of $\mathscr{M}^M$) the only distribution having μ_1, μ_2, ..., μ_n as first moments is a uniquely determined convex combination of Dirac's delta; (iii) if μ is outside $\mathscr{M}^M$, the corresponding truncated moment problem does not admit any solution.

5.2 The kernel density representation (KDR)

Let us consider a random variable X, whose pdf $p_X(x)$ with support Ω is approximated as a superposition of basis pdfs:

$$\hat{p}_X(x) = \sum_{i=1}^{N} p_i \varphi_i(x; \vartheta_i^1, \vartheta_i^2, \ldots, \vartheta_i^m) \tag{101}$$

where $\hat{p}_X(x) \cong p_X(x)$ is the approximating pdf, $\varphi_i(x; \vartheta_i^1, \vartheta_i^2, \ldots, \vartheta_i^m)$ is the i-th basis pdf with support Ω and whose parameters are $\vartheta_i^1, \vartheta_i^2, \ldots, \vartheta_i^m$, N is the number of basis pdfs used in the approximation.

It is here underlined that the superposition (101) of density functions represents a pdf if the coefficients p_i $(i = 1, 2, \ldots, n)$ satisfy the following inequalities

$$0 \leq p_i \leq 1, \qquad \sum_{i=1}^{N} p_i = 1 \tag{102}$$

Eqs.(101) and (102) remark the known property that a generic pdf may be expressed as a linear convex combination of simpler pdfs, whose coefficients have the meaning of probabilities.

Let us define the *admissible domain* $\mathcal{P}^N$ of the coefficients p_i, $(i = 1, 2, \ldots, n)$:

$$\mathcal{P}^N \equiv \left\{ p_i \mid 0 \leq p_i \leq 1, \; i = 1, 2, \ldots, n, \; \sum_{i=1}^{N} p_i = 1 \right\} \tag{103}$$

Following Athanassoulis and Gavriliadis (2002) and (2003), as a basis pdf the two-parameters kernel density function (KDF) $f_i^K(x; x_i, h_i)$ has been chosen, and then eq.(101) becomes

$$\hat{p}_X(x) = \sum_{i=1}^{N} p_i f_i^K(x; x_i, h_i) \tag{104}$$

where $f_i^K(x; x_i, h_i)$ is the i-th KDF with location parameter x_i and bandwidth h_i.

The representation (104) is really flexible, in fact each kernel is identified by its location parameter x_i and bandwidth h_i, which gives the spreading of the KDF around its location x_i. The location parameters x_i ($i = 1, 2, \ldots, N$) are points belonging to the domain Ω, they can be chosen in several different ways; however, for simplicity we chose a constant step $\Delta x = \Delta x_i = x_{i+1} - x_i$, $i = 1, 2, \ldots, N-1$. In a similar way we chose $h_i = h = b\Delta x$, $i = 1, 2, \ldots, N$, with b suitable chosen parameter, but whenever it were necessary we could choose suitable values of bandwidth in different kernel pdfs.

An interesting property of the KDF is represented by the following asymptotic limit

$$\lim_{h_i \to 0} f_i^K(x; x_i, h_i) = \delta(x - x_i) \tag{105}$$

As a consequence the Kernel Density Representation (KDR) described by (104) is able to approximate any continuous PDF as closely as required.

It is noted that several different KDFs satisfying (104) and (105) can be adopted, however here we like to mention that the choice should at least take into account of the type of support Ω of the target pdf. Consequently, if the random variable X has a bounded, semi-bounded or an unbounded support, correspondingly the kernel density functions $f_i^K(x; x_i, h_i)$ should have a bounded, semibounded or an unbounded support, as well. Therefore if the random variable X has an unbounded support, a quite natural choice for the KDFs is the Gaussian distribution; in a similar way, it appears reasonable to choose the lognormal and the beta distributions for the cases when X has a semibounded and a bounded support, respectively. A further guideline can be represented from the physical behavior of the random variables. So, if we know that X has unbounded support with low tails, the Gaussian distribution is retained as the most favourite one as a KDF; if conversely X is known to have high tails, then a better choice could be given by the Laplace distribution.

5.3 Discretization of the moment problem

The target pdf $p_X(x)$ can be approximated at any degree of accuracy by the kernel density representation (104) , so that $p_X(x) \cong \hat{p}_X(x)$. Multiplying both sides of eq.(104) by x^k, and integrating over the support, we obtain the following relation between the parameters p_i, $i = 1, 2, \ldots, N$, and the moments of the target pdf:

$$\mathbf{1}^T\mathbf{p} = 1, \qquad \mathbf{M}^K\mathbf{p} = \mu \tag{106}$$

where $\mathbf{1}$ is a unit vector of order N, while the coefficients

$$M_{ki}^K = \mu_{k,i}^K = \int_\Omega x^k f_i^K(x; x_i, h_i)dx \tag{107}$$

are the moments of order k of the i-th KDF $f_i^K(x; x_i, h_i)$, so that $\mathbf{M}^K = [M_{ki}^K]$, $k = 1, 2, \ldots, M$, $i = 1, 2, \ldots, N$, is a matrix of order $M \times N$, μ is a vector of order M collecting the assigned moments and $\mathbf{p}$ is a vectors of order N, collecting the unknown parameters p_i, $i = 1, 2, \ldots, N$.

In the analysis of the system (106) a first insight has to be focused about the parameters of the KDR. A first step is related to the choice of the "window of analysis" $\hat{\Omega} \subseteq \Omega$, which is the subset of the support Ω of X containing the "centers" x_i of the kernel pdfs and therefore such that $x_i \in \hat{\Omega}$. If X has the bounded support $\Omega \equiv [a, b]$, then the window of analysis is easily determined as $\hat{\Omega} \equiv \Omega$, while for a random variable with semibounded or unbounded support the choice of $\hat{\Omega}$ may be not immediate. In fact, from the knowledge of the sample data or from the previous experience a first choice of $\hat{\Omega}$ is made, then the whole proposed procedure is applied and the approximated pdf $\hat{p}_X(x)$ is evaluated. In order to satisfy the normalization condition the target pdf is zero toward infinity, that is $\lim_{x\to\infty} p_X(x) = 0$ or $\lim_{x\to\pm\infty} p_X(x) = 0$ for random variables with semibounded or unbounded support, respectively. Consequently, the approximated pdf $\hat{p}_X(x)$ at the boundary of $\hat{\Omega}$ has to be less than a suitable chosen tolerance level $\varepsilon > 0$. If it is not the case it is necessary to enlarge the window of analysis $\hat{\Omega}$. Note also that if the window of analysis $\hat{\Omega}$ is not adequate, then the evaluated moments through the approximated pdf $\hat{p}_X(x)$ may differ also in a substantial way from the moments of the target pdf; if it happens, it is necessary also in this case to enlarge $\hat{\Omega}$.

Once that the window of analysis is determined, extensive numerical experimentation has shown that a good choice of the location parameter is represented from an uniform discretization, that is $x_{i+1} = x_i + \Delta x$, $x_i, x_{i+1} \in \hat{\Omega}$, $i = 1, 2, \ldots, N-1$, while a good choice of the bandwidth is represented from $h = b\Delta x$, with $b = 2/3$. This choice of b from one side allows the mutual contribute of adjacent kernel pdfs, from the other side for $N \to \infty$, $\Delta x \to 0$ and also $h \to 0$, then the kernel densities converge toward Dirac's delta functions, which allow to represent any function; note that when $h \to 0$ the distinction between the different types of kernel pdfs is practically lost.

In (106) the number of moments M gives the *data information*, while the number of KDFs gives the *resolution* used for the recovery of the target pdf $p_X(x)$. Obviously, if the number of moments M increases we have more information, and then the approximation should be better; unfortunately the moment problem is strongly ill-conditioned, therefore it is not an easy task to determine good approximations of the target pdf when also the higher moments are known. In literature different solutions have been proposed for the numerical solution of the moment problem, each one with its advantages and drawbacks. In any case, in the following it will be supposed that only the lower moments ($M \leq 6$) are known with enough accuracy and therefore the question does not arise.

On the other hand, if the number N of kernel pdfs increases a better resolution is obtained, in fact: (i) if N increases , the size step Δx decreases and therefore also the bandwidth $h = b\Delta x$ decreases, and the kernel pdfs approach Dirac's delta functions which can approximate the target pdf as closely as desired; (ii) the ill-conditioning of the moment problem is due mainly to the number M of the moments, and it is slightly reduced when the number N of the kernel pdfs increases. However an excessive value of the number N of kernel pdfs increases the computational complexity of the procedure, such to be excessive if compared with the obtained improvement. Note also that eq.(106) in full rigor is valid only when $\hat{p}_X(x) = p_X(x)$, and this relationship is verified for values N not too small. Extensive numerical experimentation has shown that generally a good choice is represented by $N = 20 \div 100$.

It is of interest to observe that the eq.(106) may be written also as

$$\mu_1^K p_1 + \mu_2^K p_2 + \ldots + \mu_N^K p_N = \mu, \quad p_i \geq 0, \quad i = 1, 2, \ldots, N, \quad \sum_{i=1}^{N} p_i = 1 \tag{108}$$

that is, the moment vector μ is a vector of order M, and it may be expressed as a linear convex combination of the vectors μ_i^K, $i = 1, 2, \ldots, N$, of order M, which are the columns of the matrix $\mathbf{M}^K$. It is noted that the column vector μ_i^K collects the moments up to M-th order of the i-th kernel density function. Therefore, the proposed kernel density representation defines the "convex hull" of the vectors μ_i^K , as a truncated moment space of order M

($\mu_0 = 1$). In other words, through the KDR the effective moment space $\mathscr{M}^M$ is approximated with a polyhedral surface, obtained as the convex hull of the kernel moments of order M. Note that $\hat{\mathscr{M}}_h^{M,K} \subseteq \lim_{h\to 0} \hat{\mathscr{M}}_h^{M,K} \subseteq \mathscr{M}^M$, and that

$$\lim_{N\to\infty, h\to 0} \hat{\mathscr{M}}_h^{M,K} = \mathscr{M}^M \tag{109}$$

In virtue of the above developed considerations, it is possible to recognize that the KDR converges to any pdf compatible with the sample data only if the assigned moments belonging to the truncated moment space $\mathscr{M}^M$ belong also to the approximated moment space $\hat{\mathscr{M}}_h^{M,K}$. If this is not verified, it means that the discretization made is not adequate and therefore it is necessary refine it or reduce the bandwidth or both.

5.4 Entropy of a random variable

Entropy is the basic concept of information theory (Kapur and Kesavan, 1992). The entropy of a random variable can be interpreted as the degree of information that the observation of the variable gives. The entropy is larger for more "random" and unstructured random variables, while it is small for distributions that are clearly concentrated on certain values, i.e., when the variable is clearly clustered.

At first, let us consider the case of a discrete random variable X with support Ω and probability distribution given by

$$p_i = \Pr[X = x_i], \quad i = 1, 2, \ldots N \tag{110}$$

and whose moments are given by

$$\mu_k = \sum_{i=1}^{N} x_i^k p_i, \quad k = 1, 2, \ldots M \tag{111}$$

with $\mu_0 = \sum_{i=1}^{N} p_i = 1$. The Shannon's entropy functional (Jaynes, 1957; Kapur and Kesavan, 1992) is defined as

$$H(\hat{p}_1, \hat{p}_2, \ldots, \hat{p}_N) = -\sum_{i=1}^{N} \hat{p}_i \ln \hat{p}_i \tag{112}$$

This very well-known definition can be easily generalized for continuous-valued random variables X whose pdf is $p_X(x)$ and with moments

$$\mu_k = \int_{\Omega} x^k p_X(x) dx, \quad k = 1, 2, \ldots M \tag{113}$$

giving rise to the differential entropy

$$H[\hat{p}_X(x)] = -\int_\Omega \hat{p}_X(x)\ln[\hat{p}_X(x)]dx \tag{114}$$

According to Jaynes, the least biased density function compatible with informations is one that maximizes the entropy functional under the constraints imposed by (111) or (113) and the appropriate normalization condition. The corresponding pdf is known as the Maximum Entropy Probability Distribution, and it can be interpreted as the density that is compatible with the measurements and makes the minimum number of assumptions on the data.

A fundamental result of information theory is that a Gaussian variable has the largest entropy among all random variables of equal variance. In fact, this shows that the Gaussian distribution is the "most random" or the least structured of all distributions of equal variance. It is of interest to underline that this result holds also in multidimensional spaces, where it is seen that the Gaussian distribution has the maximum entropy among all the distribution with a given covariance matrix.

5.5 The maximum entropy probability distribution

The aim of this section is the evaluation of the maximum entropy probability distribution of a discrete random variable (110), whose the first moments are assigned through the (111) and the normalization condition $\mu_0 = \sum_{i=1}^N p_i = 1$.

By using the method of Lagrange multipliers, the extended unconstrained Lagrangian functional is expressed as

$$\mathcal{L}_\mathrm{H}(\hat{p}_1, \hat{p}_2, \ldots, \hat{p}_N, \lambda_0, \lambda_1, \ldots, \lambda_M) = -\sum_{i=1}^N \hat{p}_i \ln \hat{p}_i$$
$$-(\lambda_0 - 1)\left(\sum_{i=1}^N \hat{p}_i - \mu_0\right) - \sum_{k=1}^M \lambda_k \left(\sum_{i=1}^N x_i^k \hat{p}_i - \mu_k\right) \tag{115}$$

where $\lambda_0, \lambda_1, \ldots, \lambda_M$ are the Lagrange multipliers. Note that instead of λ_0, $\lambda_0 - 1$ is used as the first Lagrange multiplier, as a matter of mathematical convenience. The stationarity condition of the extended unconstrained functional L_H yields the Maximum Entropy (MaxEnt) distribution:

$$p_i^{ME} = \exp\left(-\lambda_0 - \sum_{k=1}^M \lambda_k x_i^k\right), \qquad i = 1, 2, \ldots, N \tag{116}$$

By substituting (116) into (115) the Lagrange multipliers can be evaluated as solution of a complicated system of non-linear equations.

However, a more efficient numerical procedure is here used. By substituting (116) into (112) one obtains

$$H(\hat{p}_1^{ME}, \hat{p}_2^{ME}, \ldots, \hat{p}_N^{ME}) = -\sum_{i=1}^{N} \hat{p}_i^{ME} \ln \hat{p}_i^{ME} = \sum_{k=0}^{M} \lambda_k \mu_k \tag{117}$$

From the normalization condition, λ_0 can be obtained as a convex function of $\lambda_0, \lambda_1, \ldots, \lambda_M$

$$\lambda_0 = \ln \left[\sum_{i=1}^{N} \exp \left(-\sum_{k=1}^{M} \lambda_k x_i^k \right) \right] \tag{118}$$

and

$$\mu_k = \frac{\sum_{i=1}^{N} x_i^k \exp\left(-\sum_{j=1}^{M} \lambda_j x_i^j\right)}{\sum_{i=1}^{N} \exp\left(-\sum_{j=1}^{M} \lambda_j x_i^j\right)}, \qquad k = 1, 2, \ldots, M \tag{119}$$

By substituting (118) into (117) one obtains

$$H^{ME} = H^{ME}(\lambda_1, \lambda_2, \ldots, \lambda_M) = \ln \left[\sum_{i=1}^{N} \exp \left(-\sum_{k=1}^{M} \lambda_k x_i^k \right) \right] + \sum_{k=1}^{M} \lambda_k \mu_k \tag{120}$$

which, taking into account of (118), is a convex function of $\lambda_1, \lambda_2, \ldots, \lambda_M$ (Kapur and Kesavan, 1992). As a consequence $H^{ME} = H^{ME}(\lambda_1, \lambda_2, \ldots, \lambda_M)$ has an unique minimum, which can be obtained through a standard convex optimization algorithm.

It is also of interest to underline that the conditions of existence of the MaxEnt distribution are identical to the ones of the MP problem. Moreover, it is possible to show that the approximant distributions p_i^{ME}, $(i = 1, 2, \ldots, N)$ presented in (116) converge in entropy to the exact p_i, $(i = 1, 2, \ldots, N)$, presented in (110), in the sense that

$$\lim_{M \to N} H[p_i^{ME}] = H[p_i], \qquad i = 1, 2, \ldots, N \tag{121}$$

Note also that

$$H[p_{i,M=1}^{ME}] \geq H[p_{i,M=2}^{ME}] \geq \ldots \geq H[p_{i,M=N}^{ME}] = H[p_i], \qquad i = 1, 2, \ldots, N \tag{122}$$

that is, adding one more moment does not necessarily imply entropy decreasing.

Finally, here we will refer some basic considerations about the numerical stability. The finite moment problem is, for its nature, an ill-posed problem. However, as previously discussed, here we consider that only a few moments are known (six at most) and therefore the question does not arise. In fact, for $M \leq 6$ the approximant p_i^{ME} and $H[p_i^{ME}]$ can be reliably estimated. In any case, in this paper the MaxEnt distribution is not obtained from the solution of a system of non-linear equations but from minimization of the MaxEnt functional (120). Following this procedure, in presence of meaningful roundoff errors one obtains coefficients p_i^{ME} , $(i = 1, 2, \ldots, N)$ in which all the calculated moments are affected by error and belong to a ball into the truncated moment space $\mathscr{M}^M$ whose center is provided by the assigned moments; as a consequence the calculated coefficients p_i^{ME} , $(i = 1, 2, \ldots, N)$, should be reliable enough also for increasing M values.

5.6 The kernel density maximum entropy (KD-ME) method

Let X be a random variable with support Ω, and let $\mu_1, \mu_2, \ldots, \mu_n$ the first assigned M moments. The task is to recover the unknown pdf of the random variable.

Step 1. First of all, it is necessary to verify that the information data are admissible, that is to verify that the sequence of numbers $\mu_1, \mu_2, \ldots, \mu_n$ is a sequence of moments of some pdf. From theoretical considerations or from the previous experience, it is established if the support Ω of the random variable is bounded, semibounded or unbounded.

Step 2. The Kernel Density Representation (KDR) is adopted, and in this way a discretization of the moment problem is obtained. As basis function the beta pdf, the lognormal pdf or the normal pdf is chosen for bounded, semibounded or unbounded support, respectively. Then the window of analysis $\hat{\Omega} \equiv [a, b]$ is chosen, together with the centers x_i of the kernel pdfs. Once that the number N of the kernel pdfs has been chosen, the step size $\Delta x = x_{i+1} - x_i$ is determined and, consequently, the bandwidth h. Then it is possible to determine the matrix $\mathbf{M}^K = [M_{ki}^K]$ and the vector μ.

Step 3. The (106) is an underdetermined system of linear equations in the unknowns p_i, so to obtain a solution it is necessary to choose a measure and to determine its minimum (or maximum). It is possible to see that the (106) represents a discrete moment problem and then a natural choice is represented from the maximization of the Discrete Entropy (DE) functional subjected to the constraints (106):

$$\mathbf{p}^{\star} = \arg\max \left\{ H[\mathbf{p}] \; s.t.{:}\mathbf{1}^T\mathbf{p} = 1, \; \mathbf{M}^K\mathbf{p} = \mu \right\} \tag{123}$$

which can be solved through a standard convex optimization algorithm. When $N \rightarrow \infty$ the DE functional converges toward the Continuous Entropy functional and therefore the obtained solution has the desired properties given from the MaxEnt principle; however the solution of (123) is obtained simply as the minimum of a convex functional without requiring the determination of complicated multidimensional integrals.

It is noted that if the optimization problem (123) does not converge to any solution, then it is necessary to increase the number N of kernels or reduce the bandwidth h or both.

Once that the coefficients p_i have been determined, it is easy to represent the target pdf $p_X(x)$ through $\hat{p}_X(x)$ by using the (104). Note that only when $M \rightarrow \infty$ the KD-ME pdf converges toward the target pdf $p_X(x)$; however, for the properties related to the MaxEnt principle only a few moments usually are enough. Moreover, it has been noted that the adoption of the moments as input data in (123) gives robustness to the non-parametric formulation (104), in the sense that $\hat{p}_X(x)$ is not wildly dependent upon the sample data.

The extension to the multivariate case is straightforward from a theoretical point of view, unfortunately the number n of discretization points grows in an exponential way with the number of dimensions. Indeed if we want to discretize the n joint random variables $X_1, X_2, \ldots, X_n$ with $N_1, N_2, \ldots, N_n$ points, respectively, then the approximation of the joint PDF requires a total amount of $N = N_1 \cdot N_2 \cdot \ldots \cdot N_n$ discretization points. Note in particular that if $N_1 = N_2 = \ldots = N_n = N_0$, then $N = N_0^n$. This shortcoming can be effectively solved by using the method of the Independent Components (IC), that we will describe in the following section.

5.7 The method of the Independent Components (IC)

Let us consider a set of n dependent and usually non Gaussian random variables $X_1, X_2, \ldots, X_n$, whose joint pdf is $p_X(x)$. Suppose now that they can be generated as a linear combination of n unknown independent and usually non Gaussian random variables $S_1, S_2, \ldots, S_n$

$$\begin{cases} X_1 = a_{11}S_1 + a_{12}S_2 + \ldots + a_{1n}S_n \\ X_2 = a_{21}S_1 + a_{22}S_2 + \ldots + a_{2n}S_n \\ \vdots \\ X_n = a_{n1}S_1 + a_{n2}S_2 + \ldots + a_{nn}S_n \end{cases} \tag{124}$$

It is often more convenient to use the compact matrix form

$$\mathbf{X} = \mathbf{AS} = \sum_{j=1}^{n} \mathbf{a}_j S_j \tag{125}$$

where $\mathbf{A} = [\ \mathbf{a}_1 \quad \mathbf{a}_2 \quad \ldots \quad \mathbf{a}_2\ ]$ is the mixing matrix of order $n \times n$, while the vectors $\mathbf{a}_j$ are its columns. It is assumed that in the IC model (124) or (125) both the independent components S_j and the mixing matrix $\mathbf{A}$ are unknown, while we have only samples of the vector $\mathbf{X}$. The fundamental hypothesis of Independent Component Analisys (ICA) is that the components $S_1, S_2, \ldots, S_n$ are statistically independent and that at least one of them follows a non Gaussian distribution. It is also assumed that $\mathbf{A}$ is a square matrix, so that an alternative formulation of (125) is

$$\mathbf{Y} = \mathbf{WX} = \sum_{j=1}^{n} \mathbf{w}_j X_j \tag{126}$$

where the random variables $Y_1, Y_2, \ldots, Y_n$ are as independent as possible, while $\mathbf{W} = \mathbf{A}^{-1}$. Through eq.(126) it is possible to find a set of independent components $Y_j \cong S_j$ through the knowledge of the matrix $\mathbf{W}$; therefore, to find a set of independent components satisfying eq.(125) is equivalent to determine the coefficients of $\mathbf{W}$, equal to the inverse of the mixing matrix.

5.8 Ambiguities of ICA

By observing the eq.(125) it is easy to recognize that the IC model has the following ambiguities:

(i) The variances of the independent components cannot be determined. The reason is that in eq.(125) for each contribute of the summation we have the product of two unknown terms, $\mathbf{a}_j$ and S_j, so that if we multiply S_j for any scalar value and we divide $\mathbf{a}_j$ for the same scalar, the contribute $\mathbf{a}_j S_j$ remains unchanged. Since S_j are random variables the most natural choice is to assume that the independent random variables S_j have zero mean and unit variance, that is $E[S_j] = 0$ and $E[S_j^2] = 1$. It is noted that in this way we have only an indetermination over the sign of S_j, fortunately this ambiguity is not significant in most applications.

(ii) The order of the independent components cannot be determined. This happens because we can change the order of the terms $\mathbf{a}_j S_j$ of the summation in eq.(125), and call any of the independent components as the first one.

5.9 Remarks about indipendence

The fundamental hypothesis of ICA is that a set of dependent non Gaussian random variables X_j may be expressed as a linear combination of independent ones S_j (or the equivalent Y_j) with zero mean and unit variance. It is here recalled that two variables are independent if information on the value of the first variable does not give any information on the second, and viceversa. In formulas one has

$$p_{Y_i|Y_1,Y_2,\ldots,Y_{i-1},Y_{i+1},\ldots,Y_n}(y_i \mid y_1, y_2, \ldots, y_{i-1}, y_{i+1}, \ldots, y_n) = p_{Y_i}(y_i) \quad (127)$$

where $p_{Y_i|Y_1,Y_2,\ldots,Y_{i-1},Y_{i+1},\ldots,Y_n}(\cdot)$ is the pdf of the variable Y_i conditioned on the remaining variables $Y_1, Y_2, \ldots, Y_{i-1}, Y_{i+1}, \ldots, Y_n$. It is known that the joint pdf of n variables is gives as

$$\begin{aligned} p_{\mathbf{Y}}(\mathbf{y}) &= p_{Y_1,Y_2,\ldots,Y_n}(y_1, y_2, \ldots, y_n) = \\ &= p_{Y_1|Y_2,Y_3,\ldots,Y_n}(y_1 \mid y_2, y_3, \ldots, y_n) \cdot \\ &\cdot p_{Y_2|Y_3,Y_4,\ldots,Y_n}(y_2 \mid y_3, y_4, \ldots, y_n) \cdot \ldots \\ &\ldots \cdot p_{Y_{n-1}|Y_n}(y_{n-1} \mid y_n) \cdot p_{Y_n}(y_n) \end{aligned} \quad (128)$$

If the variables are independent, taking account of (127) one has

$$p_{\mathbf{Y}}(\mathbf{y}) = p_{Y_1,Y_2,\ldots,Y_n}(y_1, y_2, \ldots, y_n) = p_{Y_1}(y_1) \cdot p_{Y_2}(y_2) \cdot \ldots \cdot p_{Y_n}(y_n) \quad (129)$$

From (129) it follows an interesting property of independent random variables

$$E[h_1(Y_1) \cdot h_2(Y_2) \cdot \ldots \cdot h_n(Y_n)] = E[h_1(Y_1)] \cdot E[h_2(Y_2)] \cdot \ldots \cdot E[h_n(Y_n)] \quad (130)$$

A weaker form of independence is the uncorrelatedness. Two variables Y_i and Y_j are uncorrelated if their covariance is zero: $E[Y_iY_j] - E[Y_i] \cdot E[Y_j] = 0$.

It is here underlined that the independence implies the uncorrelatedness, as it follows from (130) by setting $h_i(Y_i) = Y_i$ and $h_j(Y_j) = Y_j$. For this reason, very often the ICA methods search uncorrelated estimates of the independent components, since in this way the numerical problem is simplified and the number of free parameters is reduced. However, the converse is not true, that is the uncorrelatedness does not imply the independence.

5.10 Assumpions of ICA

The fundamental hypothesis of ICA are then the following:

(*i*) The independent components (ICs) are random variables with zero mean and unit variance, so that eq.(129) holds, and moreover

$$E[\mathbf{Y}] = \mathbf{0}, \qquad \mathbf{\Sigma_Y} = E[\mathbf{YY}^T] = \mathbf{I}_n \tag{131}$$

where $\mathbf{I}_n$ is an identity matrix of order n.

(*ii*) The mixing matrix $\mathbf{A}$ is square of order n, together with its inverse $\mathbf{W}$.

(*iii*) The observed random variables $\mathbf{X}$ are uncorrelated (but not independent) with zero-mean and unit variance:

$$E[\mathbf{X}] = \mathbf{0}, \qquad \mathbf{\Sigma_X} = E[\mathbf{XX}^T] = \mathbf{I}_n \tag{132}$$

Taking account of the eq.(125) and of the hypotheses (131) and (132) it is possible to see that the covariance matrix of $\mathbf{X}$ is $\mathbf{\Sigma_X} = \mathbf{AA}^T = \mathbf{I}_n$, from which it follows the orthogonality of the mixing matrix

$$\mathbf{WW}^T = \mathbf{I}_n, \qquad \mathbf{W} = \mathbf{A}^{-1} = \mathbf{A}^T \tag{133}$$

As a consequence, the determination of the ICs through eq.(126) requires the evaluation of only $n(n+1)/2$ free parameters.

(*iv*) The independent components must be non Gaussian. This restriction follows from the previous assumptions. Indeed, if the ICs were all Gaussian, using eq.(125) it is seen that $\mathbf{X}$ follows a multivariate standard normal distribution, independently from the mixing matrix $\mathbf{A}$. From this result it follows that the determination of independent directions needs that at least one IC follows a non Gaussian distribution.

5.11 Directions of maximum indipendence

The aim of the ICA is the evaluation of the directions $y_1, y_2, \ldots, y_n$ of maximum independence, which coincide with the directions of maximum non Gaussianity. The Central Limit Theorem states that the distribution of a sum of independent random variables tends toward a Gaussian distribution, under certain conditions. This implies that the sum of two or more independent random variables usually has a distribution that is closer to Gaussian than any of the original ones.

For sake of simplicity, assume that $\mathbf{X}$ is actually defined as a linear combination of n ICs $S_1, S_2, \ldots, S_n$ having the same distribution, i.e. $p_{S_1}(s_1) = p_{S_2}(s_2) = \ldots = p_{S_n}(s_n)$. Through ICA, starting from the data $\mathbf{X}$, we estimate a set of ICs $Y_1, Y_2, \ldots, Y_n$ likely to be close to the target ICs, the main difference being related to the sign ambiguity. According to eq.(126), the i-th IC is $Y_i = \sum_{j=1}^{n} W_{ij} X_j = \mathbf{w}_i^T \mathbf{X}$, being $\mathbf{w}_i$ a vector to be determined. If $\mathbf{w}_i$ were exactly equal to the i-th row of $\mathbf{W} = \mathbf{A}^{-1}$, then we will obtain $Y_i = S_i$. Recalling the eq.(125) and doing the coordinate transformation $\mathbf{r}_i = \mathbf{A}^T \mathbf{w}_i$, it is seen that $Y_i = \mathbf{r}_i^T \mathbf{S} = \sum_{k=1}^{n} r_{i,k} S_k$. Since a sum of two or more independent random variables is more Gaussian than the original variable, Y_i is more Gaussian than any of the S_i and it becomes least Gaussian when it in fact equals one of the S_i . This means that the directions of maximum independence are the directions of maximum non Gaussianity and that we must choose the vectors $\mathbf{w}_i$ such that the non Gaussianity of Y_i is maximized.

5.12 Measures of non Gaussianity: kurtosis and negentropy

The determination of the directions of maximum independence requires a quantitative measure of nongaussianity of the random variables Y_i. The simplest one is given from the kurtosis $\kappa(\cdot)$, which for zero mean and unit variance random variables is defined as

$$\kappa(Y_i) = E[Y_i^4] - 3 \tag{134}$$

and it is strictly related to the fourth moment of Y_i. The kurtosis is zero for Gaussian Y_i, while for non Gaussian random variables can assume positive or negative values.

Typically, the random variables with positive kurtosis (called supergaussian distributions) have "spiky" pdf around the origin and heavy tails, such as the Laplace distribution; conversely, the PDF of the random variables with negative kurtosis (known as subgaussian distributions) are "flat" around the origin and have very low values at the tails, and then

$$\kappa(Y_i) \begin{cases} > 0 & \text{supergaussian pdf} \\ = 0 & \text{gaussian pdf} \\ < 0 & \text{subgaussian pdf} \end{cases} \tag{135}$$

Consequently usually as a measure of non Gaussianity it is used not the kurtosis, but its absolute value or its square. The kurtosis has been widely used in ICA for its simplicity computational and theoretical. However, it can be very sensitive to outliers and its value can depend strongly on only a few data in the tails of the distribution, and that can be erroneous or irrelevant. In other words, kurtosis is not a robust measure of nongaussianity, when it has to be estimated from a measured sample.

A second very important measure of non Gaussianty is given by the negentropy $J(Y_i)$ (or negative entropy) of the random variable Y_i, defined as

$$J(Y_i) = H(Y_i^G) - H(Y_i) \geq 0 \tag{136}$$

where $H(\cdot)$ is the entropy operator and Y_i^G is a Gaussian random variable having the same variance of Y_i. Recalling that the Gaussian variable has the largest entropy among all random variables of equal variance, the negentropy $J(\cdot)$ is always non-negative and it is equal to zero only if Y_i follows a Gaussian distribution

$$J(Y_i) \begin{cases} = 0 & \text{gaussian pdf} \\ > 0 & \text{nongaussian pdf} \end{cases} \tag{137}$$

Moreover it is invariant with respect to an invertible linear transformation. The advantage of using the negentropy is the statistical robustness, so that it may be considered an optimal estimator as a measure of nongaussianity. From the other side, its adoption is computationally demanding, and usually simpler approximations are used. For further details about this topic, the interested reader is referred to Hyvarinen et al. (2001).

5.13 The KD-MEM applied to independent components

Let $\mathbf{X} = [\ X_1 \quad X_2 \quad \dots \quad X_n\ ]^T$ be a vector of random variables whose support Ω is defined in the n-dimensional space. The task is the evaluation of the joint pdf of $\mathbf{X}$.

Step 1. First of all, generate a relatively reduced number N_s of samples, say $1,000$. Then we can apply the ICA as described in the previous sections and evaluate the independent components $Y_1, Y_2, \dots, Y_n$ through the eq.(126).

Step 2. Let us apply the univariate KD-MEM as described previously to each IC Y_i. It is noted that it requires N_i discretization points, and that the definition of the joint pdf of the n-vector $\mathbf{Y} = [\ Y_1 \quad Y_2 \quad \dots \quad Y_n\ ]^T$

needs only $N = N_1 + N_2 + \ldots + N_n$ basis pdfs; if in particular $N_1 = N_2 = \ldots = N_n = N_0$, then $N = nN_0 \ll N_0^n$.

Step 3. Recalling that $\mathbf{A} = \mathbf{W}^{-1}$ is an orthogonal matrix whose determinant is equal to one, and that $\mathbf{X} = \mathbf{AY}$, the determination of the joint pdf of $\mathbf{X}$ is simply given as

$$p_{\mathbf{X}}(\mathbf{x}) = p_{\mathbf{Y}}(\mathbf{Wx}) = p_{Y_1}\left(\sum_{k=1}^{n} W_{1k}x_k\right) p_{Y_2}\left(\sum_{k=1}^{n} W_{2k}x_k\right) \ldots p_{Y_n}\left(\sum_{k=1}^{n} W_{nk}x_k\right) \tag{138}$$

5.14 An example: the Bouc-Wen nonlinear oscillator

As an example, consider a Bouc-Wen hysteretic oscillator defined by the following differential equations

$$\begin{aligned} m\ddot{X} + c\dot{X} + k[\alpha X + (1-\alpha)Z] = F(t) \\ \dot{Z} = -\delta\left|\dot{X}\right| |Z|^{r-1} Z - \gamma |Z|^r \dot{X} + A\dot{X} \end{aligned} \tag{139}$$

where m, c and k are the mass, damping and stiffness coefficients, α controls the degree of hysteresis, and r, A, γ and δ are parameters defining the shape of the hysteresis loop. The excitation is defined as $F(t) = -ma_g(t)$, where $a_g(t) = W(t)$ denotes the base acceleration modeled as a white-noise process of intensity $S_W = 0.0377m^2/s^3$. Let $\sigma_0 = m^2\pi S_W/(ck)$ the mean-square response of the linear system, obtained setting $\alpha = 1$. The natural frequency of the system is $\omega_0 = \sqrt{k/m} = 10rad/s$, while its damping ratio is $\varsigma_0 = c/(2m\omega_0) = 0.05$; in figure 10 we represented a sample of the hysteresis loop whose parameters are $\alpha = 0.05$, $n = A = 1$, $\gamma = \delta = 1/(2\sigma_0)$.

If we are interested to the evaluation of only a marginal pdf the univariate KD-MEM can be successfully applied. In Figure 11 we represented the pdfs of the displacement X, velocity $\dot{X}$ and the hysteretic variable Z, and we compared the solution obtained by using a crude MCS with 100,000 samples (in black) and the one given by the KD-MEM (in blue) obtained considering up to first $M = 6$ moments and using $N = 51$ kernel pdfs. The moments of the KD-MEM have been evaluated considering only 500 samples, therefore it is expected that the fourth and the sixth moment are not well estimated, however the KD-MEM pdf fits very well the target pdf.

If our aim is the evaluation of a joint pdf of two or more variables, it is necessary to use the ICA in conjunction with KD-MEM. In Figure 12 we represented the three independent components Y_1, Y_2 and Y_3 with zero mean and unit variance; it is possible to observe that the ordering of the

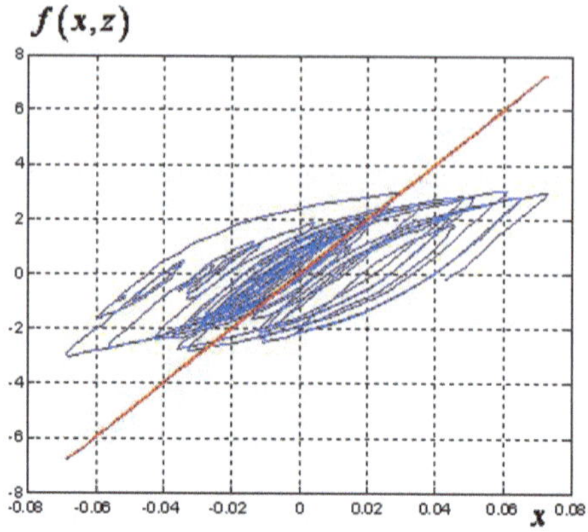

Figure 10. A sample of the hysteresis loop of the Bouc-Wen oscillator.

ICs does not correspond to the original random variables X, $\dot{X}$ and Z. As before, in blue we represented the KD-ME pdfs, while in black the target pdfs are shown, given by a crude MCS with 100,000 samples.

Starting from the knowledge of the pdfs of the ICs, using the eq.(138) it is possible to determine the joint pdfs between X, $\dot{X}$ and Z . For example purpose, in Figures 13 and 14 we represented $p_{X\dot{X}}(x,\dot{x})$ and $p_{XZ}(x,z)$.

6 Conclusions

In this work we have presented the most used methods in stochastic dynamics, analyzing in detail the Moment Equation Method, the Stochastic Linearization and the Monte Carlo Simulation. In their classical formulation, these methods suffer of known computational shortcomings, which are described along the paper. Then, we have presented some recent developments which reduce the computatonal burden and guarantee an improvement of the obtained approximation.

Bibliography

M. Alaoui-Ismaili and P. Bernard. Asymptotic analysis and linearization of the randomly perturbed two-wells duffing oscillator. *Prob. Eng. Mech.*, 12(3):171–8, 1997.

U. Alibrandi and G. Ricciardi. Efficient evaluation of the pdf of a random variable through the kernel density maximum entropy approach. *Int. J.*

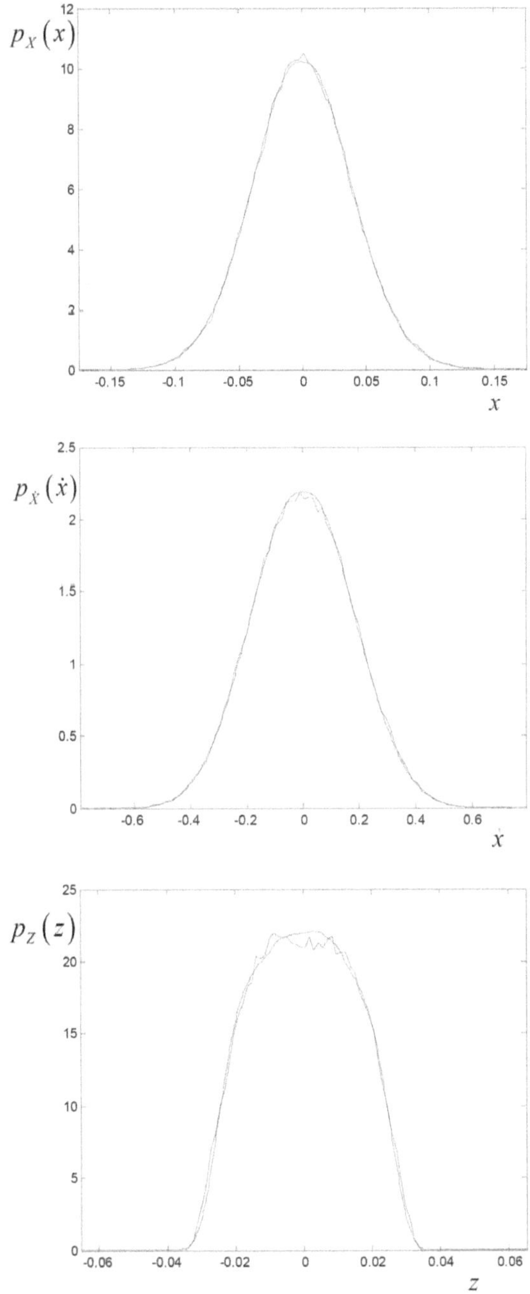

ll

Figure 11. Pdfs of the displacement X, velocity $\dot{X}$ and the hysteretic variable Z of the Bouc-Wen oscillator.

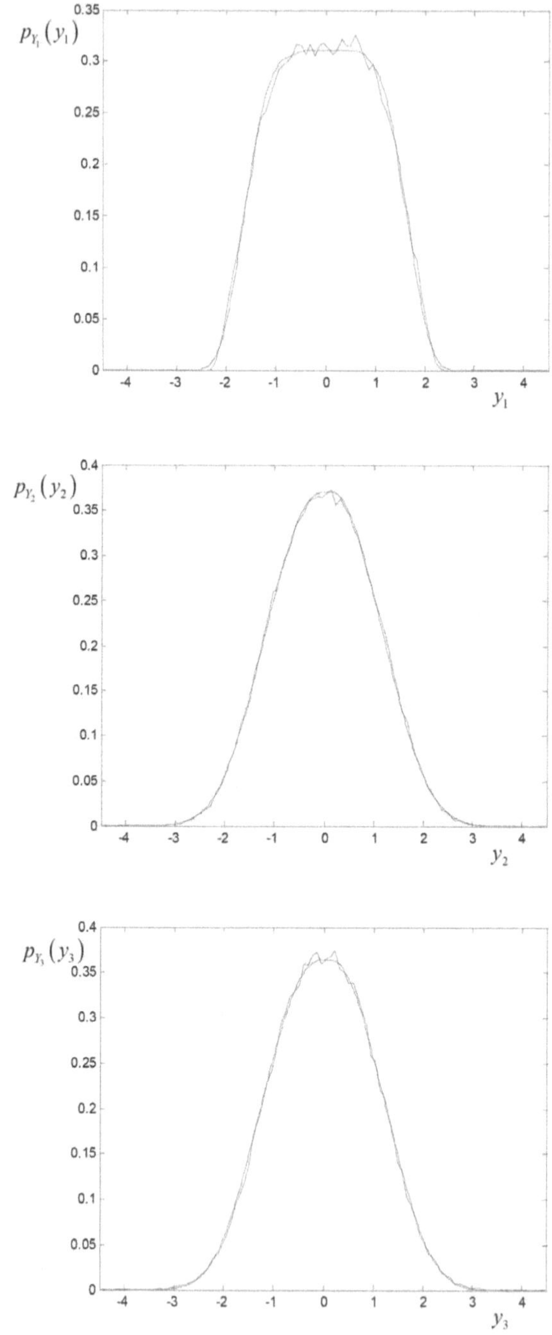

Figure 12. Pdfs of the ICs Y_1, Y_2 and Y_3 for the Bouc-Wen oscillator.

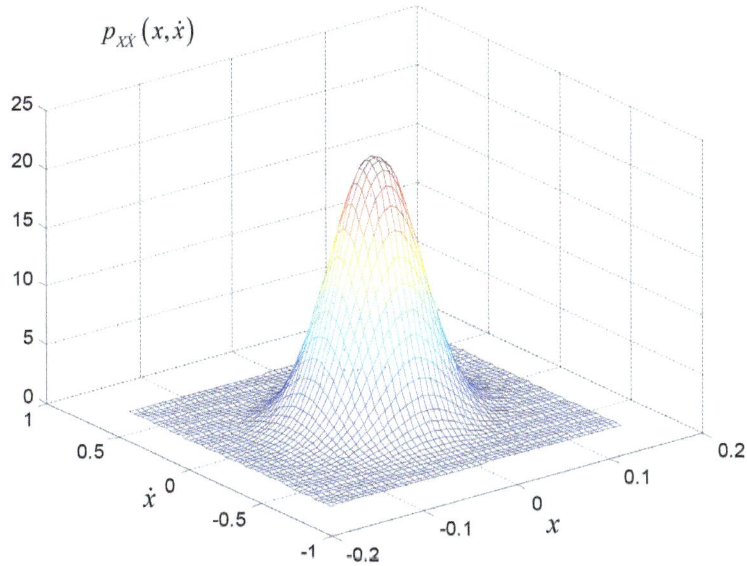

$p_{X\dot{X}}(x,\dot{x})$
25
20
15
10
5
0
1
0.5
$\dot{x}$ 0
-0.5
-1
-0.2
-0.1
0
0.1
0.2
x

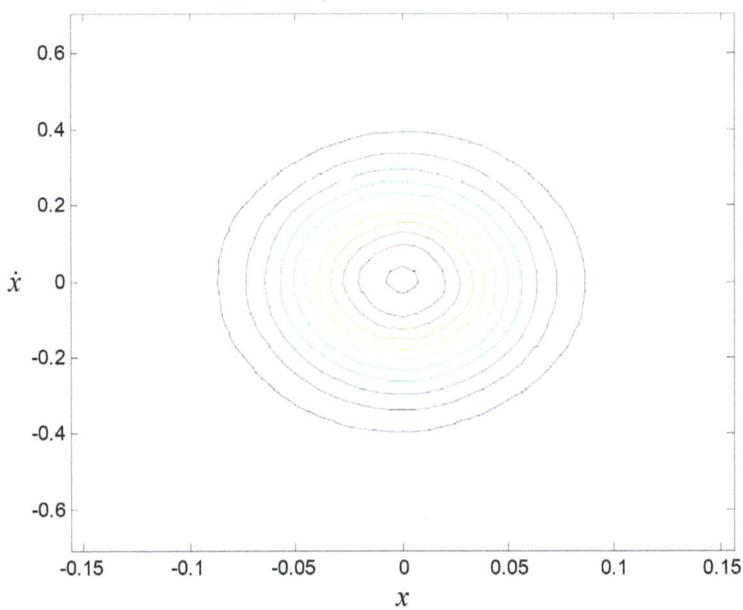

0.6
0.4
0.2
$\dot{x}$ 0
-0.2
-0.4
-0.6
-0.15
-0.1
-0.05
0
0.05
0.1
0.15
x

Figure 13.

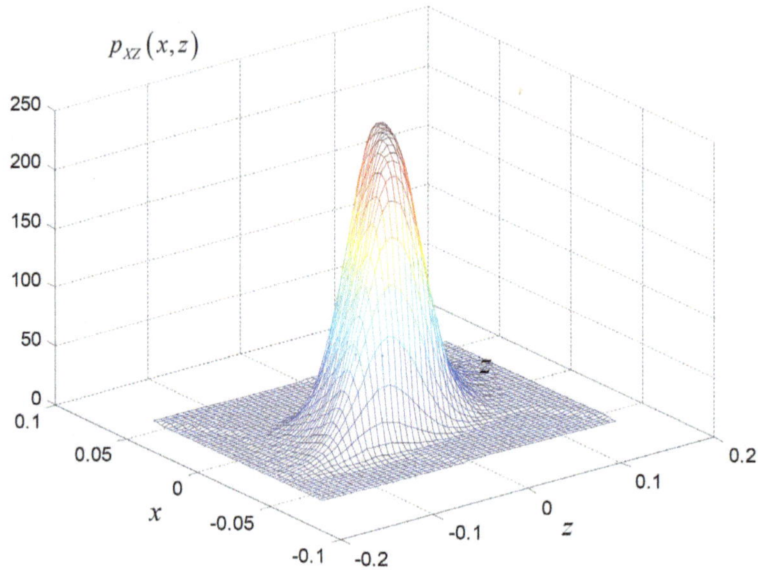
$p_{XZ}(x,z)$
250
200
150
100
50
0
0.1
0.05
0
-0.05
-0.1
x
-0.2
-0.1
0
0.1
0.2
z

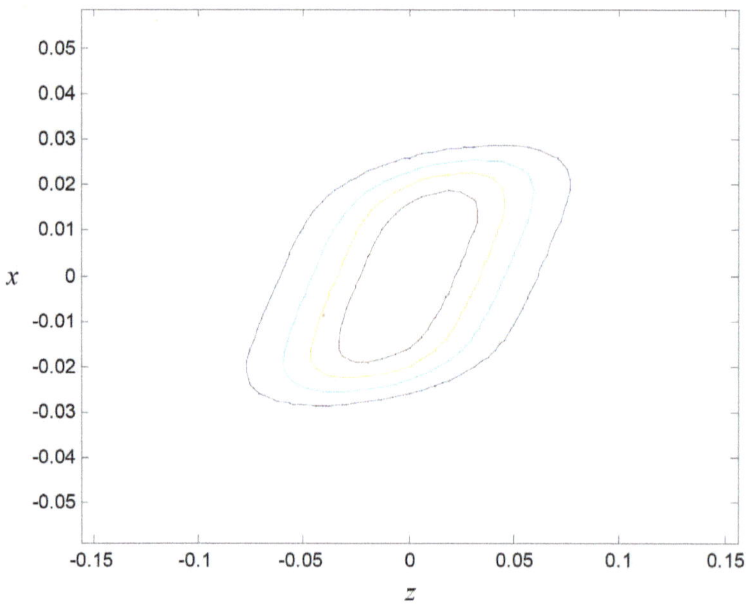
0.05
0.04
0.03
0.02
0.01
0
-0.01
-0.02
-0.03
-0.04
-0.05
x
-0.15
-0.1
-0.05
0
0.05
0.1
0.15
z

Figure 14.

Numer. Meth. Engng., 75:1511–48, 2008.

G. A. Athanassoulis and P. N. Gavriliadis. The truncated hausdorff moment problem solved by using kernel density functions. *Probabilistic Engineering Mechanics*, 17:273–91, 2002.

G. A. Athanassoulis and P. N. Gavriliadis. Moment data can be analytically completed. *Probabilistic Engineering Mechanics*, 18:329–38, 2003.

S. K. Au and J. L. Beck. First excursion probabilities for linear systems by very efficient importance sampling. *Probab. Engng. Mech.*, 6:193–207, 2001.

J. J. Beaman and J. K. Hedrick. Improved statistical linearization for analysis and control of nonlinear stochastic systems. part i: An extended statistical linearization technique. *Trans ASME, J. Dyn. Syst. Meas. Control.*, 103:14–21, 1981.

R. Bellman and J. M. Richardson. Closure and preservation of moment properties. *J. Math. Anal. Appl.*, 23:639–44, 1968.

R. C. Booton. The analysis of nonlinear control systems with random inputs. *Proceedings of the Symposium on Nonlinear Circuit Analysis,* Polytechnic Institute Brooklyn, New York, 2, 1953.

D. C. C. Bover. Moment equation methods for nonlinear stochastic system. *J. Math. Anal. Appl.*, 65:306–20, 1978.

T. K. Caughey. Response of a nonlinear string to random loading. *Trans ASME, Journal of Applied Mechanics*, 26:341–343, 1959.

R. J. Chang. Non-gaussian linearization method for stochastic parametrically and externally excited nonlinear systems. *Trans ASME, J. Dyn. Syst. Meas. Control.*, 114:20–26, 1992.

R. W. Clough and J. Penzien. *Dynamics of structures.* McGraw-Hill, 1975.

S. Crandall. Non-gaussian closure for random vibration of non-linear oscillators. *Int. J. Non-linear Mech.*, 15:303–13, 1980.

S. H. Crandall. On using non-gaussian distributions to perform statistical linearization. *Int. J. Non-Linear Mechs.*, 39:1395–1406, 2004.

M. DiPaola and G. Falsone. Stochastic dynamics of non-linear systems driven by non-normal delta correlated process. *J. Appl. Mech. ASME*, 60:1269–83, 1993.

I. Elishakoff. Stochastic linearization technique: a new interpretation and a selective review. *Shock Vibr. Dig.*, 32:179–188, 2000.

G. Falsone and G. Ricciardi. Stochastic linearization: classical approach and new developments. In A. Luongo, editor, *Recent research developments in structural dynamics*, pages 81–106. Transworld Research Network Publisher, India, 2003.

M. Grigoriu. White noise processes. *J. Eng. Mech. ASCE*, 113(5):757–65, 1987.

M. Grigoriu. Simulation of nonstationary gaussian processes by random trigonometric polynomials. *ASCE J. Engng. Mech.*, 119(2):328–43, 1993.

M. Grigoriu. Simulation of stationary non-gaussian translation processes. *ASCE J. Engng. Mech.*, 124(2):121–26, 1998.

M. Grigoriu. Moment closure by monte carlo simulation and moment sensitivity factors. *Int. J. Nonlinear Mechs.*, 34:739–48, 1999.

N. Harnpornchai, H. J. Pradlwarter, and G. I. Schueller. Stochastic analysis of dynamical systems by phase-space-controlled monte carlo simulation. *Comput. Meth. App.l Mech. Engng.*, 168:273–83, 1999.

C. S. Hsu. *Cell-to-cell mapping: a method of global analysis for nonlinear systems.* Springer, New York, 1987.

J. E. Hurtado and A. H. Barbat. Improved stochastic linearization method using mixed distribution. *Structural Safety*, 18(1):49–62, 1996.

A. Hyvarinen, J. Karhunen, and E. Oja. *Independent Component Analysis.* John Wiley and Sons, 2001.

R. A. Ibrahim. *Parametric random vibrations.* John Wiley and Sons, New York, 1985.

R. A. Ibrahim, A. Soundararajan, and H. Heo. Stochastic response of nonlinear dynamic systems based on a non-gaussian closure. *J. App.l Mech. ASME*, 52(4):965–70, 1985.

E. T. Jaynes. Information theory and statistical mechanics. *Physical Review*, 106:620–30, 1957.

K. Kanai. Semi-empirical formula for seismic characteristics of ground. *Univ. Tokio Bull. Earthquake Res. Inst.*, 35:309–25, 1957.

J. N. Kapur and H. K. Kesavan. *Entropy Optimization Principles with Applications.* Academic Press, 1992.

I. E. Kazakov. Approximate probabilistic analysis of the accuracy of operation of essentially nonlinear systems. *Automat Remote Control*, 17: 423–50, 1956.

F. Kozin. The method of statistical linearization for non-linear stochastic vibrations. In *IUTAM Symposium on Nonlinear Stochastic Dynamic Systems,* F. Ziegler and G. I. Schueller eds., Springer, Berlin, pages 45–66, 1988.

J. Lee. Improving the equivalent linearization technique for stochastic duffing oscillators. *Journal of Sound and Vibration*, 186(5):846–55, 1995.

Y. K. Lin. *Probabilistic Theory of Structural Dynamics.* McGraw-Hill, 1967.

R. C. Micaletti, A. S. Cakmak, S. R. K. Nielsen, and H. U. Koyluoglu. Error analysis of statistical linearization with gaussian closure for large-degree-of-freedom systems. *Prob. Eng. Mech.*, 13:77–84, 1998.

G. Muscolino. Response of non-linear structural systems under gaussian or non-gaussian filtered input. In F. Casciati, editor, *Dynamic Motion,*

Chaotic and Stochastic Behaviour, pages 203–299. Springer, Wien, 1993.

G. Muscolino, G. Ricciardi, and P. Cacciola. Monte carlo simulation in the stochastic analysis of non-linear systems under external stationary poisson white noise input. *Int. J. Non-Linear Mechs.*, 38:141–48, 2003.

A. Naess and J. M. Johnsen. Response statistics of nonlinear, compliant offshore structures by the path integral solution method. *Probabilistic Engineering Mechanics*, 8(2):91–106, 1993.

H. J. Pradlwarter. Non-gaussian linearization - an efficient tool to analyze nonlinear mdof-systems. *Journal of Nuclear Engineering and Design*, 128: 175–92, 1991.

G. Ricciardi. A non-gaussian stochastic linearization method. *Prob. Eng. Mech.*, 22:1–11, 2007.

G. Ricciardi. Random vibration of beam under moving loads. *J. Eng. Mech. ASCE*, 120(11):2361–80, 1994.

G. Ricciardi and I. Elishakoff. A novel local stochastic linearization method via two extremum entropy principles. *Int. J. Non-Linear Mechs.*, 37: 785–800, 2002.

H. Risken. *The Fokker-Planck equation.* Springer-Verlag, 1989.

J. B. Roberts. First passage time for randomly excited non-linear oscillators. *J. Sound. Vib.*, 109(1):33–50, 1986.

J. B. Roberts and P. D. Spanos. *Random vibration and statistical linearization.* Wiley, Chichester, 1991.

N. G. F. Sancho. On the approximate moment equations of a nonlinear stochastic differential equation. *J. Math. Anal. Appl.*, 29:384–91, 1970a.

N. G. F. Sancho. Technique for finding the moments equations of a nonlinear stochastic system. *J. Math. Phys.*, 11:771–4, 1970b.

G. I. Schueller and P. D. Spanos, editors. *Monte Carlo simulation. Proceedings of the International Conference on Monte Carlo Simulation,* Monte Carlo, 2000. A.A. Balkema, Lisse.

M. Shinozuka. Monte carlo solution of structural dynamics. *Int. J. Comput. Struct.*, 2:855–74, 1972.

M. Shinozuka and Y-K. Wen. Monte carlo solution of nonlinear vibrations. *AIAA J.*, 10(1):37–40, 1972.

L. Socha. *Linearization methods for stochastic dynamic systems.* Springer, Berlin, Heidelberg, New York, 2008.

L. Socha and T. T. Soong. Linearization in analysis of nonlinear stochastic systems. *Applied Mechanics Reviews*, 44:399–422, 1991.

P. D. Spanos and B. A. Zeldin. Monte carlo treatment of random fields: a broad perspective. *Appl. Mech. Rev.*, 51(3):219–37, 1998.

B. F. Jr. Spencer and L. A. Bergman. On the numerical solution of the fokker-planck equation for nonlinear stochastic systems. *Nonlinear Dy-*

namics, 4:357–72, 1993.

L. Sperling. Analyse stochastisch erregter nichtlinearer systeme mittels linearer differential-gleichungen fur verallgemeinerte quasimomentfunktionen. *ZAMM*, 59:169–76, 1979.

R. L. Stratonovich. *Topics in the theory of random noise.* Gordon and Breach, 1967.

H. Tajimi. A statistical method of determining the maximum response of a building structure during an earthquake. In *Proc. II World Conf. Earthquake Eng.*, Vol II, pages 781–98, 1960.

H. Tanaka. Application of an importance sampling method to time dependent reliability analyses using the girsanov transformation. In N. Shiraishi N, M. Shinozuka M, and Y. K. Wen, editors, *Structural safety and reliability*, pages 411–18. Balkema, Rotterdam, 1998.

M. F. Wehner and W. G. Wolfer. Numerical evaluation of path-integral solutions to fokker-planck equations. *Physical Review A*, 27(5):2663–70, 1983.

S.F. Wojtkiewicz, B.F. Jr. Spencer, and L.A. Bergman. On the cumulant-neglect closure methods in stochastic dynamics. *Int. J. Non-Linear Mech.*, 31(5):657–84, 1996.

E. Wong and M. Zakai. On the relation between ordinary and stochastic differential equations. *Int. J. Eng. Sci.*, 3:213–29, 1965.

W. F. Wu and Y. K. Lin. Cumulant-neglect closure for non-linear oscillators under random parametric and external excitations. *Int. J. Non-linear Mech.*, 19(4):349–62, 1984.

Stochastic Models of Uncertainties in Computational Structural Dynamics and Structural Acoustics

Christian Soize*

* Laboratoire Modélisation et Simulation Multi-Echelle (MSME UMR 8208 CNRS), Université Paris-Est, 5 bd Descartes, 77454 Marne-la-Vallée, France

1 Introduction

We present an overview concerning the main concepts, formulations and advances for the stochastic modeling of uncertainties in computational structural dynamics and structural acoustics. The *parametric* probabilistic approach, the *nonparametric* probabilistic approach and the *generalized* probabilistic approach of uncertainties are presented in the context of structural dynamics and in structural acoustics and vibration, including not only the construction of *prior* probability models but also the identification of *posterior* probability models.

2 Short overview on probabilistic modeling of uncertainties and on related topics

2.1 Uncertainty and variability

The *designed system* is used to manufacture the *real system* and to construct the nominal computational model (also called the *mean computational model* or sometimes, the mean model) using a mathematical-mechanical modeling process for which the main objective is the prediction of the responses of the real system in its environment. The real system, submitted to a given environment, can exhibit a variability in its responses due to fluctuations in the manufacturing process and due to small variations of the configuration around a nominal configuration associated with the designed system. The mean computational model which results from a mathematical-mechanical modeling process of the design system, has parameters which can be uncertain. In this case, there are *uncertainties on the computational model parameters.* In an other hand, the modeling process induces some modeling errors defined as the *model uncertainties.* Fig 1 summarizes the two types of uncertainties in a computational model and the variabilities of a real system. It is important to take into account both

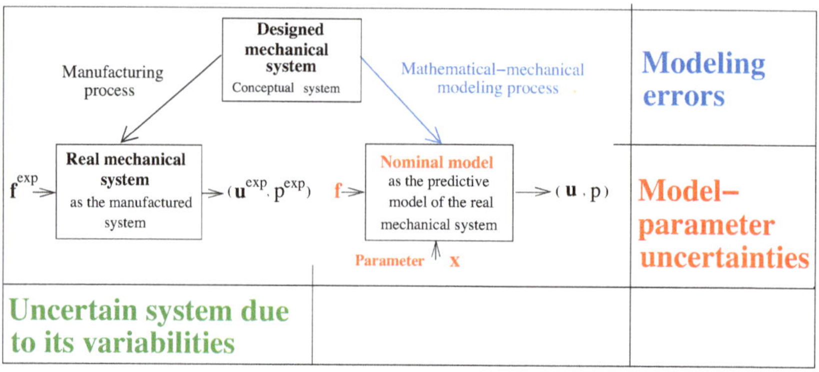

Figure 1. Variabilities and types of uncertainties in computational mechanics.

the uncertainties on the computational model parameters and the model uncertainties to improve the predictions of computational models in order to use such a computational model to carry out robust optimization, robust design and robust updating with respect to uncertainties. Today, it is well understood that, as soon as the probability theory can be used, then the probabilistic approach of uncertainties is certainly the most powerful, efficient and effective tool for modeling and for solving direct and inverse problem. The developments presented in this work are limited to the probabilistic approach.

2.2 Types of approach for stochastic modeling of uncertainties

The *parametric probabilistic approach* consists in modeling the *uncertain parameters of the computational model* by random variables and then in constructing the probabilistic model of these random variables using the available information. Such an approach is very well adapted and very efficient to take into account the uncertainties on the computational model parameters as soon as the probability theory can be used. Many works have been published and a state-of-the-art can be found, for instance, in (Mace et al., 2005; Schueller, 2005, 2007; Deodatis and Spanos, 2008; Schueller and Pradlwarter, 2009).

Concerning *model uncertainties* induced by *modeling errors*, it is today well understood that the prior and posterior probability models of the uncertain parameters of the computational model are not sufficient and do not have the capability to take into account model uncertainties in the context of computational mechanics as explained, for instance, in (Beck and Katafygiotis, 1998) and in (Soize, 2000, 2001, 2005b). Two main methods can be used to take into account model uncertainties (modeling errors).

(i) The first one consists in introducing a probabilistic model of the *output-prediction-error* which is the difference between the real system output and the computational model output. If there are no experimental data, then this method cannot really be used because there are generally no information concerning the probability model of the noise which is added to the computational model output. If experiments are available, the observed prediction error is then the difference between the measured real system output and the computational model output. A posterior probability model can then be constructed (see, for instance, (Beck and Katafygiotis, 1998; Beck and Au, 2002)) using the Bayesian method (Spall, 2003; Kaipio and Somersalo, 2005; Carlin and Louis, 2009). Such an approach is efficient but requires experimental data. It should be noted that a lot of experimental data are required in high dimension. However, with such an approach, the posterior probability model of the uncertain parameters of the computational model strongly depends on the probability model of the noise which is added to the model output and which is often unknown. In addition, for many problems, it can be necessary to take into account the modeling errors at the operators level of the mean computational model. For instance, such an approach seems to be necessary to take into account the modeling errors on the mass and the stiffness operators of a computational dynamical model in order to analyze the generalized eigenvalue problem. It is also the case for the robust design optimization performed with an uncertain computational model for which the design parameters of the computational model are not fixed but run through an admissible set of values.

(ii) The second one is based on the *nonparametric probabilistic approach* of model uncertainties induced by modeling errors, which has been proposed in (Soize, 2000) as an alternative method to the output-prediction-error method in order to take into account modeling errors at the operators level by introducing random operators and not at the model output level by introducing an additive noise. It should be noted that such an approach allows a prior probability model of model uncertainties to be constructed even if no experimental data are available. The nonparametric probabilistic approach is based on the use of a reduced-order model and the random matrix theory. It consists in directly constructing the stochastic modeling of the operators of the mean computational model. The random matrix theory (Mehta, 1991) and its developments in the context of dynamics, vibration and acoustics (Soize, 2000, 2001, 2005b, 2010b; Wright and Weaver, 2010) is used to construct the prior probability distribution of the random matrices modeling the uncertain operators of the mean computational model. This prior probability distribution is constructed by using the Maximum Entropy Principle (Jaynes, 1957), in the context of Information Theory (Shannon, 1948), for which the constraints are defined by the available information (Soize, 2000, 2001, 2003a, 2005a,b, 2010b). Since the paper (Soize, 2000), many works have been published in order: (a) to validate the nonparametric probabilistic approach of both the computational-model-parameter uncertainties and the model uncertainties induced by modeling errors, with experimental results (Chebli and Soize, 2004; Soize, 2005b; Chen et al., 2006; Duchereau and Soize, 2006; Soize et al., 2008a; Durand et al., 2008; Fernandez et al., 2009, 2010), (b) to extend the applicability of the theory to other areas (Soize, 2003b; Soize and Chebli, 2003; Capiez-Lernout and Soize, 2004; Desceliers et al., 2004; Capiez-Lernout et al., 2005; Cottereau et al., 2007; Soize, 2008b; Das and Ghanem, 2009; Kassem et al., 2009), (c) to extend the theory to new ensembles of positive-definite random matrices yielding a more flexible description of the dispersion levels (Mignolet and Soize, 2008a), (d) to apply the theory for the analysis of complex dynamical systems in the medium-frequency range, including structural-acoustic systems, (Ghanem and Sarkar, 2003; Soize, 2003b; Chebli and Soize, 2004; Capiez-Lernout et al., 2006; Duchereau and Soize, 2006; Arnst et al., 2006; Cottereau et al., 2008; Durand et al., 2008; Pellissetti et al., 2008; Desceliers et al., 2009; Fernandez et al., 2009, 2010; Kassem et al., 2011), and (e) to analyze nonlinear dynamical systems (i) with local nonlinear elements (Desceliers et al., 2004; Sampaio and Soize, 2007a,b; Batou and Soize, 2009a,b; Ritto et al., 2009, 2010; Wang et al., 2010) and (ii) with nonlinear geometrical effects (Mignolet and Soize, 2008b).

Concerning the coupling of the parametric probabilistic approach of uncertain computational model parameters, with the nonparametric probabilistic approach of model uncertainties induced by modeling errors, a methodology has been proposed in (Soize, 2010a; Batou et al., 2011). This *generalized probabilistic approach* of uncertainties in computational dynamics uses the random matrix theory. The proposed approach allows the prior probability model of each type of uncertainties (uncertainties on the computational model parameters and model uncertainties) to be separately constructed and identified. The modeling errors are not taken into account with the usual output-prediction-error method but with the nonparametric probabilistic approach.

2.3 Types of representation for stochastic modeling of uncertainties

A fundamental question is related to the construction of the prior probability model of uncertain computational model parameters but also to the construction of the prior probability model of model uncertainties induced by modeling errors. Such a prior probability model can then be used to study the propagation of uncertainties through the mechanical system which is analyzed. If experimental data are available for outputs of the mechanical system, then these experimental data can be used (1) to identify the parameters of the prior probability model using, for instance, the maximum likelihood method (Serfling, 1980; Spall, 2003) or (2) to construct a posterior probability model using the Bayesian method (Spall, 2003; Carlin and Louis, 2009).

Two main methods are available to construct the prior probability model of a random vector $\theta \mapsto \mathbf{X}(\theta) = (X_1(\theta), \ldots, X_{n_p}(\theta))$ defined on a probability space $(\Theta, \mathcal{T}, \mathcal{P})$, with values in $\mathbb{R}^{n_p}$, and whose probability distribution on $\mathbb{R}^{n_p}$ is denoted by $P_{\mathbf{X}}(d\mathbf{x})$ (this random vector can be the approximation in finite dimension of a stochastic process or a random field). It should be noted that these two methods can also be used for random matrices, random fields, etc. We introduce the space $\mathcal{L}^2_{n_p}$ of all the second-order random variables $\mathbf{X}$ such that

$$E\{\|\mathbf{X}\|^2\} = \int_{\mathbb{R}^{n_p}} \|\mathbf{x}\|^2 \, P_{\mathbf{X}}(d\mathbf{x}) < +\infty \quad ,$$

in which E is the mathematical expectation, where $\mathbf{x} = (x_1, \ldots, x_{n_p})$ and $d\mathbf{x} = (dx_1 \ldots dx_{n_p})$, and where $\|\mathbf{x}\|^2 = x_1^2 + \ldots + x_{n_p}^2$ is the square of the Euclidean norm of vector $\mathbf{x}$.

(i)- The first method is a direct approach which consists in directly constructing the probability distribution $P_{\mathbf{X}}(d\mathbf{x})$ on $\mathbb{R}^{n_p}$ in using, for instance, the maximum entropy principle (see Section 2.4).

(ii)- The second one is an indirect approach which consists in introducing a representation $\mathbf{X} = \mathbf{h}(\mathbf{\Xi})$ for which $\mathbf{X}$ is the transformation by a deterministic nonlinear (measurable) mapping $\mathbf{h}$ (which has to be constructed) of

a $\mathbb{R}^{N_g}$-valued random variable $\boldsymbol{\Xi} = (\Xi_1, \ldots, \Xi_{N_g})$ whose probability distribution $P_{\boldsymbol{\Xi}}(d\boldsymbol{\xi})$ is given and then is known. Then $P_{\mathbf{X}}$ is the transformation of $P_{\boldsymbol{\Xi}}$ by mapping $\mathbf{h}$. Two main types of methods can be used.

(ii.1) - The first one corresponds to the Polynomial Chaos representation (Ghanem and Spanos, 1990, 1991, 2003) based on the Wiener mathematical works (Wiener, 1938). The Polynomial Chaos representation of random variable $\mathbf{X}$ in $\mathcal{L}^2_{n_p}$ allows the effective construction of mapping $\mathbf{h}$ to be carried out as

$$\mathbf{X} = \Sigma_{j_1=0}^{+\infty} \ldots \Sigma_{j_{N_g}=0}^{+\infty} \mathbf{a}_{j_1,\ldots,j_{N_g}} \psi_{j_1}(\Xi_1) \times \ldots \times \psi_{j_{N_g}}(\Xi_{N_g}) \quad ,$$

in which ψ_{j_k} are given real polynomials related to the probability distribution of $\boldsymbol{\Xi}$ and where $\mathbf{a}_{j_1,\ldots,j_{N_g}}$ are deterministic coefficients in $\mathbb{R}^{n_p}$. Introducing the multi-index $\boldsymbol{\alpha} = (j_1, \ldots, j_{N_g})$ and the multi-dimensional polynomials $\psi_{\boldsymbol{\alpha}}(\boldsymbol{\Xi}) = \psi_{j_1}(\Xi_1) \times \ldots \times \psi_{j_{N_g}}(\Xi_{N_g})$ of random vector $\boldsymbol{\Xi}$, the previous polynomial chaos decomposition can be rewritten as

$$\mathbf{X} = \Sigma_{\boldsymbol{\alpha}}\, \mathbf{a}_{\boldsymbol{\alpha}}\, \psi_{\boldsymbol{\alpha}}(\boldsymbol{\Xi}) \quad .$$

The multi-dimensional polynomials $\psi_{\boldsymbol{\alpha}}(\boldsymbol{\Xi})$ are orthogonal and normalized,

$$E\{\psi_{\boldsymbol{\alpha}}(\boldsymbol{\Xi})\, \psi_{\boldsymbol{\beta}}(\boldsymbol{\Xi})\} = \int_{\mathbb{R}^{N_g}} \psi_{\boldsymbol{\alpha}}(\boldsymbol{\xi})\, \psi_{\boldsymbol{\beta}}(\boldsymbol{\xi})\, P_{\boldsymbol{\Xi}}(d\boldsymbol{\xi}) = \delta_{\boldsymbol{\alpha}\boldsymbol{\beta}} \quad ,$$

in which $\delta_{\boldsymbol{\alpha}\boldsymbol{\alpha}} = 1$ and $\delta_{\boldsymbol{\alpha}\boldsymbol{\beta}} = 0$. The coefficients $\mathbf{a}_{\boldsymbol{\alpha}} = \mathbf{a}_{j_1,\ldots,j_{N_g}}$ are vectors in $\mathbb{R}^{n_p}$ which completely define mapping $\mathbf{h}$ and which are given by

$$\mathbf{a}_{\boldsymbol{\alpha}} = E\{\mathbf{X}\, \psi_{\boldsymbol{\alpha}}(\boldsymbol{\Xi})\} \quad .$$

The construction of $\mathbf{h}$ then consists in identifying the vector-valued coefficients $\mathbf{a}_{\boldsymbol{\alpha}}$. If $\boldsymbol{\Xi}$ is a normalized Gaussian random vector, then the polynomials are the normalized Hermite polynomials.

The Polynomial Chaos representation can be applied in infinite dimension for stochastic processes and random fields (Ghanem and Kruger, 1996; Sakamoto and Ghanem, 2002; Xiu and Karniadakis, 2002a; Debusschere et al., 2004; LeMaitre et al., 2004b,a; Lucor et al., 2004; Soize and Ghanem, 2004; Wan and Karniadakis, 2006; LeMaitre et al., 2007; Nouy, 2007; Arnst and Ghanem, 2008; Das et al., 2008; Ghosh and Ghanem, 2008; Das et al., 2009; Nouy, 2009; Soize and Ghanem, 2009; LeMaitre and Knio, 2010; Soize and Desceliers, 2010).

Today, many applications using Polynomial Chaos representations have been carried out for modeling and solving direct and inverse problems in the domains of computational fluid mechanics and solid mechanics (Ghanem and Spanos, 1990, 1991; Ghanem and Kruger, 1996; Ghanem and Dham, 1998; Ghanem and Red-Horse, 1999; Ghanem, 1999; Ghanem and Pellissetti, 2002; LeMaitre et al., 2002; Xiu and Karniadakis, 2002b, 2003; Ghanem et al., 2005; Wan and Karniadakis, 2005; Ghanem and Doostan, 2006; Berveiller et al., 2006; Desceliers et al., 2006; Faverjon and Ghanem, 2006; Knio and Maitre, 2006; Wan and Karniadakis, 2006; Blatman and Sudret, 2007; Ghanem and Ghosh, 2007; Desceliers et al., 2007; LeMaitre et al., 2007; Lucor et al., 2007; Marzouk et al., 2007; Rupert and Miller, 2007; Arnst and Ghanem, 2008; Arnst et al., 2008; Ghanem et al., 2008; Guilleminot et al., 2008, 2009; Marzouk and Najm, 2009; Najm, 2009; Stefanou et al., 2009; LeMaitre and Knio, 2010; Soize, 2010c). We also refer the reader to Section 2.6 for a short overview concerning the analysis of the propagation of uncertainties and to Section 2.7 concerning the identification and inverse stochastic methods.

Sometimes, it can be interesting to replace the deterministic coefficients of the polynomial chaos decomposition by random coefficients. The theory of the polynomial chaos expansion with random coefficients of random vector $\mathbf{X}$ can be found in (Soize and Ghanem, 2009) and is written as

$$\mathbf{X} = \Sigma_\alpha \, \mathbf{A}_\alpha \, \psi_\alpha(\Xi) \quad ,$$

in which $\{\mathbf{A}_\alpha\}_\alpha$ are random coefficients with values in $\mathbb{R}^{n_p}$. Such a polynomial chaos representation with random coefficients is used in (Das et al., 2008; Ghanem et al., 2008; Soize and Ghanem, 2009; Arnst et al., 2010).

(ii.2) - The second one consists in introducing a prior algebraic representation $\mathbf{X} = \mathbf{h}(\Xi, \mathbf{s})$ in which $\mathbf{s}$ is a vector parameter which has a small dimension and which must be identified, where Ξ is a vector-valued random variable with a given probability distribution P_Ξ and where $\mathbf{h}$ is a given nonlinear mapping. For instance, tensor-valued random fields representations constructed with such an approach can be found in (Soize, 2006, 2008a; Guilleminot et al., 2011; Guilleminot and Soize, 2011).

In theory, method (ii.1) allows any random vector $\mathbf{X}$ in $\mathcal{L}^2_{n_p}$ to be represented by a Polynomial Chaos expansion. In practice, the representation can require a very large number of coefficients to get convergence yielding very difficult problems in high dimension for the identification problem and then requires adapted methodologies and adapted methods (Soize and Desceliers, 2010; Soize, 2010c, 2011). In general, method (ii.2) does not allow any random vector $\mathbf{X}$ in $\mathcal{L}^2_{n_p}$ to be represented but allows a family of representations to be constructed in a subset of $\mathcal{L}^2_{n_p}$ when $\mathbf{s}$ runs through its admissible set (but, in counter part, the identification of $\mathbf{s}$ is effective and efficient).

2.4 Construction of prior probability models using the Maximum Entropy Principle

The measure of uncertainties using the entropy has been introduced by (Shannon, 1948) in the framework of Information Theory. The maximum entropy principle (that is to say the maximization of the level of uncertainties) has been introduced by (Jaynes, 1957) to construct the prior probability model of a random variable under the constraints defined by the available information. This principle appears as a major tool to construct the prior probability model (1) of the uncertain parameters of the computational model using the parametric probabilistic approach, (2) of both the computational-model-parameter uncertainties and the model uncertainties, using the nonparametric probabilistic approach and (3) of the generalized approach of uncertainties corresponding to a full coupling of the parametric and the nonparametric probabilistic approaches.

Let $\mathbf{x} = (x_1, \ldots, x_{n_p})$ be a real vector and let $\mathbf{X} = (X_1, \ldots, X_{n_p})$ be a second-order random variable with values in $\mathbb{R}^{n_p}$ whose probability distribution $P_{\mathbf{X}}$ is defined by a probability density function $\mathbf{x} \mapsto p_{\mathbf{X}}(\mathbf{x})$ on $\mathbb{R}^{n_p}$ with respect to $d\mathbf{x} = dx_1 \ldots dx_{n_p}$ and which verifies the normalization condition $\int_{\mathbb{R}^{n_p}} p_{\mathbf{X}}(\mathbf{x})\, d\mathbf{x} = 1$. It is assumed that $\mathbf{X}$ is with values in any bounded or unbounded part $\mathcal{X}$ of $\mathbb{R}^{n_p}$ and consequently, the support of $p_{\mathbf{X}}$ is $\mathcal{X}$. The available information defines a constraint equation on $\mathbb{R}^{\mu}$ written as

$$E\{\mathbf{g}(\mathbf{X})\} = \mathbf{f} \quad ,$$

in which E is the mathematical expectation, $\mathbf{f}$ is a given vector in $\mathbb{R}^{\mu}$ and where $\mathbf{x} \mapsto \mathbf{g}(\mathbf{x})$ is a given function (measurable) from $\mathbb{R}^{n_p}$ into $\mathbb{R}^{\mu}$. Since $p_{\mathbf{X}}$ is unknown and has to be constructed, we introduce the set $\mathcal{C}$ of all the probability density functions $\mathbf{x} \mapsto p(\mathbf{x})$ defined on $\mathbb{R}^{n_p}$, with values in $\mathbb{R}^+$, with support $\mathcal{X}$, verifying the normalization condition and the constraint equation $E\{\mathbf{g}(\mathbf{X})\} = \mathbf{f}$, that is to say,

$$\mathcal{C} = \left\{ \mathbf{x} \mapsto p(\mathbf{x}) \geq 0 \,,\, \operatorname{Supp} p = \mathcal{X} \,, \int_{\mathcal{X}} p(\mathbf{x})\, d\mathbf{x} = 1 \,, \int_{\mathcal{X}} \mathbf{g}(\mathbf{x})\, p(\mathbf{x})\, d\mathbf{x} = \mathbf{f} \right\}.$$

The maximum entropy principle consists in finding $p_{\mathbf{X}}$ in $\mathcal{C}$ which maximizes the entropy (that is to say which maximizes the uncertainties),

$$p_{\mathbf{X}} = \arg\max_{p\in C} S(p) \quad , \quad S(p) = -\int_{\mathbb{R}^N} p(\mathbf{x}) \log(p(\mathbf{x}))\, d\mathbf{x} \quad ,$$

in which $S(p)$ is the entropy of the probability density function p. Introducing the Lagrange multiplier $\boldsymbol{\lambda} \in \mathcal{L}_\mu \subset \mathbb{R}^\mu$ (associated with the constraint), where $\mathcal{L}_\mu$ is the subset of $\mathbb{R}^\mu$ of all the admissible values for $\boldsymbol{\lambda}$, it can easily be seen that the solution of the optimization problem can be written as

$$p_{\mathbf{X}}(\mathbf{x}) = c_0\, \mathbb{1}_{\mathcal{X}}(\mathbf{x}) \exp(- < \boldsymbol{\lambda}, \mathbf{g}(\mathbf{x}) >) \quad , \quad \forall \mathbf{x} \in \mathbb{R}^{n_p}$$

in which $< \mathbf{x}, \mathbf{y} >= x_1 y_1 + \ldots + x_\mu y_\mu$ and where $\mathbb{1}_{\mathcal{X}}$ is the indicator function of set $\mathcal{X}$. The normalization constant c_0 and the Lagrange multiplier $\boldsymbol{\lambda}$ are calculated in solving the following nonlinear vectorial algebraic equations

$$c_0 \int_{\mathcal{X}} \exp(- < \boldsymbol{\lambda}, \mathbf{g}(\mathbf{x}) >)\, d\mathbf{x} = 1 \quad ,$$

$$c_0 \int_{\mathcal{X}} \mathbf{g}(\mathbf{x}) \exp(- < \boldsymbol{\lambda}, \mathbf{g}(\mathbf{x}) >)\, d\mathbf{x} = \mathbf{f} \quad .$$

These algebraic equations can be solved using appropriated algorithms. Then, it is necessary to construct a generator of independent realizations of random variable $\mathbf{X}$ whose probability density function is the one which has been built. In small dimension (n_p is a few units), there is no difficulty. In high dimension (n_p several hundreds or several thousands), there are two major difficulties. The first one is related to the calculation of an integral in high dimension of the type $c_0 \int_{\mathcal{X}} \mathbf{g}(\mathbf{x}) \exp(- < \boldsymbol{\lambda}, \mathbf{g}(\mathbf{x}) >)\, d\mathbf{x}$. Such a calculation is necessary to implement the algorithm for computing c_0 and $\boldsymbol{\lambda}$. The second one is the construction of the generator once c_0 et $\boldsymbol{\lambda}$ have been calculated. These two aspects can be solved using the Markov Chain Monte Carlo methods (MCMC) (Spall, 2003; Kaipio and Somersalo, 2005; Rubinstein and Kroese, 2008). The transition kernel of the homogeneous (stationary) Markov chain of the MCMC method can be constructed using the Metropolis-Hastings algorithm (Metropolis and Ulam, 1949; Hastings, 1970; Casella and George, 1992) or the Gibbs algorithm (Geman and Geman, 1984) which is a slightly different algorithm for which the kernel is directly derived from the transition probability density function and for which the Gibbs realizations are always accepted. These two algorithms construct the transition kernel for which the invariant measure is $P_{\mathbf{X}}$. In general, these algorithms are effective but can not be when there are attraction regions which do not correspond to the invariant measure. These situations can not be easily detected and are time consuming. An alternative approach (Soize, 2008b) of the class of the Gibbs method has been proposed to avoid these difficulties. This approach is based on the introduction of an Itô stochastic differential equation whose unique invariant measure is $P_{\mathbf{X}}$. This invariant measure is then the explicit solution of a Fokker-Planck equation (Soize, 1994). The algorithm is then obtained by the discretization of the Itô equation.

2.5 Random matrix theory

The random matrix theory were introduced and developed in mathematical statistics by Wishart and others in the 1930s and was intensively studied by physicists and mathematicians in the context of nuclear physics. These works began with Wigner in the 1950s and received an important effort in the 1960s by Wigner, Dyson, Mehta and others. In 1965, Poter published a volume of important papers in this field, followed, in 1967 by the first edition of the Mehta book whose second edition (Mehta, 1991) published in 1991 is an excellent synthesis of the random matrix theory. We refer the reader to (Wright and Weaver, 2010) and (Soize, 2010b) for an introduction to random matrix theory presented in the context of Mechanics. Concerning multivariate statistical analysis and statistics of random matrices, the reader will find additional developments in Fougeaud and Fuchs (Fougeaud and Fuchs, 1967) and Anderson (Anderson, 1958).

Why the Gaussian orthogonal ensemble cannot be used if positiveness property is required. The Gaussian Orthogonal Ensemble (GOE), for which the mean value is the unity matrix $[I_n]$ (Soize, 2003a), is the set of the random matrices $[\mathbf{G}^{\rm GOE}]$) which can be written as $[\mathbf{G}^{\rm GOE}] = [I_n] + [\mathbf{B}^{\rm GOE}]$ in which $[\mathbf{B}^{\rm GOE}]$ belongs to the GOE (Mehta, 1991), for which the mean value is the zero matrix, $[0]$, and consequently, would be a second-order centered random matrix with values in $\mathbb{M}_n^S(\mathbb{R})$ such that $E\{[\mathbf{B}^{\rm GOE}]\} = [\,0\,]$ and $E\{\|[\mathbf{B}^{\rm GOE}]\|_F^2\} < +\infty$, in which $\|A\|_F^2 = \text{tr}\{[A]^T[A]\} = \sum_{j=1}^n \sum_{k=1}^n [A]_{jk}^2$ is the square of the Frobenius norm of the symmetric real matrix $[A]$. The pdf of random matrix $[\mathbf{B}^{\rm GOE}]$, with respect to $\widetilde{d}[B] = 2^{n(n-1)/4}\,\Pi_{1\leq i\leq j\leq n}\, d[B]_{ij}$, is written as

$$p_{[\mathbf{B}^{\rm GOE}]}([B]) = C_n \times \exp\left\{-\frac{(n+1)}{4\delta^2}\text{tr}\{[B]^2\}\right\} \quad .$$

The constant C_n of normalization can easily be calculated and δ is the coefficient of variation of the random matrix $[\mathbf{G}^{\rm GOE}]$ which is such that $\delta^2 = n^{-1}E\{\|\,[\mathbf{G}^{\rm GOE}] - [I_n]\,\|_F^2\}$ because $\|\,[I_n]\,\|_F^2 = n$. The real-valued random variables $\{[\mathbf{B}^{\rm GOE}]_{jk}, j \leq k\}$ are statistically independent, second order, centered and Gaussian. It can be seen that $[\mathbf{G}^{\text{GOE}}]$ is with values in $\mathbb{M}_n^S(\mathbb{R})$ but is not positive definite. In addition,

$$E\{\|[\mathbf{G}^{\text{GOE}}]^{-1}\|^2\} = +\infty \quad .$$

Consequently, $[\mathbf{G}^{\text{GOE}}]$ is not acceptable if positiveness property and integrability of the inverse are required.

Ensemble SG_0^+ of random matrices. The GOE cannot be used when positiveness property and integrability of the inverse are required. Consequently, we need new ensembles of random matrices which will be used to develop the nonparametric probabilistic approach of uncertainties in computational solid and fluid mechanics, and which differ from the GOE and from the other known ensembles of the random matrix theory.

The objective of this section is then to summarize the construction given in (Soize, 2000, 2001, 2003a, 2005b) of the ensemble SG_0^+ of random matrices $[\mathbf{G}_0]$ defined on the probability space $(\Theta, \mathcal{T}, \mathcal{P})$, with values in the set $\mathbb{M}_n^+(\mathbb{R})$ of all the positive definite symmetric $(n \times n)$ real matrices and such that

$$E\{[\mathbf{G}_0]\} = [I_n] \quad , \quad E\{\log(\det[\mathbf{G}_0])\} = C \quad , \quad |C| < +\infty \quad .$$

The probability distribution $P_{[\mathbf{G}_0]} = p_{[\mathbf{G}_0]}([G])\, \tilde{d}G$ is defined by a probability density function $[G] \mapsto p_{[\mathbf{G}_0]}([G])$ from $\mathbb{M}_n^+(\mathbb{R})$ into $\mathbb{R}^+$ with respect to the volume element $\tilde{d}G$ on the set $\mathbb{M}_n^S(\mathbb{R})$ of all the symmetric $(n \times n)$ real matrices, which is such that $\tilde{d}G = 2^{n(n-1)/4}\, \Pi_{1 \leq j \leq k \leq n}\, dG_{jk}$. This probability density function can then verify the normalization condition,

$$\int_{\mathbb{M}_n^+(\mathbb{R})} p_{[\mathbf{G}_0]}([G])\, \tilde{d}G = 1 \quad .$$

Let δ be the positive real number defined by

$$\delta = \left\{ \frac{E\{\|\, [\mathbf{G}_0] - E\{[\mathbf{G}_0]\}\, \|_F^2\}}{\|\, E\{[\mathbf{G}_0]\}\, \|_F^2} \right\}^{1/2} = \left\{ \frac{1}{n} E\{\|\, [\mathbf{G}_0] - [I_n]\, \|_F^2\} \right\}^{1/2} \quad ,$$

which will allow the dispersion of the probability model of random matrix $[\mathbf{G}_0]$ to be controlled. For δ such that $0 < \delta < (n+1)^{1/2}(n+5)^{-1/2}$, the use of the maximum entropy principle under the constraints defined by the above available information yields the following algebraic expression of the probability density function of random matrix $[\mathbf{G}_0]$,

$$p_{[\mathbf{G}_0]}([G]) = \mathbb{1}_{\mathbb{M}_n^+(\mathbb{R})}([G]) \times C_{\mathbf{G}_0} \times \left(\det [G]\right)^{(n+1)\frac{(1-\delta^2)}{2\delta^2}} \times e^{-\frac{(n+1)}{2\delta^2}\,\mathrm{tr}[G]} \quad ,$$

in which $\mathbb{1}_{\mathbb{M}_n^+(\mathbb{R})}([G])$ is equal to 1 if $[G] \in \mathbb{M}_n^+(\mathbb{R})$ and is equal to zero if $[G] \notin \mathbb{M}_n^+(\mathbb{R})$, where $\text{tr}[G]$ is the trace of matrix $[G]$, where $\det[G]$ is the determinant of matrix $[G]$ and where the positive constant $C_{\mathbf{G}_0}$ is such that

$$C_{\mathbf{G}_0} = (2\pi)^{-n(n-1)/4} \left(\frac{n+1}{2\delta^2}\right)^{n(n+1)(2\delta^2)^{-1}} \left\{\Pi_{j=1}^n \Gamma\left(\frac{n+1}{2\delta^2} + \frac{1-j}{2}\right)\right\}^{-1} ,$$

and where, for all $z > 0$, $\Gamma(z) = \int_0^{+\infty} t^{z-1}\, e^{-t}\, dt$. Note that $\{[\mathbf{G}_0]_{jk}, 1 \leq j \leq k \leq n\}$ are dependent random variables. If $(n+1)/\delta^2$ is an integer, then this probability density function coincides with the Wishart probability distribution (Anderson, 1958; Fougeaud and Fuchs, 1967). If $(n+1)/\delta^2$ is not an integer, then this probability density function can be viewed as a particular case of the Wishart distribution, in infinite dimension, for stochastic processes (?).

Let $\|G\| = \sup_{\|\mathbf{x}\| \leq 1} \|[G]\,\mathbf{x}\|$ be the operator norm of matrix $[G]$ which is such that $\|[G]\,\mathbf{x}\| \leq \|G\|\,\|\mathbf{x}\|$ for all $\mathbf{x}$ in $\mathbb{R}^n$. Let $\|G\|_F$ be the Frobenius norm of $[G]$ which is defined by $\|G\|_F^2 = \text{tr}\{[G]^T[G]\} = \sum_{j=1}^n \sum_{k=1}^n [G]_{jk}^2$ and which is such that $\|G\| \leq \|G\|_F \leq \sqrt{n}\,\|G\|$. It is proven (Soize, 2001) that

$$E\{\|[\mathbf{G}_0]^{-1}\|^2\} \leq E\{\|[\mathbf{G}_0]^{-1}\|_F^2\} < +\infty \quad .$$

In general, the above equation does not imply that $n \mapsto E\{\|[\mathbf{G}_0]^{-1}\|^2\}$ is a bounded function with respect to n, but, in the present case, we have the following fundamental property,

$$\forall n \geq 2 \quad , \quad E\{\|[\mathbf{G}_0]^{-1}\|^2\} \leq C_\delta < +\infty \quad ,$$

in which C_δ is a positive finite constant which is independent of n but which depends on δ. The above equation means that $n \mapsto E\{\|[\mathbf{G}_0]^{-1}\|^2\}$ is a bounded function from $\{n \geq 2\}$ into $\mathbb{R}^+$.

The generator of independent realizations (which is required to solve the random equations with the Monte Carlo method) can be constructed (Soize, 2000, 2001) using the following algebraic representation. Random matrix $[\mathbf{G}_0]$ can be written (Cholesky decomposition) as $[\mathbf{G}_0] = [\mathbf{L}]^T\,[\mathbf{L}]$ in which $[\mathbf{L}]$ is an upper triangular $(n \times n)$ random matrix such that:

(1) random variables $\{[\mathbf{L}]_{jj'}, j \leq j'\}$ are independent;

(2) for $j < j'$, the real-valued random variable $[\mathbf{L}]_{jj'}$ is written as $[\mathbf{L}]_{jj'} = \sigma_n U_{jj'}$ in which $\sigma_n = \delta(n+1)^{-1/2}$ and where $U_{jj'}$ is a real-valued

Gaussian random variable with zero mean and variance equal to 1;

(3) for $j = j'$, the positive-valued random variable $[\mathbf{L}]_{jj}$ is written as $[\mathbf{L}]_{jj} = \sigma_n \sqrt{2V_j}$ in which σ_n is defined above and where V_j is a positive-valued gamma random variable whose probability density function is $p_{V_j}(v) = \mathbb{1}_{\mathbb{R}^+}(v) \frac{1}{\Gamma(a_j)}\, v^{a_j - 1}\, e^{-v}$, in which $a_j = \frac{n+1}{2\delta^2} + \frac{1-j}{2}$. It should be noted that the probability density function of each diagonal element $[\mathbf{L}]_{jj}$ of the random matrix $[\mathbf{L}]$ depends on the rank j of the element.

Ensemble $\mathrm{SG}_\varepsilon^+$ of random matrices. Let $\mathcal{L}_n^2$ be the set of all the second-order random variables, defined on the probability space $(\Theta, \mathcal{T}, P)$, with values in $\mathbb{R}^n$, equipped with the inner product $\ll \mathbf{X}, \mathbf{Y} \gg = E\{< \mathbf{X}\,, \mathbf{Y} >\}$ and with the associated norm $|||\mathbf{X}||| = \ll \mathbf{X}, \mathbf{X} \gg^{1/2}$. Let $0 \leq \varepsilon \ll 1$ be a positive number as small as one wants. The ensemble $\mathrm{SG}_\varepsilon^+$ is defined as the ensemble of all the random matrices such that

$$[\mathbf{G}] = \frac{1}{1+\varepsilon} \{[\mathbf{G}_0] + \varepsilon\, [I_n]\} \quad ,$$

in which $[\mathbf{G}_0]$ is a random matrix which belongs to ensemble SG_0^+. Let $[\mathbf{G}]$ be in $\mathrm{SG}_\varepsilon^+$ with $\varepsilon \geq 0$ fixed as small as one wants (possibly, ε can be equal to zero and in such a case, $\mathrm{SG}_\varepsilon^+ = \mathrm{SG}_0^+$ and then, $[\mathbf{G}] = [\mathbf{G}_0]$). It can easily be seen that

$$E\{[\mathbf{G}]\} = [I_n] \quad .$$

Let $\mathbb{b}(\mathbf{X}, \mathbf{Y})$ be the bilinear form on $\mathcal{L}_n^2 \times \mathcal{L}_n^2$ such that $\mathbb{b}(\mathbf{X}, \mathbf{Y}) = \ll [\mathbf{G}]\, \mathbf{X}\,, \mathbf{Y} \gg$. For all $\mathbf{X}$ in $\mathcal{L}_n^2$, we have

$$\mathbb{b}(\mathbf{X}, \mathbf{X}) \geq c_\varepsilon |||\mathbf{X}|||^2$$

in which $c_\varepsilon = \varepsilon/(1+\varepsilon)$. The proof can easily be obtained. We have $\mathbb{b}(\mathbf{X}, \mathbf{X}) = 1/(1+\varepsilon) \ll [\mathbf{G}_0]\, \mathbf{X}\,, \mathbf{X} \gg + \varepsilon/(1+\varepsilon) |||\mathbf{X}|||^2 \geq c_\varepsilon |||\mathbf{X}|||^2$, because, for all $[\mathbf{G}_0]$ in SG_0^+, and for all $\mathbf{X}$ in $\mathcal{L}_n^2$, we have $\ll [\mathbf{G}_0]\, \mathbf{X}\,, \mathbf{X} \gg \geq 0$. Finally, for all $\varepsilon \geq 0$, it can be proven that

$$E\{\|[\mathbf{G}]^{-1}\|_F^2\} < +\infty \quad .$$

2.6 Propagation of uncertainties and stochastic solvers

Concerning the methods and formulations to analyze the propagation of uncertainties in the computational model, the choice of a specific method depends on the desired accuracy on the model output and on the nature of the expected probabilistic information. These last two decades, a growing interest has been devoted to spectral stochastic methods, pioneered by Roger Ghanem in 1990-1991 (Ghanem and Spanos, 1990, 1991, 2003), and which provide an explicit representation of the random model output as a function of the basic random parameters modeling the input uncertainties (Ghanem and Spanos, 1991; Ghanem and Kruger, 1996; Ghanem and Red-Horse, 1999; Ghanem, 1999; Xiu and Karniadakis, 2002a; Ghanem and Spanos, 2003; Ghosh and Ghanem, 2008; Matthies, 2008; Nouy, 2009; LeMaitre and Knio, 2010). An approximation of the random model output is sought on suitable functional approximation bases.

There are two distinct classes of techniques for the definition and computation of this approximation.

(1) The first class of techniques rely on Galerkin-type projections of weak solutions of models involving differential or partial differential equations (Ghanem and Spanos, 1991, 2003; Deb et al., 2001; LeMaitre et al., 2004b,a; Babuska et al., 2005; Frauenfelder et al., 2005; Matthies and Keese, 2005; Wan and Karniadakis, 2005, 2006; Mathelin and LeMaitre, 2007; Wan and Karniadakis, 2009). Alternative methods, based on the construction of optimal separated representations of the solution, have been performed to reduce computational requirements. They consist in representing the solution on optimal reduced bases of deterministic functions and stochastic functions (scalar-valued random variables). Several strategies are proposed for the construction of reduced bases using approximate Karhunen-Loeve expansions (or classical spectral decompositions) of the solution (Matthies and Keese, 2005; Doostan et al., 2007). Another method which is called Generalized Spectral Decomposition method, can be used to construct such representations without knowing *a priori* the solution nor an approximation of it (Nouy, 2007, 2008; Nouy and Maitre, 2009). An advantage of these algorithms is that they allow a separation of deterministic problems for the computation of deterministic functions and stochastic algebraic equations for the computation of stochastic functions. In that sense, they can be considered as partly non-intrusive techniques for computing Galerkin-type projections. Similarly, multidimensional extensions of separated representations techniques have been proposed (Doostan and Iaccarino, 2009; Nouy, 2010). These methods exploit the tensor product structure of the solution function space, resulting from the product structure of the probability space defined by input random parameters. However, it does not circumvent the curse of dimensionality associated with the dramatic increase in the dimension of stochastic approximation spaces when dealing with high stochastic dimension and high approximation resolution along each stochastic dimension.

(2) The second class is composed of methods based on a direct simulation (Matthies and Keese, 2005; Berveiller et al., 2006; Babuska et al., 2007; Blatman and Sudret, 2007; Ganapathysubramanian and Zabaras, 2007; Webster et al., 2007; Zabaras and Ganapathysubramanian, 2008; Ma and Zabaras, 2009). These methods are often called non-intrusive since they offer the advantage of only requiring the availability of classical deterministic codes. Finally, the direct Monte Carlo numerical simulation method (see for instance (Fishman, 1996; Rubinstein and Kroese, 2008) is a very effective and efficient method because this method (i) is non-intrusive, (ii) is adapted to massively parallel computation without any software developments, (iii) is such that its convergence can be controlled during the computation, and (iv) the speed of convergence is independent of the dimension. The speed of convergence of the Monte Carlo method can be improved using advanced Monte Carlo simulation procedures (Papadrakakis and Papadopoulos, 1996; Pradlwarter and Schueller, 1997; Papadrakakis and Kotsopulos, 1999; Schueller, 2009), subset simulation technics (Au and Beck, 2003a), important sampling for high dimension problems (Au and Beck, 2003b), local domain Monte Carlo Simulation (Pradlwarter and Schueller, 2010).

2.7 Identification of the prior and posterior probability models of uncertainties

The identification of the parameters of the probability model of uncertainties (parametric and nonparametric probabilistic approaches) is a problem belonging to the class of the statistical inverse problems, see for instance (Kaipio and Somersalo, 2005).

Let us assume that experimental data are available for observations related to the random computational model output. The experimental identification of the parameters of the prior probability distributions of the random variables which model the uncertain parameters in the computational model can be performed, either in minimizing a distance between the experimental observations and the random model observations (such as least-square method), or in using the maximum likelihood method (Serfling, 1980; Spall, 2003). Such an approach is described in Sections 3.5, 5.3 and 6.5, and many applications can be found in the literature. In the domain of structural dynamics, vibrations and structural acoustics, we refer the reader to (Peters and Roeck, 2001; Ghanem et al., 2005; Arnst et al., 2008; Goller et al., 2009) for the stochastic system identification in structural dynamics and vibration, to (Faverjon and Ghanem, 2006) for the stochastic inversion in acoustic scattering, to (Cataldo et al., 2009) for the probabilistic modeling of a nonlinear dynamical system used for producing voice and to (Leissing et al., 2010) for the computational model for long-range non-linear propagation over urban cities. With such an identification, we then obtained an optimal prior probability distribution. The Bayesian method then allows a posterior probability distribution to be constructed from the optimal prior probability distribution and from the experimental data. Many works have been published in the literature: see, for instance, the textbooks on the Bayesian method such as (Bernardo and Smith, 2000; Kaipio and Somersalo, 2005; Congdon, 2007; Carlin and Louis, 2009) and papers devoted to the use of the Bayesian method in the context of uncertain mechanical and dynamical systems such as (Beck and Katafygiotis, 1998; Katafygiotis and Beck, 1998; Beck and Au, 2002; Ching et al., 2006; Cheung and Beck, 2009; Beck, 2010; Cheung and Beck, 2010; Tan et al., 2010). We will use such a Bayesian approach in Sections 3.6 and 5.4.

General developments concerning the experimental identification of the parameters of the prior probability distributions of random matrices used to model the uncertainties in computational mechanics can be found in (Soize, 2005b). Many works have been published in this field, such as (Chebli and Soize, 2004; Arnst et al., 2006, 2008; Chen et al., 2006; Duchereau and Soize, 2006; Soize et al., 2008a; Batou et al., 2011) for the experimental identification of the nonparametric probabilistic model of uncertainties in structural dynamics, as (Durand et al., 2008; Fernandez et al., 2010) in structural acoustics for the low- and medium-frequency ranges and as (Batou and Soize, 2009a,b) for nonlinear structural dynamical systems. The identification of the generalized probabilistic approach of uncertainties can be found in (Soize, 2010a) and also below in Sections 5.3.

2.8 Robust updating of computational models and robust design with uncertain computational models

Robust updating or robust design optimization consists in updating a computational model or in optimizing the design of a mechanical system with a computational model, in taking into account the uncertainties in the computational model parameters and the modeling uncertainties. An overview on computational methods in optimization considering uncertainties can be found in (Schueller and Jensen, 2008). Robust updating and robust design developments with uncertainties on the computational model parameters are developed in (Beck et al., 1999; Papadimitriou et al., 2000, 2001; Papadrakakis and Lagaros, 2002; Taflanidis and Beck, 2008; Goller et al., 2009) while robust updating or robust design optimization with modeling uncertainties can be found in (Capiez-Lernout and Soize, 2008a,b,c; Soize et al., 2008b; Ritto et al., 2010).

3 Parametric probabilistic approach of uncertainties in computational structural dynamics

3.1 Introduction of the mean computational model in computational structural dynamics

The developments are presented for computational models in structural dynamics. The dynamical system is a damped fixed structure around a static equilibrium configuration considered as a natural state without prestresses and subjected to an external load. For given nominal values of the parameters of the dynamical system, the basic finite element model is called the *mean computational model.* In addition, it is assumed that a set of model parameters has been identified as the uncertain model parameters. These model parameters are the components of a vector $\mathbf{x} = (x_1, \ldots, x_{n_p})$ belonging to an admissible set $\mathcal{C}_{\text{par}}$ which is a subset of $\mathbb{R}^{n_p}$. Using the usual finite element method to construct a finite approximation of the weak formulation of the boundary value problem yields the dynamical equation of the mean computational model which is then written as

$$[\mathbb{M}(\mathbf{x})]\,\ddot{\mathbb{y}}(t) + [\mathbb{D}(\mathbf{x})]\,\dot{\mathbb{y}}(t) + [\mathbb{K}(\mathbf{x})]\,\mathbb{y}(t) + \mathbb{f}_{\text{NL}}(\mathbb{y}(t), \dot{\mathbb{y}}(t); \mathbf{x}) = \mathbb{f}(t; \mathbf{x}) \quad , \tag{1}$$

in which $\mathbb{y}(t)$ is the unknown time response vector of the m degrees of freedom (DOF) (displacements and/or rotations); $\dot{\mathbb{y}}(t)$ and $\ddot{\mathbb{y}}(t)$ are the velocity and acceleration vectors respectively; $\mathbb{f}(t; \mathbf{x})$ is the known external load vector of the m inputs (forces and/or moments); $[\mathbb{M}(\mathbf{x})]$, $[\mathbb{D}(\mathbf{x})]$ and $[\mathbb{K}(\mathbf{x})]$ are the mass, damping and stiffness matrices of the mean computational model, which belong to the set $\mathbb{M}_m^+(\mathbb{R})$ of all the positive-definite symmetric $(m \times m)$ real matrices; $(\mathbf{y}, \mathbf{z}) \mapsto \mathbb{f}_{\text{NL}}(\mathbf{y}, \mathbf{z}; \mathbf{x})$ is the nonlinear mapping modeling local nonlinear forces.

3.2 Introduction of the reduced mean computational model

In the context of the parametric probabilistic approach of model-parameter uncertainties, parameter $\mathbf{x}$ is modeled by a random variable $\mathbf{X}$. The mean value of $\mathbf{X}$ is chosen as the nominal value $\underline{\mathbf{x}} = (\underline{x}_1, \ldots, \underline{x}_{n_p})$ of the uncertain model parameter $\mathbf{x}$. The support of the probability distribution of $\mathbf{X}$ on $\mathbb{R}^{n_p}$ is $\mathcal{C}_{\text{par}} \subset \mathbb{R}^{n_p}$.

For all $\mathbf{x}$ fixed in $\mathcal{C}_{\text{par}}$, let $\{\phi_1(\mathbf{x}), \ldots, \phi_m(\mathbf{x})\}$ be an algebraic basis of $\mathbb{R}^m$ constructed, for instance, either with the elastic modes of the linearized system, either with the elastic modes of the underlying linear system, or with the POD (Proper Orthogonal Decomposition) modes of the nonlinear system). It should be noted that the POD modes depend on the external load vector,$\mathbb{f}(t;\mathbf{x})$. Below, it is assumed that the elastic modes of the underlying linear system are used (Sampaio and Soize, 2007b). In such a framework, there are two main possibilities to construct the reduced mean computational model.

(1) The first one consists in solving the generalized eigenvalue problem associated with the mean mass and stiffness matrices for $\mathbf{x}$ fixed to its nominal value $\underline{\mathbf{x}}$. We then obtain the elastic modes of the nominal mean computational model which are independent of $\mathbf{x}$ and which depend only on $\underline{\mathbf{x}}$ which is fixed. In this case, when $\mathbf{x}$ runs through $\mathcal{C}_{\text{par}}$, matrices $[\mathbb{M}(\mathbf{x})]$ and $[\mathbb{K}(\mathbf{x})]$ have to be projected on the subspace spanned by the elastic modes of the nominal mean computational model. For very large computational model (m can be several tens of millions) such an operation is not easy to perform with usual commercial softwares which often are black boxes.

(2) The second one consists in solving the generalized eigenvalue problem associated with the mean mass and stiffness matrices for each required $\mathbf{x}$ belonging to $\mathcal{C}_{\text{par}}$. In this case, the elastic modes of the mean computational model depend on $\mathbf{x}$. In the context of the parametric probabilistic approach of model-parameter uncertainties, we then have to solve a random generalized eigenvalue problem and such an approach is better adapted to usual commercial softwares and allows a fast convergence to be obtained with respect to the dimension of the reduced-order model. In this context, some algorithms have been developed for random eigenvalue problems of large systems (Szekely and Schuller, 2001; Pradlwarter et al., 2002). In order to limit the developments, we will focus the presentation using this second approach. The extension to the first approach is straightforward from a theoretical point of view (see for instance (Soize, 2000)). Finally, it should be noted that the random generalized eigenvalue problem can also be considered in a polynomial chaos decomposition for which an efficient approach has been proposed (Ghanem and Ghosh, 2007). Such an ingredient can be added without difficulty in the developments presented below but would induce an addition degree of difficulty in the understanding which could mask the main ideas of the parametric probabilistic approach of model-parameter uncertainties.

For each value of $\mathbf{x}$ given in $\mathcal{C}_{\text{par}}$, the generalized eigenvalue problem associated with the mean mass and stiffness matrices is written as

$$[\mathbb{K}(\mathbf{x})]\,\boldsymbol{\phi}(\mathbf{x}) = \lambda(\mathbf{x})\,[\mathbb{M}(\mathbf{x})]\,\boldsymbol{\phi}(\mathbf{x}) \quad , \tag{2}$$

for which the eigenvalues $0 < \lambda_1(\mathbf{x}) \leq \lambda_2(\mathbf{x}) \leq \ldots \leq \lambda_m(\mathbf{x})$ and the associated elastic modes $\{\boldsymbol{\phi}_1(\mathbf{x}), \boldsymbol{\phi}_2(\mathbf{x}), \ldots\}$ are such that

$$<[\mathbb{M}(\mathbf{x})]\,\boldsymbol{\phi}_\alpha(\mathbf{x})\,,\boldsymbol{\phi}_\beta(\mathbf{x})>= \mu_\alpha(\mathbf{x})\,\delta_{\alpha\beta} \quad , \tag{3}$$

$$<[\mathbb{K}(\mathbf{x})]\,\boldsymbol{\phi}_\alpha(\mathbf{x})\,,\boldsymbol{\phi}_\beta(\mathbf{x})>= \mu_\alpha(\mathbf{x})\,\omega_\alpha(\mathbf{x})^2\,\delta_{\alpha\beta} \quad , \tag{4}$$

in which $\omega_\alpha(\mathbf{x}) = \sqrt{\lambda_\alpha(\mathbf{x})}$ is the eigenfrequency of elastic mode $\boldsymbol{\phi}_\alpha(\mathbf{x})$ whose normalization is defined by the generalized mass $\mu_\alpha(\mathbf{x})$ and where $< \mathbf{u}\,,\mathbf{v} >= \sum_j u_j v_j$ is the Euclidean inner product of the vectors $\mathbf{u}$ and $\mathbf{v}$. For each value of $\mathbf{x}$ given in $\mathcal{C}_{\text{par}}$, the reduced mean computational model of the dynamical system is obtained in constructing the projection of the mean computational model on the subspace $\mathbb{H}_n$ of $\mathbb{R}^m$ spanned by $\{\boldsymbol{\phi}_1(\mathbf{x}), \ldots, \boldsymbol{\phi}_n(\mathbf{x})\}$ with $n \ll m$. Let $[\phi(\mathbf{x})]$ be the $(m \times n)$ real matrix whose columns are vectors $\{\boldsymbol{\phi}_1(\mathbf{x}), \ldots, \boldsymbol{\phi}_n(\mathbf{x})\}$. The generalized force $\mathbf{f}(t;\mathbf{x})$ is a $\mathbb{R}^n$-vector such that $\mathbf{f}(t;\mathbf{x}) = [\phi(\mathbf{x})]^T\,\mathbb{f}(t;\mathbf{x})$. For all $\mathbf{x}$ in $\mathcal{C}_{\text{par}}$, the generalized mass, damping and stiffness matrices $[M(\mathbf{x})]$, $[D(\mathbf{x})]$ and $[K(\mathbf{x})]$ belong to the set $\mathbb{M}_n^+(\mathbb{R})$ of all the positive-definite symmetric $(n \times n)$ real matrices, and are defined by

$$[M(\mathbf{x})]_{\alpha\beta} = \mu_\alpha(\mathbf{x})\,\delta_{\alpha\beta} \quad , \quad [D(\mathbf{x})]_{\alpha\beta} =<[\mathbb{D}(\mathbf{x})]\,\boldsymbol{\phi}_\beta(\mathbf{x})\,,\boldsymbol{\phi}_\alpha(\mathbf{x})> \quad , \tag{5}$$

$$[K(\mathbf{x})]_{\alpha\beta} = \mu_\alpha(\mathbf{x})\,\omega_\alpha(\mathbf{x})^2\,\delta_{\alpha\beta} \quad , \tag{6}$$

in which, generally, $[D(\mathbf{x})]$ is a full matrix. Consequently, for all $\mathbf{x}$ in $\mathcal{C}_{\text{par}}$ and for all fixed t, the reduced mean computational model of the dynamical system is written as the projection $\mathbf{y}^n(t)$ of $\mathbb{y}(t)$ on $\mathbb{H}_n$ which can be written as $\mathbf{y}^n(t) = [\phi(\mathbf{x})]\,\mathbf{q}(t)$ in which $\mathbf{q}(t)$ is the vector in $\mathbb{R}^n$ of the generalized coordinates. The reduced-order model is then written as

$$\mathbf{y}^n(t) = [\phi(\mathbf{x})]\,\mathbf{q}(t) \quad ,$$

$$[M(\mathbf{x})]\,\ddot{\mathbf{q}}(t) + [D(\mathbf{x})]\,\dot{\mathbf{q}}(t) + [K(\mathbf{x})]\,\mathbf{q}(t) + \mathbf{F}_{\text{NL}}(\mathbf{q}(t), \dot{\mathbf{q}}(t);\mathbf{x}) = \mathbf{f}(t;\mathbf{x}) \quad , \tag{7}$$

in which $\mathbf{F}_{\text{NL}}(\mathbf{q}(t), \dot{\mathbf{q}}(t);\mathbf{x}) = [\phi(\mathbf{x})]^T\,\mathbb{f}_{\text{NL}}([\phi(\mathbf{x})]\,\mathbf{q}(t), [\phi(\mathbf{x})]\,\dot{\mathbf{q}}(t);\mathbf{x})$. In the particular case for which $\mathbb{f}_{\text{NL}} = 0$, then the corresponding equation in the frequency domain is written as

$$-\omega^2[M(\mathbf{x})]\,\mathbf{q}(\omega) + i\omega[D(\mathbf{x})]\,\mathbf{q}(\omega) + [K(\mathbf{x})]\,\mathbf{q}(\omega) = \mathbf{f}(\omega;\mathbf{x}) \quad ,$$

in which $\mathbf{q}(\omega) = \int_{\mathbb{R}} e^{-i\omega t}\mathbf{q}(t)\, dt$ and $\mathbf{f}(\omega; \mathbf{x}) = \int_{\mathbb{R}} e^{-i\omega t}\mathbf{f}(t; \mathbf{x})\, dt$.

Convergence. Below, we will denote by n_0 the value of n for which the response $\mathbf{y}^n$ is converged to $\mathbb{y}$, with a given accuracy, for all the values of $\mathbf{x}$ in $\mathcal{C}_{\text{par}}$.

3.3 Methodology for the parametric probabilistic approach of model-parameter uncertainties

The value of n is fixed to the value n_0 previously defined. The methodology for the parametric probabilistic approach of model-parameter uncertainties consists in modeling the uncertain model parameter $\mathbf{x}$ (whose nominal value is $\underline{\mathbf{x}}$) by a random variable $\mathbf{X}$ defined on a probability space $(\Theta, \mathcal{T}, \mathcal{P})$, with values in $\mathbb{R}^{n_p}$. Consequently, the generalized matrices in Eq. (7) become random matrices $[M(\mathbf{X})]$, $[D(\mathbf{X})]$ and $[K(\mathbf{X})]$ and, for all fixed t, the generalized external force $\mathbf{f}(t; \mathbf{x})$ becomes a random vector $\mathbf{f}(t; \mathbf{X})$. The mean values of these random matrices are denoted by $[\underline{M}]$, $[\underline{D}]$, $[\underline{K}]$, and are such that

$$E\{[M(\mathbf{X})]\} = [\underline{M}] \;,\; E\{[D(\mathbf{X})]\} = [\underline{D}] \;,\; E\{[K(\mathbf{X})]\} = [\underline{K}] \;, \tag{8}$$

in which E is the mathematical expectation. It should be noted that the mean matrices $[\underline{M}]$, $[\underline{D}]$ and $[\underline{K}]$ are different from the matrices $[M(\underline{\mathbf{x}})]$, $[D(\underline{\mathbf{x}})]$ and $[K(\underline{\mathbf{x}})]$ of the mean (nominal) computational model. The parametric probabilistic approach of uncertainties then consists in replacing the mean computational model by the following stochastic reduced-order computational model,

$$\mathbf{Y}(t) = [\,\phi(\mathbf{X})]\, \mathbf{Q}(t) \quad , \tag{9}$$

$$[M(\mathbf{X})]\, \ddot{\mathbf{Q}}(t) + [D(\mathbf{X})]\, \dot{\mathbf{Q}}(t) + [K(\mathbf{X})]\, \mathbf{Q}(t) + \mathbf{F}_{\text{NL}}(\mathbf{Q}(t), \dot{\mathbf{Q}}(t); \mathbf{X}) = \mathbf{f}(t; \mathbf{X}) \quad , \tag{10}$$

in which for all fixed t, $\mathbf{Y}(t)$ is a $\mathbb{R}^m$-valued random vector and $\mathbf{Q}(t)$ is a $\mathbb{R}^n$-valued random vector. As soon as the probability model of random vector $\mathbf{X}$ is constructed, the stochastic computational model defined by Eqs. (9) and (10) can be solved using the methods presented in Section 2.6. In particular, the direct Monte Carlo method is efficient for a reduced-order stochastic computational model and its convergence rate is independent of the value of n_p. Consequently, such a method can be used in high dimension.

3.4 Construction of the prior probability model of model-parameter uncertainties

The unknown probability distribution of $\mathbf{X}$ is assumed to be defined by a probability density function $\mathbf{x} \mapsto p_{\mathbf{X}}(\mathbf{x})$ from $\mathbb{R}^{n_p}$ into $\mathbb{R}^+ = [0\,, +\infty[$. Under the assumption that no experimental data are available to construct $p_{\mathbf{X}}$, the prior model can be constructed using the maximum entropy principle (see Section 2.4). For such a construction, the available information has to be defined. Since $\mathbf{x}$ belongs to $\mathcal{C}_{\text{par}}$, the support of $p_{\mathbf{X}}$ must be $\mathcal{C}_{\text{par}}$ and the normalization condition must be verified. We then have,

$$\text{supp}\, p_{\mathbf{X}} = \mathcal{C}_{\text{par}} \subset \mathbb{R}^{n_p} \quad , \quad \int_{\mathbb{R}^{n_p}} p_{\mathbf{X}}(\mathbf{x})\, d\mathbf{x} = \int_{\mathcal{C}_{\text{par}}} p_{\mathbf{X}}(\mathbf{x})\, d\mathbf{x} = 1 \quad . \tag{11}$$

Since the nominal value of $\mathbf{x}$ is $\underline{\mathbf{x}} \in \mathcal{C}_{\text{par}}$, an additional available information consists in writing that the mean value $E\{\mathbf{X}\}$ of $\mathbf{X}$ is equal to $\underline{\mathbf{x}}$ which yields the following constraint equation,

$$\int_{\mathbb{R}^{n_p}} \mathbf{x}\, p_{\mathbf{X}}(\mathbf{x})\, d\mathbf{x} = \underline{\mathbf{x}} \quad . \tag{12}$$

In general, an additional available information can be deduced from the analysis of the mathematical properties of the solution of the stochastic computational model under construction. The random solution $\mathbf{Q}$ of the stochastic computational model defined by Eq. (10) must be a second-order vector-valued stochastic process (because the dynamical system is stable) which means that, for all t, we must have $E\{\|\mathbf{Q}(t)\|^2\} < +\infty$. In order that such a property be verified, it is necessary to introduce a constraint which can always be written as the equation $E\{\mathbf{h}(\mathbf{X})\} = \boldsymbol{\gamma}$ on $\mathbb{R}^{\mu_X}$, in which $\boldsymbol{\gamma} = (\gamma_1, \ldots, \gamma_{\mu_X})$ is a given vector in $\mathbb{R}^{\mu_X}$ with $\mu_X \geq 1$ and where $\mathbf{x} \mapsto \mathbf{h}(\mathbf{x}) = (h_1(\mathbf{x}), \ldots, h_{\mu_X}(\mathbf{x}))$ is a given (measurable) mapping from $\mathbb{R}^{n_p}$ into $\mathbb{R}^{\mu_X}$. Consequently, this constraint is defined as follows,

$$\int_{\mathbb{R}^{n_p}} \mathbf{h}(\mathbf{x})\, p_{\mathbf{X}}(\mathbf{x})\, d\mathbf{x} = \boldsymbol{\gamma} \quad . \tag{13}$$

Let $\mathcal{C}$ be the set of all the probability density functions $p_{\mathbf{X}}$ defined on $\mathbb{R}^{n_p}$ with values in $\mathbb{R}^+$ such that Eqs. (11) to (13) hold. The prior model $p_{\mathbf{X}} \in \mathcal{C}$ can then be constructed using the maximum entropy principle (see Section 2.4). Since Eq. (13) is a vectorial equation of dimension μ_X, the solution $p_{\mathbf{X}}$ of the maximum entropy principle depends on the free $\mathbb{R}^{\mu_X}$-valued parameter $\boldsymbol{\gamma}$. However, parameter $\boldsymbol{\gamma}$ has generally no physical meaning and it is better to express $\boldsymbol{\gamma}$ in terms of a $\mathbb{R}^{\mu_X}$-valued parameter $\boldsymbol{\delta}_X$ which corresponds to a well defined statistical quantity for random variable $\mathbf{X}$. In general, $\boldsymbol{\delta}_X$ does not run through $\mathbb{R}^{\mu_X}$ but must belong to an admissible set $\mathcal{C}_X$ which is a subset of $\mathbb{R}^{\mu_X}$. Consequently, $p_{\mathbf{X}}$ depends on $\underline{\mathbf{x}}$ and $\boldsymbol{\delta}_X$ and we will write this probability density function as

$$\mathbf{x} \mapsto p_{\mathbf{X}}(\mathbf{x}; \underline{\mathbf{x}}, \boldsymbol{\delta}_X) \quad \text{with} \quad (\underline{\mathbf{x}}, \boldsymbol{\delta}_X) \in \mathcal{C}_{\text{par}} \times \mathcal{C}_X \subset \mathbb{R}^{n_p} \times \mathbb{R}^{\mu_X} \quad . \tag{14}$$

3.5 Estimation of the parameters of the prior probability model of the uncertain model parameter

The value of n is fixed to the value n_0 for which, for all values of $\mathbf{x}$ in $\mathcal{C}_{\text{par}}$, the response $\mathbf{y}^n$ is converged with respect to n (with a given accuracy). The prior probability model $p_{\mathbf{X}}(\mathbf{x}; \underline{\mathbf{x}}, \boldsymbol{\delta}_X)$ of random variable $\mathbf{X}$ relative to the model-parameter uncertainties, depends on parameters $\underline{\mathbf{x}}$ and $\boldsymbol{\delta}_X$ belonging to admissible sets $\mathcal{C}_{\text{par}}$ and $\mathcal{C}_X$. If no experimental data are available, then $\underline{\mathbf{x}}$ is fixed to the nominal value and $\boldsymbol{\delta}_X$ must be considered as a parameter to perform a sensitivity analysis of the stochastic solution. Such a prior probability model of model-parameter uncertainties then allows the robustness of the solution to be analyzed in function of the level of model-parameter uncertainties controlled by $\boldsymbol{\delta}_X$.

For the particular case for which a few experimental data exist, $\underline{\mathbf{x}}$ can be updated and $\boldsymbol{\delta}_X$ can be estimated with the experimental data. The updating of $\underline{\mathbf{x}}$ and the estimation of $\boldsymbol{\delta}_X$ must then be performed with observations of the systems for which experimental data are available. Let $\mathbf{W}$ be the random vector which is observed and which is with values in $\mathbb{R}^{m_{\text{obs}}}$. It is assumed that $\mathbf{W}$ is independent of t but depends on $\{\mathbf{Y}(t), t \in \mathcal{J}\}$ in which $\mathcal{J}$ is any part of $\mathbb{R}$. For all $(\underline{\mathbf{x}}, \boldsymbol{\delta}_X) \in \mathcal{C}_{\text{par}} \times \mathcal{C}_X$, and for all $\mathbf{w}$ fixed, the probability density function of $\mathbf{W}$ is denoted by $p_{\mathbf{W}}(\mathbf{w}; \underline{\mathbf{x}}, \boldsymbol{\delta}_X)$. On the other hand, it is assumed that ν_{exp} independent experimental values $\mathbf{w}_1^{\text{exp}}, \ldots, \mathbf{w}_{\nu_{\text{exp}}}^{\text{exp}}$ are available. The optimal value $(\underline{\mathbf{x}}^{\text{opt}}, \boldsymbol{\delta}_X^{\text{opt}})$ of $(\underline{\mathbf{x}}, \boldsymbol{\delta}_X)$ can be estimated by maximizing the logarithm of the likelihood function (maximum likelihood method (Serfling, 1980; Spall, 2003)),

$$(\underline{\mathbf{x}}^{\text{opt}}, \boldsymbol{\delta}_X^{\text{opt}}) = \arg \max_{(\underline{\mathbf{x}}, \boldsymbol{\delta}_X) \in \mathcal{C}_{\text{par}} \times \mathcal{C}_X} \{ \sum_{r=1}^{\nu_{\text{exp}}} \log p_{\mathbf{W}}(\mathbf{w}_r^{\text{exp}}; \underline{\mathbf{x}}, \boldsymbol{\delta}_X) \} \quad . \tag{15}$$

For all r, the quantity $p_{\mathbf{W}}(\mathbf{w}_r^{\text{exp}}; \underline{\mathbf{x}}, \boldsymbol{\delta}_X)$ is estimated with the stochastic computational model (using, for instance, the independent realizations of $\mathbf{W}$ calculated with the Monte Carlo method and the multivariate Gaussian kernel density estimation method.

3.6 Posterior probability model of uncertainties using output-prediction-error method and the Bayesian method

Let $p_{\mathbf{X}}^{\text{prior}}(\mathbf{x}) = p_{\mathbf{X}}(\mathbf{x}; \underline{\mathbf{x}}^{\text{opt}}, \boldsymbol{\delta}_X^{\text{opt}})$ be the optimal prior probability density function of $\mathbf{X}$, constructed with the optimal value $(\underline{\mathbf{x}}^{\text{opt}}, \boldsymbol{\delta}_X^{\text{opt}})$ of $(\underline{\mathbf{x}}, \boldsymbol{\delta}_X)$ which has been calculated in Section 3.5 solving the optimization problem defined by Eq. (15). The objective of this section is to estimate the posterior probability density function, $p_{\mathbf{X}}^{\text{post}}(\mathbf{x})$, of $\mathbf{X}$ using the experimental data associated with observation $\mathbf{W}$ introduced in Section 3.5 and using the Bayesian method.

In order to apply the Bayesian method (Serfling, 1980; Bernardo and Smith, 2000; Spall, 2003; Congdon, 2007; Carlin and Louis, 2009) to estimate the posterior probability density function $p_{\mathbf{X}}^{\text{post}}(\mathbf{x})$ of $\mathbf{X}$, the output-prediction-error method (Walter and Pronzato, 1997; Beck and Katafygiotis, 1998; Kaipio and Somersalo, 2005) is used and then, an additive noise $\mathbf{B}$ is added to the observation as explained in Section 2.2-(i). In this condition, the random observed output $\mathbf{W}^{\text{out}}$ with values in $\mathbb{R}^{m_{\text{obs}}}$, for which experimental data $\mathbf{w}_1^{\text{exp}}, \ldots, \mathbf{w}_{\nu_{\text{exp}}}^{\text{exp}}$ are available (see Section 3.5), is written as

$$\mathbf{W}^{\text{out}} = \mathbf{W} + \mathbf{B} \quad , \tag{16}$$

in which $\mathbf{W}$ is the computational model output with values in $\mathbb{R}^{m_{\text{obs}}}$. The noise $\mathbf{B}$ is a $\mathbb{R}^{m_{\text{obs}}}$-valued random vector, defined on probability space $(\Theta, \mathcal{T}, \mathcal{P})$, which is assumed to be independent of $\mathbf{X}$, and consequently, which is independent of $\mathbf{W}$. It should be noted that this hypothesis concerning the independence of $\mathbf{W}$ and $\mathbf{B}$ could easily be removed. It is also assumed that the probability density function $p_{\mathbf{B}}(\mathbf{b})$ of random vector $\mathbf{B}$, with respect to $d\mathbf{b}$, is known. For instance, it is often assumed that $\mathbf{B}$ is a centered Gaussian random vector for which the covariance matrix $[C_{\mathbf{B}}] = E\{\mathbf{B}\mathbf{B}^T\}$ is assumed to be invertible. In such a case, we would have

$$p_{\mathbf{B}}(\mathbf{b}) = (2\pi)^{-m_{\text{obs}}/2} \,(\det[C_{\mathbf{B}}])^{-1/2} \exp(-\frac{1}{2} < [C_{\mathbf{B}}]^{-1}\mathbf{b}\,, \mathbf{b} >) \quad . \tag{17}$$

The posterior probability density function $p_{\mathbf{X}}^{\text{post}}(\mathbf{x})$ is then calculated by the Bayes formula,

$$p_{\mathbf{X}}^{\text{post}}(\mathbf{x}) = \mathcal{L}(\mathbf{x})\, p_{\mathbf{X}}^{\text{prior}}(\mathbf{x}) \quad , \tag{18}$$

in which $\mathbf{x} \mapsto \mathcal{L}(\mathbf{x})$ is the likelihood function defined on $\mathbb{R}^{n_p}$, with values in $\mathbb{R}^+$, such that

$$\mathcal{L}(\mathbf{x}) = \frac{\Pi_{r=1}^{\nu_{\text{exp}}} p_{\mathbf{W}^{\text{out}}|\mathbf{X}}(\mathbf{w}_r^{\text{exp}}|\mathbf{x})}{E\{\Pi_{r=1}^{\nu_{\text{exp}}} p_{\mathbf{W}^{\text{out}}|\mathbf{X}}(\mathbf{w}_r^{\text{exp}}|\mathbf{X}^{\text{prior}})\}} \quad . \tag{19}$$

In Eq. (19), $p_{\mathbf{W}^{\text{out}}|\mathbf{X}}(\mathbf{w}_\ell^{\text{exp}}|\mathbf{x})$ is the experimental value of the conditional probability density function $\mathbf{w} \mapsto p_{\mathbf{W}^{\text{out}}|\mathbf{X}}(\mathbf{w}|\mathbf{x})$ of the random observed output $\mathbf{W}^{\text{out}}$ given $\mathbf{X} = \mathbf{x}$ in $\mathcal{C}_{\text{par}}$. Taking into account Eq. (16) and that $\mathbf{B}$ and $\mathbf{W}$ are assumed to be independent, it can easily be deduced that

$$p_{\mathbf{W}^{\text{out}}|\mathbf{X}}(\mathbf{w}_r^{\text{exp}}|\mathbf{x}) = p_{\mathbf{B}}(\mathbf{w}_r^{\text{exp}} - \mathbb{h}(\mathbf{x})) \quad , \tag{20}$$

in which $\mathbf{w} = \mathbb{h}(\mathbf{x})$ is the model observation depending on $\{\mathbf{y}^n(t), t \in \mathcal{J}\}$ which is calculated using Eq. (7). It should be noted that the posterior probability density function strongly depends on the choice of the probability model of the output additive noise $\mathbf{B}$.

4 Nonparametric probabilistic approach of uncertainties in computational structural dynamics

The nonparametric probabilistic approach of uncertainties has been introduced in (Soize, 2000, 2001, 2005b) to take into account both the model-parameter uncertainties and the model uncertainties induced by modeling errors, without separating the contribution of each one of these two types of uncertainties. Below, we summarized this approach which allows the uncertainties to be taken into account in the computational model.

4.1 Methodology to take into account both the model-parameter uncertainties and the model uncertainties (modeling errors)

Let $(\Theta', \mathcal{T}', \mathcal{P}')$ be another probability space (we recall that probability space $(\Theta, \mathcal{T}, \mathcal{P})$ is devoted to the probabilistic model of model-parameter uncertainties using the parametric probabilistic approach (Section 3). Let $\underline{\mathbf{x}}$ be the nominal value of model parameter $\mathbf{x}$, which is fixed in $\mathcal{C}_{\text{par}}$. Let $\underline{\mathbf{y}}^n$ be the response of the mean reduced-order computational model at order n for the nominal value $\underline{\mathbf{x}}$ of parameter $\mathbf{x}$, and which is such that (see Eq. (7)),

$$\underline{\mathbf{y}}^n(t) = [\,\phi(\underline{\mathbf{x}})]\,\underline{\mathbf{q}}(t) \quad , \tag{21}$$

$$[M(\underline{\mathbf{x}})]\ddot{\underline{\mathbf{q}}}(t)+[D(\underline{\mathbf{x}})]]\dot{\underline{\mathbf{q}}}(t)+[K(\underline{\mathbf{x}})]]\underline{\mathbf{q}}(t)+\mathbf{F}_{\text{NL}}(\underline{\mathbf{q}}(t),\dot{\underline{\mathbf{q}}}(t);\underline{\mathbf{x}})=\mathbf{f}(t;\underline{\mathbf{x}}). \tag{22}$$

The value of n is fixed to the value n_0 defined at the end of Section 3.2, for which, for all $\underline{\mathbf{x}}$ fixed in $\mathcal{C}_{\text{par}}$, the response $\underline{\mathbf{y}}^n$ is converged with respect to n. For n fixed to the value n_0, the nonparametric probabilistic approach of uncertainties then consists in replacing, in Eq. (22), the matrices $[M(\underline{\mathbf{x}})]$, $[D(\underline{\mathbf{x}})]$ and $[K(\underline{\mathbf{x}})]$ by the random matrices

$$[\mathbf{M}(\underline{\mathbf{x}})] = \{\theta' \mapsto [\mathbf{M}(\theta';\underline{\mathbf{x}})]\} \quad ,$$

$$[\mathbf{D}(\underline{\mathbf{x}})] = \{\theta' \mapsto [\mathbf{D}(\theta';\underline{\mathbf{x}})]\} \quad , \tag{23}$$

$$[\mathbf{K}(\underline{\mathbf{x}})] = \{\theta' \mapsto [\mathbf{K}(\theta';\underline{\mathbf{x}})]\} \quad , \tag{24}$$

defined on probability space $(\Theta', \mathcal{T}', \mathcal{P}')$, depending on the nominal value $\underline{\mathbf{x}}$ of parameter $\mathbf{x}$ and for which the probability distributions will be defined in Section 4.2. Taking into account such a construction, it can be proven that

$$E\{[\mathbf{M}(\underline{\mathbf{x}})]\}=[M(\underline{\mathbf{x}})]\ ,\ E\{[\mathbf{D}(\underline{\mathbf{x}})]\}=[D(\underline{\mathbf{x}})]\ ,\ E\{[\mathbf{K}(\underline{\mathbf{x}})]\}=[K(\underline{\mathbf{x}})]\ . \tag{25}$$

The deterministic reduced computational model defined by Eqs. (21) and (22), is then replaced by the following stochastic reduced computational model,

$$\mathbf{Y}(t) = [\,\phi(\underline{\mathbf{x}})]\,\mathbf{Q}(t) \quad , \tag{26}$$

$$[\mathbf{M}(\underline{\mathbf{x}})]\,\ddot{\mathbf{Q}}(t) + [\mathbf{D}(\underline{\mathbf{x}})]\,\dot{\mathbf{Q}}(t) + [\mathbf{K}(\underline{\mathbf{x}})]\,\mathbf{Q}(t) + \mathbf{F}_{\mathrm{NL}}(\mathbf{Q}(t),\dot{\mathbf{Q}}(t);\underline{\mathbf{x}}) = \mathbf{f}(t;\underline{\mathbf{x}}) \quad , \tag{27}$$

in which for all fixed t, $\mathbf{Y}(t) = \{\theta' \mapsto \mathbf{Y}(\theta';t)\}$ and $\mathbf{Q}(t) = \{\theta' \mapsto \mathbf{Q}(\theta';t)\}$ are $\mathbb{R}^m$- and $\mathbb{R}^n$-valued random vectors defined on Θ'. The realizations $\mathbf{Y}(\theta';t)$ and $\mathbf{Q}(\theta';t)$ of the random variables $\mathbf{Y}(t)$ and $\mathbf{Q}(t)$ verify the deterministic equations

$$\mathbf{Y}(\theta';t) = [\,\phi(\underline{\mathbf{x}})]\,\mathbf{Q}(\theta';t) \quad , \tag{28}$$

$$\begin{aligned}[\mathbf{M}(\theta';\underline{\mathbf{x}})]\,\ddot{\mathbf{Q}}(\theta';t) + [\mathbf{D}(\theta';\underline{\mathbf{x}})]\,\dot{\mathbf{Q}}(\theta';t) + [\mathbf{K}(\theta';\underline{\mathbf{x}})]\,\mathbf{Q}(\theta';t)\\ +\mathbf{F}_{\mathrm{NL}}(\mathbf{Q}(\theta';t),\dot{\mathbf{Q}}(\theta';t);\underline{\mathbf{x}}) = \mathbf{f}(t;\underline{\mathbf{x}})\,.\end{aligned} \tag{29}$$

4.2 Construction of the prior probability model of the random matrices

For n fixed to the value n_0 and as explained in (Soize, 2000, 2001, 2005b), the random matrices $[\mathbf{M}(\underline{\mathbf{x}})]$, $[\mathbf{D}(\underline{\mathbf{x}})]$ and $[\mathbf{K}(\underline{\mathbf{x}})]$ introduced in Eq. (27) are written as

$$\begin{aligned}[\mathbf{M}(\underline{\mathbf{x}})] &= [L_M(\underline{\mathbf{x}})]^T\,[\mathbf{G}_M]\,[L_M(\underline{\mathbf{x}})] \quad ,\\ [\mathbf{D}(\underline{\mathbf{x}})] &= [L_D(\underline{\mathbf{x}})]^T\,[\mathbf{G}_D]\,[L_D(\underline{\mathbf{x}})] \quad ,\\ [\mathbf{K}(\underline{\mathbf{x}})] &= [L_K(\underline{\mathbf{x}})]^T\,[\mathbf{G}_K]\,[L_K(\underline{\mathbf{x}})] \quad ,\end{aligned} \tag{30}$$

in which $[L_M(\underline{\mathbf{x}})]$, $[L_D(\underline{\mathbf{x}})]$ and $[L_K(\underline{\mathbf{x}})]$ are the upper triangular matrices such that $[M(\underline{\mathbf{x}})] = [L_M(\underline{\mathbf{x}})]^T\,[L_M(\underline{\mathbf{x}})]$, $[D(\underline{\mathbf{x}})] = [L_D(\underline{\mathbf{x}})]^T\,[L_D(\underline{\mathbf{x}})]$ and $[K(\underline{\mathbf{x}})] = [L_K(\underline{\mathbf{x}})]^T\,[L_K(\underline{\mathbf{x}})]$.

In Eq. (30), $[\mathbf{G}_M]$, $[\mathbf{G}_D]$ and $[\mathbf{G}_K]$ are random matrices which are defined on probability space $(\Theta',\mathcal{T}',\mathcal{P}')$ with values in $\mathbb{M}_n^+(\mathbb{R})$. The joint probability density function of these random matrices $[\mathbf{G}_M]$, $[\mathbf{G}_D]$ and $[\mathbf{G}_K]$ is constructed using the maximum entropy principle under the constraints defined by the available information. Taking into account the available information introduced in the nonparametric probabilistic approach (Soize,

2000, 2001, 2005b), it is proven that these three random matrices are statistically independent and each one belongs to ensemble $\mathrm{SG}_{\varepsilon}^{+}$ of random matrices defined in Section 2.5. Consequently, these three random matrices depend on the free positive real dispersion parameters δ_M, δ_D and δ_K which allow the level of the statistical fluctuations, that is to say the level of uncertainties, to be controlled. Let $\boldsymbol{\delta}_G = (\delta_M, \delta_D, \delta_K)$ be the vector of the dispersion parameters, which belongs to an admissible set $\mathcal{C}_G \subset \mathbb{R}^3$. Consequently, the joint probability density function of the random matrices $[\mathbf{G}_M]$, $[\mathbf{G}_D]$ and $[\mathbf{G}_K]$ is written as

$$
\begin{aligned}
([G_M],[G_D],[G_K]) \mapsto p_{\mathbf{G}}([G_M],[G_D],[G_K];\boldsymbol{\delta}_G) = p_{[\mathbf{G}_M]]}([G_M];\delta_M) \\
\times\, p_{[\mathbf{G}_D]]}([G_D];\delta_D) \times p_{[\mathbf{G}_K]]}([G_K];\delta_K) \quad , \quad \boldsymbol{\delta}_G \in \mathcal{C}_G \subset \mathbb{R}^3 \quad . \qquad (31)
\end{aligned}
$$

The probability distributions of random matrices $[\mathbf{G}_M]$, $[\mathbf{G}_D]$ and $[\mathbf{G}_K]$ and their algebraic representations, which are useful for generating independent realizations $[\mathbf{G}_M(\theta')]$, $[\mathbf{G}_D(\theta')]$ and $[\mathbf{G}_K(\theta')]$, are explicitly defined in Section 2.5. From Eq. (30), for θ' in Θ', it can then be deduced the realizations $[\mathbf{M}(\theta',\underline{\mathbf{x}})]$, $[\mathbf{D}(\theta',\underline{\mathbf{x}})]$ and $[\mathbf{K}(\theta',\underline{\mathbf{x}})]$ which are given by

$$
\begin{aligned}
[\mathbf{M}(\theta';\underline{\mathbf{x}})] &= [L_M(\underline{\mathbf{x}})]^T\,[\mathbf{G}_M(\theta')]\,[L_M(\underline{\mathbf{x}})] \quad , \\
[\mathbf{D}(\theta';\underline{\mathbf{x}})] &= [L_D(\underline{\mathbf{x}})]^T\,[\mathbf{G}_D(\theta')]\,[L_D(\underline{\mathbf{x}})] \quad , \\
[\mathbf{K}(\theta';\underline{\mathbf{x}})] &= [L_K(\underline{\mathbf{x}})]^T\,[\mathbf{G}_K(\theta')]\,[L_K(\underline{\mathbf{x}})] \quad .
\end{aligned}
$$

4.3 Estimation of the parameters of the prior probability model of uncertainties

As explained above, in the nonparametric probabilistic approach of uncertainties, n is fixed to the value n_0 for which the response of the mean reduced-order computational model is converged with respect to n. The prior probability model of uncertainties then depends on parameters $\underline{\mathbf{x}}$ and $\boldsymbol{\delta}_G$ belonging to the admissible sets $\mathcal{C}_{\text{par}}$ and $\mathcal{C}_G$. If no experimental data are available, then $\underline{\mathbf{x}}$ is fixed to its nominal value and $\boldsymbol{\delta}_G$ must be considered as a vector-valued parameter for performing a sensitivity analysis of the stochastic solution with respect to the level of uncertainties. Such a prior probability model of both the model-parameter uncertainties and the model uncertainties then allows the robustness of the solution to be analyzed as a function of the level of uncertainties which is controlled by $\boldsymbol{\delta}_G$.

If a few experimental data are available, the methodology presented in Section 3.5 can then be used to update the nominal value $\underline{\mathbf{x}}$ and to estimate $\boldsymbol{\delta}_G$. As previously, let $\mathbf{W}$ be the random vector which is observed, which is independent of t, but which depends on $\{\mathbf{Y}(t), t \in \mathcal{J}\}$, in which $\mathcal{J}$ is any part of

$\mathbb{R}$ and where $\mathbf{Y}$ is the vector-valued stochastic process which is the second-order random solution of Eqs. (26) and (27). For all $(\underline{\mathbf{x}}, \boldsymbol{\delta}_G)$ in $\mathcal{C}_{\text{par}} \times \mathcal{C}_G$, the probability density function of $\mathbf{W}$ is denoted by $\mathbf{w} \mapsto p_{\mathbf{W}}(\mathbf{w}; \underline{\mathbf{x}}, \boldsymbol{\delta}_G)$. The optimal value $(\underline{\mathbf{x}}^{\text{opt}}, \boldsymbol{\delta}_G^{\text{opt}})$ of $(\underline{\mathbf{x}}, \boldsymbol{\delta}_G)$ can be estimated by maximizing the logarithm of the likelihood function,

$$(\underline{\mathbf{x}}^{\text{opt}}, \boldsymbol{\delta}_G^{\text{opt}}) = \arg \max_{(\underline{\mathbf{x}}, \boldsymbol{\delta}_G) \in \mathcal{C}_{\text{par}} \times \mathcal{C}_G} \{ \sum_{r=1}^{\nu_{\text{exp}}} \log p_{\mathbf{W}}(\mathbf{w}_r^{\text{exp}}; \underline{\mathbf{x}}, \boldsymbol{\delta}_G \}, \tag{32}$$

in which $\mathbf{w}_1^{\text{exp}}, \ldots, \mathbf{w}_{\nu_{\text{exp}}}^{\text{exp}}$ are ν_{exp} independent experimental data corresponding to $\mathbf{W}$. Similarly to Section 3.5, for all r, $p_{\mathbf{W}}(\mathbf{w}_r^{\text{exp}}; \underline{\mathbf{x}}, \boldsymbol{\delta}_G)$ is estimated with the stochastic computational model defined by Eqs. (26) and (27) (using, for instance, the independent realizations of $\mathbf{W}$ calculated with the Monte Carlo method and the multivariate Gaussian kernel density estimation method.

4.4 Comments about the applications and the validations

Concerning the applications in linear and nonlinear structural dynamics, we refer the reader to (Soize, 2000, 2001, 2003a,b; Soize and Chebli, 2003; Capiez-Lernout and Soize, 2004; Desceliers et al., 2004; Capiez-Lernout et al., 2005; Arnst et al., 2006; Capiez-Lernout et al., 2006; Cottereau et al., 2007; Sampaio and Soize, 2007a; Cottereau et al., 2008; Mignolet and Soize, 2008a; Pellissetti et al., 2008; Ritto et al., 2009). Identification methodologies, applied to nonlinear structural dynamical problems, can be found in (Soize et al., 2008a; Batou and Soize, 2009a,b). Experimental validations are presented in (Chebli and Soize, 2004; Chen et al., 2006; Duchereau and Soize, 2006; Durand et al., 2008; Desceliers et al., 2009). Some applications devoted to robust optimization problems, that is to say to optimization problems using uncertain computational models, can be found in (Capiez-Lernout and Soize, 2008a,c; Soize et al., 2008b; Ritto et al., 2010).

5 Generalized probabilistic approach of uncertainties in computational structural dynamics

In Section 4, we have presented the nonparametric probabilistic approach of uncertainties which has been introduced in (Soize, 2000, 2001, 2005b) to take into account both the model-parameter uncertainties and the model uncertainties induced by modeling errors, without separating the contribution of each one of these two types of uncertainties. An extension of the nonparametric approach of uncertainties, called the generalized probabilistic approach of uncertainties, has been proposed (Soize, 2010a) and allows the prior probability model of each type of uncertainties (model-parameter uncertainties and model uncertainties induced by modeling errors) to be separately constructed. Below, we summarized this approach which is based on the use of the parametric probabilistic approach of model-parameter uncertainties, presented in Section 3, and on the use of the nonparametric probabilistic approach of model uncertainties induced by modeling errors, introduced in Section 4.

5.1 Methodology of the generalized probabilistic approach

Probability space $(\Theta, \mathcal{T}, \mathcal{P})$, introduced in Section 3, is devoted to the probability model of model-parameter uncertainties using the parametric probabilistic approach, while $(\Theta', \mathcal{T}', \mathcal{P}')$, introduced in Section 4, is devoted to the probability model of model uncertainties (modeling errors) using the nonparametric probabilistic approach. Similarly to the nonparametric probabilistic approach of uncertainties introduced in Section 4, the value of n is fixed to the value n_0 defined at the end of Section 3.2. For all $\mathbf{x}$ in $\mathcal{C}_{\text{par}}$, the response $\mathbf{y}^n$ computed with Eq. (7), is converged with respect to n (with a given accuracy). For n fixed to the value n_0, the generalized probabilistic approach of uncertainties then consists in replacing, for all $\mathbf{x}$ in $\mathcal{C}_{\text{par}}$, the matrices $[M(\mathbf{x})]$, $[D(\mathbf{x})]$ and $[K(\mathbf{x})]$ by the random matrices

$$[\mathbf{M}(\mathbf{x})] = \{\theta' \mapsto [\mathbf{M}(\theta'; \mathbf{x})]\}\,,$$

$$[\mathbf{D}(\mathbf{x})] = \{\theta' \mapsto [\mathbf{D}(\theta'; \mathbf{x})]\}\,,$$

$$[\mathbf{K}(\mathbf{x})] = \{\theta' \mapsto [\mathbf{K}(\theta'; \mathbf{x})]\}\,,$$

on probability space $(\Theta', \mathcal{T}', \mathcal{P}')$ and which will be defined in Section 5.2. The generalized probabilistic approach of uncertainties then consists in replacing in Eq. (10), in order to take into account modeling errors, the dependent random matrices $[M(\mathbf{X})]$, $[D(\mathbf{X})]$ and $[K(\mathbf{X})]$ by the dependent random matrices

$$[\mathbf{M}(\mathbf{X})] = \{(\theta, \theta') \mapsto [\mathbf{M}(\theta'; \mathbf{X}(\theta))]\} \quad ,$$

$$[\mathbf{D}(\mathbf{X})] = \{(\theta, \theta') \mapsto [\mathbf{D}(\theta'; \mathbf{X}(\theta))]\} \quad ,$$

$$[\mathbf{K}(\mathbf{X})] = \{(\theta, \theta') \mapsto [\mathbf{K}(\theta'; \mathbf{X}(\theta))]\} \quad ,$$

defined on $\Theta \times \Theta'$. Taking into account the construction which will be given in Section 5.2, it can easily be proven that

$$E\{[\mathbf{M}(\mathbf{X})]\} = [\underline{M}] \ , \ E\{[\mathbf{D}(\mathbf{X})]\} = [\underline{D}] \ , \ E\{[\mathbf{K}(\mathbf{X})]\} = [\underline{K}] \ , \qquad (33)$$

in which the matrices $[\underline{M}]$, $[\underline{D}]$ and $[\underline{K}]$ are the deterministic matrices introduced in Section 3.3. The stochastic reduced computational model, defined by Eqs. (9) and (10), is then replaced by the following stochastic reduced computational model,

$$\mathbf{Y}(t) = [\,\phi(\mathbf{X})]\,\mathbf{Q}(t) \quad , \tag{34}$$

$$[\mathbf{M}(\mathbf{X})]\,\ddot{\mathbf{Q}}(t) + [\mathbf{D}(\mathbf{X})]\,\dot{\mathbf{Q}}(t) + [\mathbf{K}(\mathbf{X})]\,\mathbf{Q}(t) + \mathbf{F}_{\mathrm{NL}}(\mathbf{Q}(t), \dot{\mathbf{Q}}(t); \mathbf{X}) = \mathbf{f}(t; \mathbf{X}) \quad , \tag{35}$$

in which for all fixed t, $\mathbf{Y}(t) = \{(\theta, \theta') \mapsto \mathbf{Y}(\theta, \theta'; t)\}$ and $\mathbf{Q}(t) = \{(\theta, \theta') \mapsto \mathbf{Q}(\theta, \theta'; t)\}$ are $\mathbb{R}^m$- and $\mathbb{R}^n$-valued random vectors defined on $\Theta \times \Theta'$. Let $\mathbf{X}(\theta_\ell)$ be any realization of random variable $\mathbf{X}$ for θ_ℓ in Θ. For all $\mathbf{x}$ in $\mathcal{C}_{\mathrm{par}}$, let $[\mathbf{M}(\theta'_{\ell'}; \mathbf{x})]$, $[\mathbf{D}(\theta'_{\ell'}; \mathbf{x})]$ and $[\mathbf{K}(\theta'_{\ell'}; \mathbf{x})]$ be any realizations of random matrices $[\mathbf{M}(\mathbf{x})]$, $[\mathbf{D}(\mathbf{x})]$, $[\mathbf{K}(\mathbf{x})]$ for $\theta'_{\ell'}$ in Θ'. The realizations $\mathbf{Y}(\theta_\ell, \theta'_{\ell'}; t)$ and $\mathbf{Q}(\theta_\ell, \theta'_{\ell'}; t)$ of random variables $\mathbf{Y}(t)$ and $\mathbf{Q}(t)$ verify the following deterministic equations

$$\mathbf{Y}(\theta_\ell, \theta'_{\ell'}; t) = [\,\phi(\mathbf{X}(\theta_\ell))]\,\mathbf{Q}(\theta_\ell, \theta'_{\ell'}; t) \quad , \tag{36}$$

$$\begin{aligned} &[\mathbf{M}(\theta'_{\ell'}; \mathbf{X}(\theta_\ell))]\,\ddot{\mathbf{Q}}(\theta_\ell, \theta'_{\ell'}; t) + [\mathbf{D}(\theta'_{\ell'}; \mathbf{X}(\theta_\ell))]\,\dot{\mathbf{Q}}(\theta_\ell, \theta'_{\ell'}; t) \\ &+ [\mathbf{K}(\theta'_{\ell'}; \mathbf{X}(\theta_\ell))]\,\mathbf{Q}(\theta_\ell, \theta'_{\ell'}; t) + \mathbf{F}_{\mathrm{NL}}(\mathbf{Q}(\theta_\ell, \theta'_{\ell'}; t), \dot{\mathbf{Q}}(\theta_\ell, \theta'_{\ell'}; t); \mathbf{X}(\theta_\ell)) \\ &= \mathbf{f}(t; \mathbf{X}(\theta_\ell)) \quad . \end{aligned} \tag{37}$$

5.2 Construction of the prior probability model of the random matrices

As explained in (Soize, 2010a), the dependent random matrices $[\mathbf{M}(\mathbf{X})]$, $[\mathbf{D}(\mathbf{X})]$ and $[\mathbf{K}(\mathbf{X})]$, introduced in Eq. (35), are written as

$$\begin{aligned} [\mathbf{M}(\mathbf{X})] &= [L_M(\mathbf{X})]^T\,[\mathbf{G}_M]\,[L_M(\mathbf{X})] \quad , \\ [\mathbf{D}(\mathbf{X})] &= [L_D(\mathbf{X})]^T\,[\mathbf{G}_D]\,[L_D(\mathbf{X})] \quad , \\ [\mathbf{K}(\mathbf{X})] &= [L_K(\mathbf{X})]^T\,[\mathbf{G}_K]\,[L_K(\mathbf{X})] \quad , \end{aligned} \tag{38}$$

in which, for all $\mathbf{x}$ in $\mathcal{C}_{\mathrm{par}}$, $[L_M(\mathbf{x})]$, $[L_D(\mathbf{x})]$ and $[L_K(\mathbf{x})]$ are the upper triangular matrices such that $[M(\mathbf{x})] = [L_M(\mathbf{x})]^T\,[L_M(\mathbf{x})]$, $[D(\mathbf{x})] = [L_D(\mathbf{x})]^T\,[L_D(\mathbf{x})]$ and $[K(\mathbf{x})] = [L_K(\mathbf{x})]^T\,[L_K(\mathbf{x})]$. In Eq. (38), $[\mathbf{G}_M]$, $[\mathbf{G}_D]$ and $[\mathbf{G}_K]$ are the random matrices on probability space $(\Theta', \mathcal{T}', \mathcal{P}')$, with values in $\mathbb{M}_n^+(\mathbb{R})$, defined in Section 4.2. These three random matrices are statistically independent and each one belongs to ensemble $\mathrm{SG}_\varepsilon^+$ of random matrices defined in Section 2.5. Consequently, these three random matrices depend on free positive real dispersion parameters δ_M, δ_D and δ_K which allow the level of model uncertainties to be controlled. As previously, let $\boldsymbol{\delta}_G = (\delta_M, \delta_D, \delta_K)$ be the vector of the dispersion parameters, which belongs to an admissible set $\mathcal{C}_G \subset \mathbb{R}^3$. The algebraic representation of each random matrix $[\mathbf{G}_M]$, $[\mathbf{G}_D]$ or $[\mathbf{G}_K]$, given in Section 2.5, is useful for generating independent realizations $[\mathbf{G}_M(\theta')]$, $[\mathbf{G}_D(\theta')]$ and $[\mathbf{G}_K(\theta')]$. From Eq. (38), for (θ, θ') in $\Theta \times \Theta'$, it can then be deduced the realization $[\mathbf{M}(\theta'; \mathbf{X}(\theta))]$, $[\mathbf{D}(\theta'; \mathbf{X}(\theta))]$ and $[\mathbf{K}(\theta'; \mathbf{X}(\theta))]$ which are given by

$$
\begin{aligned}
[\mathbf{M}(\theta'; \mathbf{X}(\theta))] &= [L_M(\mathbf{X}(\theta))]^T \, [\mathbf{G}_M(\theta')] \, [L_M(\mathbf{X}(\theta))] \quad , \\
[\mathbf{D}(\theta'; \mathbf{X}(\theta))] &= [L_D(\mathbf{X}(\theta))]^T \, [\mathbf{G}_D(\theta')] \, [L_D(\mathbf{X}(\theta))] \quad , \\
[\mathbf{K}(\theta'; \mathbf{X}(\theta))] &= [L_K(\mathbf{X}(\theta))]^T \, [\mathbf{G}_K(\theta')] \, [L_K(\mathbf{X}(\theta))] \quad .
\end{aligned}
$$

5.3 Estimation of the parameters of the prior probability model of uncertainties

The value of n is fixed to the value n_0 defined in Section 5.1. The prior probability model of uncertainties then depends on parameters $\underline{\mathbf{x}}$, $\boldsymbol{\delta}_X$ and $\boldsymbol{\delta}_G$ belonging to the admissible sets $\mathcal{C}_{\text{par}}$, $\mathcal{C}_X$ and $\mathcal{C}_G$. If no experimental data are available, then $\underline{\mathbf{x}}$ is fixed to the nominal value $\mathbf{x}_0$ and, $\boldsymbol{\delta}_X$ and $\boldsymbol{\delta}_G$ must be considered as parameters to perform a sensitivity analysis of the stochastic solution. Such a prior probability model of the model-parameter uncertainties and the model uncertainties then allows the robustness of the solution to be analyzed in function of the level of model-parameter uncertainties controlled by $\boldsymbol{\delta}_X$ and in function of model uncertainties controlled by $\boldsymbol{\delta}_G$.

If a few experimental data are available, the methodology presented in Sections 3.5 and 4.3, can then be used to update $\underline{\mathbf{x}}$ and to estimate $\boldsymbol{\delta}_X$ and $\boldsymbol{\delta}_G$. As previously, let $\mathbf{W}$ be the random vector which is observed, which is independent of t, but which depends on $\{\mathbf{Y}(t), t \in \mathcal{J}\}$, in which $\mathcal{J}$ is any part of $\mathbb{R}$ and where $\mathbf{Y}$ is the vector-valued stochastic process which is the second-order random solution of Eqs. (34) and (35). For all $(\underline{\mathbf{x}}, \boldsymbol{\delta}_X, \boldsymbol{\delta}_G) \in \mathcal{C}_{\text{par}} \times \mathcal{C}_X \times \mathcal{C}_G$, the probability density function of $\mathbf{W}$ is denoted by $\mathbf{w} \mapsto p_{\mathbf{W}}(\mathbf{w}; \underline{\mathbf{x}}, \boldsymbol{\delta}_X, \boldsymbol{\delta}_G)$. The optimal value $(\underline{\mathbf{x}}^{\text{opt}}, \boldsymbol{\delta}_X^{\text{opt}}, \boldsymbol{\delta}_G^{\text{opt}})$ of $(\underline{\mathbf{x}}, \boldsymbol{\delta}_X, \boldsymbol{\delta}_G)$ can be estimated by maximizing the logarithm of the likelihood function,

$$
(\underline{\mathbf{x}}^{\text{opt}}, \boldsymbol{\delta}_X^{\text{opt}}, \boldsymbol{\delta}_G^{\text{opt}}) = \arg \max_{(\underline{\mathbf{x}}, \boldsymbol{\delta}_X, \boldsymbol{\delta}_G) \in \mathcal{C}_{\text{par}} \times \mathcal{C}_X \times \mathcal{C}_G} \{ \sum_{r=1}^{\nu_{\text{exp}}} \log p_{\mathbf{W}}(\mathbf{w}_r^{\text{exp}}; \underline{\mathbf{x}}, \boldsymbol{\delta}_X, \boldsymbol{\delta}_G \}, \tag{39}
$$

in which $\mathbf{w}_1^{\text{exp}}, \ldots, \mathbf{w}_{\nu_{\text{exp}}}^{\text{exp}}$ are ν_{exp} independent experimental data corresponding to $\mathbf{W}$. Similarly to Sections 3.5 and 4.3, for all r, the quantity $p_{\mathbf{W}}(\mathbf{w}_r^{\text{exp}};$

$\underline{\mathbf{x}}$, $\boldsymbol{\delta}_X, \boldsymbol{\delta}_G$) is estimated with the stochastic computational model defined by Eqs. (34) and (35) (using, for instance, the independent realizations of $\mathbf{W}$ calculated with the Monte Carlo method and the multivariate Gaussian kernel density estimation method.

5.4 Posterior probability model of uncertainties using the Bayesian method

Let $p_{\mathbf{X}}^{\rm prior}(\mathbf{x}) = p_{\mathbf{X}}(\mathbf{x}; \underline{\mathbf{x}}^{\rm opt}, \boldsymbol{\delta}_X^{\rm opt})$ and $p_{\mathbf{G}}([G_M],[G_D],[G_K];\boldsymbol{\delta}_G^{\rm opt})$ be the optimal prior probability density functions of random vector $\mathbf{X}$ and of random matrices $[\mathbf{G}_M]$, $[\mathbf{G}_D]$, $[\mathbf{G}_K]$, constructed with the optimal value ($\underline{\mathbf{x}}^{\rm opt}$, $\boldsymbol{\delta}_X^{\rm opt}$, $\boldsymbol{\delta}_G^{\rm opt}$) of ($\underline{\mathbf{x}}$, $\boldsymbol{\delta}_X, \boldsymbol{\delta}_G$) which has been calculated in Section 5.3. The objective of this section is the following. For this given optimal prior probability model $p_{\mathbf{G}}([G_M],[G_D],[G_K];\boldsymbol{\delta}_G^{\rm opt})$ of model uncertainties induced by modeling errors, the posterior probability density function $p_{\mathbf{X}}^{\rm post}(\mathbf{x})$ of model-parameter uncertainties is estimated using the experimental data associated with observation $\mathbf{W}$ and using the Bayesian method.

Posterior probability density function of the model-parameter uncertainties. Let $\mathbf{w}_1^{\rm exp}, \ldots, \mathbf{w}_{\nu_{\rm exp}}^{\rm exp}$ be the $\nu_{\rm exp}$ independent experimental data corresponding to observation $\mathbf{W}$ introduced in Section 5.3. The Bayesian method allows the posterior probability density function $p_{\mathbf{X}}^{\rm post}(\mathbf{x})$ to be calculated by

$$p_{\mathbf{X}}^{\rm post}(\mathbf{x}) = \mathcal{L}(\mathbf{x})\, p_{\mathbf{X}}^{\rm prior}(\mathbf{x}) \quad , \tag{40}$$

in which $\mathbf{x} \mapsto \mathcal{L}(\mathbf{x})$ is the likelihood function defined on $\mathbb{R}^{n_p}$, with values in $\mathbb{R}^+$, such that

$$\mathcal{L}(\mathbf{x}) = \frac{\Pi_{r=1}^{\nu_{\rm exp}}\, p_{\mathbf{W}|\mathbf{X}}(\mathbf{w}_r^{\rm exp}|\mathbf{x})}{E\{\Pi_{r=1}^{\nu_{\rm exp}}\, p_{\mathbf{W}|\mathbf{X}}(\mathbf{w}_r^{\rm exp}|\mathbf{X}^{\rm prior})\}} \quad . \tag{41}$$

In Eq. (41), $p_{\mathbf{W}|\mathbf{X}}(\mathbf{w}_\ell^{\rm exp}|\mathbf{x})$ is the experimental value of the conditional probability density function $\mathbf{w} \mapsto p_{\mathbf{W}|\mathbf{X}}(\mathbf{w}|\mathbf{x})$ of the random observation $\mathbf{W}$ given $\mathbf{X} = \mathbf{x}$ in $\mathcal{C}_{\rm par}$. Equation (41) shows that likelihood function $\mathcal{L}$ must verify the following equation,

$$E\{\mathcal{L}(\mathbf{X}^{\rm prior})\} = \int_{\mathbb{R}^{n_p}} \mathcal{L}(\mathbf{x})\, p_{\mathbf{X}}^{\rm prior}(\mathbf{x})\, d\mathbf{x} = 1 \quad . \tag{42}$$

Posterior probability density functions of the responses. Let $\mathbf{U} = (\mathbf{V}, \mathbf{W})$ be the random response vector in which $\mathbf{W}$ is the random vector which is experimentally observed and where $\mathbf{V}$ is another random response vector whose components are quantities which are not experimentally observed. As previously, $\mathbf{U}$ is independent of t, but depends on $\{\mathbf{Y}(t), t \in \mathcal{J}\}$, in which $\mathcal{J}$ is any part of $\mathbb{R}$ and where $\mathbf{Y}$ is the vector-valued stochastic process which is the second-order random solution of Eqs. (34) and (35). The probability density function $\mathbf{u} \mapsto p_{\mathbf{U}^{\text{post}}}(\mathbf{u})$ of the posterior random response vector $\mathbf{U}^{\text{post}}$ is then given by

$$p_{\mathbf{U}^{\text{post}}}(\mathbf{u}) = \int_{\mathbb{R}^{n_p}} p_{\mathbf{U}|\mathbf{X}}(\mathbf{u}|\mathbf{x})\, p_{\mathbf{X}}^{\text{post}}(\mathbf{x})\, d\mathbf{x} \quad , \tag{43}$$

in which the conditional probability density function $p_{\mathbf{U}|\mathbf{X}}(\mathbf{u}|\mathbf{x})$ of $\mathbf{U}$, given $\mathbf{X} = \mathbf{x}$, is constructed using the stochastic computational model defined by Eqs. (34) and (35) with $\mathbf{X} = \mathbf{x}$. From Eq. (40), it can be deduced that

$$p_{\mathbf{U}^{\text{post}}}(\mathbf{u}) = E\{\mathcal{L}(\mathbf{X}^{\text{prior}})\, p_{\mathbf{U}|\mathbf{X}}(\mathbf{u}|\mathbf{X}^{\text{prior}})\} \quad .$$

Let U_k^{post} be any component of random vector $\mathbf{U}^{\text{post}}$. The probability density function $u_k \mapsto p_{U_k^{\text{post}}}(u_k)$ on $\mathbb{R}$ of the posterior random variable U_k^{post} is then given by

$$p_{U_k^{\text{post}}}(u_k) = E\{\mathcal{L}(\mathbf{X}^{\text{prior}})\, p_{U_k|\mathbf{X}}(u_k|\mathbf{X}^{\text{prior}})\} \quad , \tag{44}$$

in which $u_k \mapsto p_{U_k|\mathbf{X}}(u_k|\mathbf{x})$ is the conditional probability density function of the real valued random variable U_k given $\mathbf{X} = \mathbf{x}$ and which is also constructed using the stochastic computational model defined by Eqs. (34) and (35) with $\mathbf{X} = \mathbf{x}$.

Computational aspects. We use the notation introduced in Sections 5.1 and 5.2 concerning the realizations. Let $\mathbf{X}^{\text{prior}}(\theta_1), \ldots, \mathbf{X}^{\text{prior}}(\theta_\nu)$ be ν independent realizations of $\mathbf{X}^{\text{prior}}$ whose probability density function is $\mathbf{x} \mapsto p_{\mathbf{X}}^{\text{prior}}(\mathbf{x})$. For ν sufficiently large, the right-hand side of Eq. (44) can be estimated by

$$p_{U_k^{\text{post}}}(u_k) \simeq \frac{1}{\nu} \sum_{\ell=1}^{\nu} \mathcal{L}(\mathbf{X}^{\text{prior}}(\theta_\ell))\, p_{U_k|\mathbf{X}}(u_k|\mathbf{X}^{\text{prior}}(\theta_\ell)) \quad . \tag{45}$$

Let

$$([\mathbf{G}_M(\theta_1')], [\mathbf{G}_D(\theta_1')], [\mathbf{G}_K(\theta_1')]), \ldots, [\mathbf{G}_M(\theta_{\nu'}')], [\mathbf{G}_D(\theta_{\nu'}')], [\mathbf{G}_K(\theta_{\nu'}')])$$

be ν' independent realizations of $([\mathbf{G}_M], [\mathbf{G}_D], [\mathbf{G}_K])$ whose probability density function is $([G_M], [G_D], [G_K]) \mapsto p_{\mathbf{G}}([G_M], [G_D], [G_K]; \delta_G^{\text{opt}})$. For fixed θ_ℓ, the computational model defined by Eqs. (36) and (37) with $\mathbf{X}(\theta_\ell) = \mathbf{X}^{\text{prior}}(\theta_\ell) = \mathbf{x}$ is used to calculate the ν' independent realizations $\mathbf{U}(\theta_1'|\mathbf{x})$, $\ldots, \mathbf{U}(\theta_{\nu'}'|\mathbf{x})$. It can then be deduced $\mathbf{W}(\theta_1'|\mathbf{X}^{\text{prior}}(\theta_\ell)), \ldots, \mathbf{W}(\theta_{\nu'}'|\mathbf{X}^{\text{prior}}(\theta_\ell))$

and, for all fixed k, we can compute $U_k(\theta'_1|\mathbf{X}^{\text{prior}}(\theta_\ell)), \ldots, U_k(\theta'_{\nu'}|\mathbf{X}^{\text{prior}}(\theta_\ell))$.

(1) Using independent realizations $\mathbf{W}(\theta'_1|\mathbf{X}^{\text{prior}}(\theta_\ell)), \ldots, \mathbf{W}(\theta'_{\nu'}|\mathbf{X}^{\text{prior}}(\theta_\ell))$ and the multivariate Gaussian kernel density estimation method (see Section ??), for $r = 1, \ldots, \nu_{\text{exp}}$, we can estimate $p_{\mathbf{W}|\mathbf{X}}(\mathbf{w}_r^{\text{exp}}|\mathbf{X}^{\text{prior}}(\theta_\ell))$ and then, for $\ell = 1, \ldots, \nu$, we can deduce $\mathcal{L}(\mathbf{X}^{\text{prior}}(\theta_\ell))$ using Eq. (41).

(2) For all fixed k, $p_{U_k|\mathbf{X}}(u_k|\mathbf{X}^{\text{prior}}(\theta_\ell))$ is estimated using the independent realizations $U_k(\theta'_1|\mathbf{X}^{\text{prior}}(\theta_\ell)), \ldots, U_k(\theta'_{\nu'}|\mathbf{X}^{\text{prior}}(\theta_\ell))$ and the kernel estimation method (Bowman and Azzalini, 1997). Using Eq. (45), it can then be deduced $p_{U_k^{\text{post}}}(u_k)$.

6 Nonparametric probabilistic approach of uncertainties in structural-acoustic models for the low- and medium-frequency ranges

This section is devoted to the predictions of complex structural-acoustic systems in the low- and medium-frequency ranges for which computational structural-acoustic models are required. The presentation is limited to a bounded structure coupled with a bounded internal acoustic cavity. In order to simplify the presentation, the acoustic coupling of the structure with the unbounded external acoustic fluid is not considered here but can be taken into account without any difficulties (Ohayon and Soize, 1998). For complex structural-acoustic systems, the mean computational structural-acoustic model is not sufficient to predict the responses of the real system due to the presence of both the model-parameter uncertainties and the model uncertainties induced by modeling errors in the computational model and by the variability of the real system with respect to the design system. In this section, we show how the nonparametric probabilistic approach of uncertainties presented in Section 4 can be used to model the uncertainties in the structural-acoustic computational model. We could present such an approach in the context of the generalized probabilistic approach presented in Section 5. However, we will limit the presentation to the context of the nonparametric probabilistic approach in order to simplify the presentation. Such an approach has been used in (Durand et al., 2008; Fernandez et al., 2009, 2010; Kassem et al., 2009, 2011) for structural-acoustic modeling of automotive vehicles in presence of uncertainties. The extension to the generalized probabilistic approach is straightforward and can easily be constructed in following the methodology introduced in Section 5.

6.1 Reduced mean structural-acoustic model

Mean boundary value problem. The structural-acoustic model is developed in the context of the 3D linear elastoacoustics for a structural-acoustic system made up of a damped elastic structure coupled with a closed internal acoustic cavity filled with a dissipative acoustic fluid. The linear responses of the structural-acoustic system are studied around a static equilibrium state which is taken as natural state at rest. The external acoustic fluid is a gas and its effects on the structural-acoustic system are assumed to be negligible in the frequency band of analysis. The mean computational structural-acoustic system is derived from the mean boundary value problem using the finite element method.

The damped structure occupies a bounded domain, is fixed (no rigid body displacements) and is subjected to external loads. This structure is made up of a linear non homogeneous anisotropic viscoelastic material. In the frequency domain, the real part of the viscoelastic tensor of the constitutive equation is assumed to be independent of the frequency but the imaginary part depends on the frequency. Concerning the internal dissipative acoustic fluid, the formulation in pressure is used. The mass density of the acoustic fluid is taken at equilibrium, the speed of sound is assumed to be constant and the coefficient due to the viscosity of the fluid depends on frequency (the coefficient due to thermal conduction is neglected). There is an acoustic source density inside the acoustic cavity. The corresponding equations and the associated boundary value problem can be found in (Ohayon and Soize, 1998).

The frequency band of analysis is denoted by $\mathcal{F} = [\omega_{min}\,, \omega_{max}]$ rad/s.

Mean computational structural-acoustic model. The finite element method is used to approximate the boundary value problem. A finite element mesh of the structure and the internal acoustic fluid is then introduced. Let $\mathbb{u}(\omega) = (\mathbb{u}_1(\omega), \ldots, \mathbb{u}_{m_s}(\omega))$ be the complex vector of the m_s degrees-of-freedom of the structure according to the finite element discretization of the displacement field. Let $\mathbb{p}(\omega) = (\mathbb{p}_1(\omega), \ldots, \mathbb{p}_{m_a}(\omega))$ be the complex vector of the m_a degrees-of-freedom of the acoustic fluid according to the finite element discretization of the pressure field. The finite element discretization of the boundary value problem yields the following mean computational structural-acoustic model,

$$\begin{bmatrix} [\mathbb{A}^s(\omega)] & [\mathbb{H}] \\ \omega^2 [\mathbb{H}]^T & [\mathbb{A}^a(\omega)] \end{bmatrix} \begin{bmatrix} \mathbb{u}(\omega) \\ \mathbb{p}(\omega) \end{bmatrix} = \begin{bmatrix} \mathbb{f}^s(\omega) \\ \mathbb{f}^a(\omega) \end{bmatrix} \quad , \tag{46}$$

where $[\mathbb{A}^s(\omega)]$ is the dynamical stiffness matrix of the damped structure *in vacuo* which is a symmetric $(m_s \times m_s)$ complex matrix such that

$$[\mathbb{A}^s(\omega)] = -\omega^2[\mathbb{M}^s] + i\omega[\mathbb{D}^s(\omega)] + [\mathbb{K}^s] \quad , \tag{47}$$

in which $[\mathbb{M}^s]$, $[\mathbb{D}^s(\omega)]$ and $[\mathbb{K}^s]$ are the mass, damping and stiffness matrices of the structure which belong to the set $\mathbb{M}^+_{m_s}(\mathbb{R})$ of all the positive-definite symmetric $(m_s \times m_s)$ real matrices. In Eq. (46), $[\mathbb{A}^a(\omega)]$ is the dynamical stiffness matrix of the dissipative acoustic fluid which is a symmetric $(m_a \times m_a)$ complex matrix such that

$$[\mathbb{A}^a(\omega)] = -\omega^2[\mathbb{M}^a] + i\omega[\mathbb{D}^a(\omega)] + [\mathbb{K}^a] \quad , \tag{48}$$

in which $[\mathbb{M}^a]$, $[\mathbb{D}^a(\omega)]$ and $[\mathbb{K}^a]$ are the "mass", "damping" and "stiffness" matrices of the acoustic cavity with fixed coupling interface. The matrix $[\mathbb{M}^a]$ belongs to $\mathbb{M}^+_{m_a}(\mathbb{R})$ and, the matrices $[\mathbb{D}^a(\omega)]$ and $[\mathbb{K}^a]$ belong to the set $\mathbb{M}^{+0}_{m_a}(\mathbb{R})$ of all the semi-positive definite symmetric $(m_a \times m_a)$ real matrices for which the rank is $m_a - 1$. The matrix $[\mathbb{H}]$ is the structural-acoustic coupling matrix which belongs to the set $\mathbb{M}_{m_s,m_a}(\mathbb{R})$ of all the $(m_s \times m_a)$ real matrices.

Reduced mean computational structural-acoustic model. The projection of the mean computational structural-acoustic equation (46) on the structural modes *in vacuo* and on the acoustic modes of the acoustic cavity with fixed coupling interface yields the reduced mean computational structural-acoustic model (Ohayon and Soize, 1998). The structural modes *in vacuo* and the acoustic modes of the cavity with fixed coupling interface are calculated by solving the two generalized eigenvalue problems

$$[\mathbb{K}^s]\,\boldsymbol{\phi}^s = \lambda^s[\mathbb{M}^s]\,\boldsymbol{\phi}^s \quad , \quad [\mathbb{K}^a]\,\boldsymbol{\phi}^a = \lambda^a[\mathbb{M}^a]\,\boldsymbol{\phi}^a \quad . \tag{49}$$

Let $[\Phi^s]$ be the $(m_s \times n)$ real matrix whose columns are the n structural modes $\boldsymbol{\phi}^s$ associated with the n first positive eigenvalues $0 < \lambda^s_1 \leq \lambda^s_2 \leq \ldots \leq \lambda^s_n$ (the n first structural eigenfrequencies are $\omega^s_\alpha = \sqrt{\lambda^s_\alpha}$). Let $[\Phi^a]$ be the $(m_a \times m)$ real matrix whose columns are constituted (1) of the constant pressure mode associated with the zero eigenvalue $\lambda_1 = 0$ and (2) of the acoustic modes $\boldsymbol{\phi}^a$ associated with the positive eigenvalues $0 < \lambda^a_2 \leq \lambda^a_3 \leq \ldots \leq \lambda^a_m$ (the $m-1$ first acoustical eigenfrequencies are $\omega^a_\beta = \sqrt{\lambda^a_\beta}$). It should be noted that the constant pressure mode is kept in order to model the quasi-static variation of the internal fluid pressure induced by the deformation of the coupling interface (Ohayon and Soize, 1998). The eigenvectors satisfied the orthogonality conditions

$$[\Phi^s]^T[\mathbb{M}^s]\,[\Phi^s] = [I_n] \quad , \quad [\Phi^s]^T[\mathbb{K}^s]\,[\Phi^s] = [\lambda^s] \quad , \tag{50}$$

$$[\Phi^a]^T [\mathbb{M}^a] [\Phi^a] = [I_m] \quad , \quad [\Phi^a]^T [\mathbb{K}^a] [\Phi^a] = [\lambda^a] \quad , \tag{51}$$

in which $[\lambda^s]_{\alpha\alpha'} = \lambda^s_\alpha \, \delta_{\alpha\alpha'}$ and $[\lambda^a]_{\beta\beta'} = \lambda^a_\beta \, \delta_{\beta\beta'}$ are diagonal matrices, where $[I_n]$ and $[I_m]$ are the unity matrices of dimension n and m respectively, and where $\delta_{\alpha\alpha} = 1$ and $\delta_{\alpha\alpha'} = 0$ for $\alpha \neq \alpha'$. Using Eqs. (50) and (51), the projection of Eq. (46) yields the reduced mean matrix model of the structural-acoustic system (also called the reduced mean computational structural-acoustic model). For all ω in the frequency band $\mathcal{F}$, the structural displacement $\mathbb{u}^n(\omega)$ and the internal acoustic pressure $\mathbb{p}^m(\omega)$ are written as

$$\mathbb{u}^n(\omega) = [\Phi^s] \, \mathbf{q}^s(\omega) \quad , \quad \mathbb{p}^m(\omega) = [\Phi^a] \, \mathbf{q}^a(\omega) \quad , \tag{52}$$

in which the complex vectors $\mathbf{q}^s(\omega)$ and $\mathbf{q}^a(\omega)$ are such that

$$\begin{bmatrix} [A^s(\omega)] & [H] \\ \omega^2 [H]^T & [A^a(\omega)] \end{bmatrix} \begin{bmatrix} \mathbf{q}^s(\omega) \\ \mathbf{q}^a(\omega) \end{bmatrix} = \begin{bmatrix} \mathbf{f}^s(\omega) \\ \mathbf{f}^a(\omega) \end{bmatrix} \quad , \tag{53}$$

in which $[H] = [\Phi^s]^T \, [\mathbb{H}] \, [\Phi^a]$ is in $\mathbb{M}_{n,m}(\mathbb{R})$ and where $\mathbf{f}^s(\omega) = [\Phi^s]^T \, \mathbb{f}^s(\omega)$ and $\mathbf{f}^a(\omega) = [\Phi^a]^T \, \mathbb{f}^a(\omega)$ are complex vectors. The generalized dynamical stiffness matrix $[A^s(\omega)]$ of the damped structure is the $(n \times n)$ complex matrix written as

$$[A^s(\omega)] = -\omega^2 [M^s] + i\omega [D^s(\omega)] + [K^s] \quad , \tag{54}$$

in which $[M^s] = [\Phi^s]^T \, [\mathbb{M}^s] \, [\Phi^s] = [I_n]$ and $[K^s] = [\Phi^s]^T \, [\mathbb{K}^s] \, [\Phi^s] = [\lambda^s]$ are diagonal matrices belonging to $\mathbb{M}_n^+(\mathbb{R})$. The generalized damping matrix $[D^s(\omega)] = [\Phi^s]^T \, [\mathbb{D}^s(\omega)] \, [\Phi^s]$ is a full matrix belonging to $\mathbb{M}_n^+(\mathbb{R})$. The generalized dynamical stiffness matrix $[A^a(\omega)]$ of the dissipative acoustic fluid is the $(m \times m)$ complex matrix written as

$$[A^a(\omega)] = -\omega^2 [M^a] + i\omega [D^a(\omega)] + [K^a] \quad , \tag{55}$$

in which $[M^a] = [\Phi^a]^T \, [\mathbb{M}^a] \, [\Phi^a] = [I_m]$ is a diagonal matrix belonging to $\mathbb{M}_m^+(\mathbb{R})$ and where $[D^a(\omega)] = [\Phi^a]^T \, [\mathbb{D}^a(\omega)] \, [\Phi^a] = \tau(\omega) \, [\lambda^a]$ and $[K^a] = [\Phi^a]^T \, [\mathbb{K}^a] \, [\Phi^a] = [\lambda^a]$ are diagonal matrices belonging to $\mathbb{M}_m^{+0}(\mathbb{R})$ of rank $m - 1$.

Convergence. Below, we will denote by n_0 and m_0 the values of n and m for which the responses $\mathbb{u}^n$ and $\mathbb{p}^m$ are converged to $\mathbb{u}$ and $\mathbb{p}$, with a given accuracy.

6.2 Stochastic reduced-order model of the computational structural-acoustic model using the nonparametric probabilistic approach of uncertainties

For n and m fixed to the values n_0 and m_0 which have been previously defined, we use the nonparametric probabilistic approach of both the model-parameter uncertainties and the model uncertainties following the methodology presented in Section 4. For all ω in $\mathcal{F}$, the random vector $\mathbf{U}^n(\omega)$ of the structural displacements and the random vector $\mathbf{P}^m(\omega)$ of the acoustic pressures are given by

$$\mathbf{U}^n(\omega) = [\Phi^s]\,\mathbf{Q}^s(\omega) \quad , \quad \mathbf{P}^m(\omega) = [\Phi^a]\,\mathbf{Q}^a(\omega) \quad , \tag{56}$$

in which the random vectors $\mathbf{Q}^s(\omega)$ and $\mathbf{Q}^a(\omega)$ are such that

$$\begin{bmatrix} [\mathbf{A}^s(\omega)] & [\mathbf{H}] \\ \omega^2[\mathbf{H}]^T & [\mathbf{A}^a(\omega)] \end{bmatrix} \begin{bmatrix} \mathbf{Q}^s(\omega) \\ \mathbf{Q}^a(\omega) \end{bmatrix} = \begin{bmatrix} \mathbf{f}^s(\omega) \\ \mathbf{f}^a(\omega) \end{bmatrix} \quad , \tag{57}$$

where $[\mathbf{H}]$ is a random matrix with values in the set $\mathbb{M}_{n,m}(\mathbb{R})$. The random generalized dynamical stiffness matrix of the damped structure is written as

$$[\mathbf{A}^s(\omega)] = -\omega^2[\mathbf{M}^s] + i\omega[\mathbf{D}^s(\omega)] + [\mathbf{K}^s] \quad , \tag{58}$$

in which $[\mathbf{M}^s]$, $[\mathbf{D}^s(\omega)]$ and $[\mathbf{K}^s]$ are random matrices with values in $\mathbb{M}_n^+(\mathbb{R})$. The random generalized dynamical stiffness matrix of the dissipative acoustic fluid is written as

$$[\mathbf{A}^a(\omega)] = -\omega^2[\mathbf{M}^a] + i\omega[\mathbf{D}^a(\omega)] + [\mathbf{K}^a] \quad , \tag{59}$$

in which $[\mathbf{M}^a]$ is a random matrix with values in $\mathbb{M}_m^+(\mathbb{R})$ and where $[\mathbf{D}^a(\omega)]$ and $[\mathbf{K}^a]$ are random matrices with values in $\mathbb{M}_m^{+0}(\mathbb{R})$. The rank of the two random matrices $[\mathbf{D}^a(\omega)]$ and $[\mathbf{K}^a]$ is fixed and is equal to $m-1$.

The development of the nonparametric probabilistic approach then requires to precisely define the available information for the random matrices above in order to construct their probability distributions using the maximum entropy principle which allows to maximize uncertainties. For all ω fixed in $\mathcal{F}$, this available information is defined as follows:

(i) The random matrices $[\mathbf{H}]$, $[\mathbf{M}^s]$, $[\mathbf{D}^s(\omega)]$, $[\mathbf{K}^s]$, $[\mathbf{M}^a]$, $[\mathbf{D}^a(\omega)]$ and $[\mathbf{K}^a]$ are defined on a probability space $(\Theta', \mathcal{T}', \mathcal{P}')$, and for θ' in Θ', we have

$$[\mathbf{H}(\theta')] \ \text{in}\ \mathbb{M}_{n,m}(\mathbb{R}) \quad , \tag{60}$$

$$[\mathbf{M}^s(\theta')]\,,\ [\mathbf{D}^s(\omega,\theta')] \ \text{and}\ [\mathbf{K}^s(\theta')] \ \text{in}\ \mathbb{M}_n^+(\mathbb{R}) \quad , \tag{61}$$

$$[\mathbf{M}^a(\theta')] \ \text{in}\ \mathbb{M}_m^+(\mathbb{R}) \quad ; \quad [\mathbf{D}^a(\omega,\theta')] \ \text{and}\ [\mathbf{K}^a(\theta')] \ \text{in}\ \mathbb{M}_m^{+0}(\mathbb{R}) \quad . \tag{62}$$

(ii) The mean value of these random matrices must be equal to the matrices of the reduced mean computational structural-acoustic model defined in Section 6.1:

$$E\{[\mathbf{H}]\} = [H] \quad , \tag{63}$$

$$E\{[\mathbf{M}^s]\} = [M^s]\,,\; E\{[\mathbf{D}^s(\omega)]\} = [D^s(\omega)]\,,\; E\{[\mathbf{K}^s]\} = [K^s] \quad , \tag{64}$$

$$E\{[\mathbf{M}^a]\} = [M^a]\,,\; E\{[\mathbf{D}^a(\omega)]\} = [D^a(\omega)]\,,\; E\{[\mathbf{K}^a]\} = [K^a] \quad , \tag{65}$$

in which $E\{[\mathbf{B}]\}$ denotes the mathematical expectation of random matrix $[\mathbf{B}]$.

(iii) For all ω fixed in $\mathcal{F}$, the probability distributions of the random matrices $[\mathbf{H}]$, $[\mathbf{M}^s]$, $[\mathbf{D}^s(\omega)]$, $[\mathbf{K}^s]$, $[\mathbf{M}^a]$, $[\mathbf{D}^a(\omega)]$ and $[\mathbf{K}^a]$ must be such that the random solution $(\mathbf{U}^n(\omega), \mathbf{P}^m(\omega))$ of Eqs. (56) to (57) is a second-order random vector,

$$E\{\|\mathbf{U}^n(\omega)\|^2\} < +\infty \quad , \quad E\{\|\mathbf{P}^m(\omega)\|^2\} < +\infty \quad . \tag{66}$$

6.3 Construction of the prior probability model of uncertainties

(i) Let us assumed that $n = n_0$ and $m = m_0$ in which n_0 and m_0 are defined in Section 6.1, with $n \geq m$. The elastic modes of the structure and the acoustic modes of the internal acoustic cavity are selected in order for that the matrix $[H]$ which belongs to $\mathbb{M}_{n,m}(\mathbb{R})$ be such that $[H]\,\mathbf{q}^a = 0$ implies $\mathbf{q}^a = 0$. It should be noted that if $m \geq n$ the following construction must be applied to $[H]^T$ instead of $[H]$. Then, using the singular value decomposition of the rectangular matrix $[H]$, one can write $[H] = [C]\,[T]$ in which the rectangular matrix $[C]$ in $\mathbb{M}_{n,m}(\mathbb{R})$ is such that $[C]^T\,[C] = [\,I_m]$ and where the symmetric square matrix $[T]$ belongs to $\mathbb{M}_m^+(\mathbb{R})$ and can then be written (using the Cholesky decomposition) as $[T] = [L_T]^T\,[L_T]$ in which $[L_T]$ is an upper triangular matrix in $\mathbb{M}_m(\mathbb{R})$. Since $[T]$ belongs to $\mathbb{M}_m^+(\mathbb{R})$, we introduce the random matrix $[\mathbf{T}]$ with values in $\mathbb{M}_m^+(\mathbb{R})$ such that $E\{[\mathbf{T}]\} = [T]$. The random matrix $[\mathbf{H}]$ with values in $\mathbb{M}_{n,m}(\mathbb{R})$ can then be written (Soize, 2005b) as

$$[\mathbf{H}] = [C]\,[\mathbf{T}] \quad , \quad [\mathbf{T}] = [L_T]^T\,[\mathbf{G}_H]\,[L_T] \quad , \tag{67}$$

with $[\mathbf{G}_H]$ a random matrix with values in $\mathbb{M}_m^+(\mathbb{R})$ such that $E\{[\mathbf{G}_H]\} = [I_m]$.

(ii) For all ω in $\mathcal{F}$, it can be seen that the positive-definite matrices $[M^s]$, $[K^s]$, $[D^s(\omega)]$, $M^a]$ and the semi-positive definite matrices $[K^a]$, $[D^a(\omega)]$ can be written as

$$[M^s] = [L_{M^s}]^T\,[L_{M^s}] \quad , \quad [K^s] = [L_{K^s}]^T\,[L_{K^s}] \quad , \tag{68}$$

$$[D^s(\omega)] = [L_{D^s}(\omega)]^T \, [L_{D^s}(\omega)] \quad , \quad [M^a] = [L_{M^a}]^T \, [L_{M^a}] \quad , \qquad (69)$$

$$[K^a] = [L_{K^a}]^T \, [L_{K^a}] \quad , \quad [D^a(\omega)] = [L_{D^a}(\omega)]^T \, [L_{D^a}(\omega)] \quad , \qquad (70)$$

in which $[L_{M^s}]$ is the unity matrix $[I_n]$, $[L_{K^s}]$ is the diagonal matrix $[\lambda^s]^{1/2}$ in $\mathbb{M}_n(\mathbb{R})$, $[L_{D^s}(\omega)]$ is an upper triangular matrix in $\mathbb{M}_n(\mathbb{R})$, $[L_{M^a}]$ is the unity matrix $[I_m]$, $[L_{K^a}]$ is a rectangular matrix in $\mathbb{M}_{m-1,m}(\mathbb{R})$ such that, for all $\beta = 1, \ldots, m-1$, we have $[L_{K^a}]_{\beta 1} = 0$ and for $\beta' = 2, \ldots, m$, we have $[L_{K^a}]_{\beta\beta'} = \sqrt{\lambda^a_{\beta'}}\, \delta_{(\beta+1)\beta'}$. Finally, $[L_{D^a}(\omega)] = \sqrt{\tau(\omega)}\, [L_{K^a}]$ is a rectangular matrix in $\mathbb{M}_{m-1,m}(\mathbb{R})$. The random matrices $[\mathbf{M}^s]$, $[\mathbf{K}^s]$ and $[\mathbf{D}^s(\omega)]$ are written as

$$[\mathbf{M}^s] = [L_{M^s}]^T \, [\mathbf{G}_{M^s}] \, [L_{M^s}] \quad , \quad [\mathbf{K}^s] = [L_{K^s}]^T \, [\mathbf{G}_{K^s}] \, [L_{K^s}] \quad , \qquad (71)$$

$$[\mathbf{D}^s(\omega)] = [L_{D^s}(\omega)]^T \, [\mathbf{G}_{D^s}] \, [L_{D^s}(\omega)] \quad , \qquad (72)$$

in which $[\mathbf{G}_{M^s}]$, $[\mathbf{G}_{K^s}]$ and $[\mathbf{G}_{D^s}]$ are random matrices with values in $\mathbb{M}_n^+(\mathbb{R})$. These three random matrices are independent of ω and such that $E\{[\mathbf{G}_{M^s}]\} = E\{[\mathbf{G}_{K^s}]\} = E\{[\mathbf{G}_{D^s}]\} = [I_n]$. Similarly, the random matrices $[\mathbf{M}^a]$, $[\mathbf{K}^a]$ and $[\mathbf{D}^a(\omega)]$ are written as

$$[\mathbf{M}^a] = [L_{M^a}]^T \, [\mathbf{G}_{M^a}] \, [L_{M^a}] \quad , \quad [\mathbf{K}^a] = [L_{K^a}]^T \, [\mathbf{G}_{K^a}] \, [L_{K^a}] \quad , \qquad (73)$$

$$[\mathbf{D}^a(\omega)] = [L_{D^a}(\omega)]^T \, [\mathbf{G}_{D^a}] \, [L_{D^a}(\omega)] \quad , \qquad (74)$$

in which $[\mathbf{G}_{M^a}]$ is a random matrix with values in $\mathbb{M}_m^+(\mathbb{R})$, independent of ω and such that $E\{[\mathbf{G}_{M^a}]\} = [I_m]$ and where $[\mathbf{G}_{K^a}]$ and $[\mathbf{G}_{D^a}]$ are random matrices with values in $\mathbb{M}_{m-1}^+(\mathbb{R})$, independent of ω and such that $E\{[\mathbf{G}_{K^a}]\} = E\{[\mathbf{G}_{D^a}]\} = [I_{m-1}]$.

(iii) Taking into account Eqs. (67) to (74), the stochastic reduced computational model is completely defined as soon as the joint probability distribution of the random matrices $[\mathbf{G}_H]$, $[\mathbf{G}_{M^s}]$, $[\mathbf{G}_{D^s}]$, $[G_{K^s}]$, $[\mathbf{G}_{M^a}]$, $[\mathbf{G}_{D^a}]$ and $[\mathbf{G}_{K^a}]$ are constructed. Let $[\mathbf{G}]$ be any one of these random matrices with values in $\mathbb{M}_\mu^+(\mathbb{R})$ for which μ is equal to n, m or $m-1$. We then have $E\{[\mathbf{G}]\} = [I_\mu]$ and it has been proven that $E\{\|[\mathbf{G}]^{-1}\|_F^2\}$ must be finite in order that Eq. (66) be verified. Consequently, random matrix $[\mathbf{G}]$ must belong to ensemble $\mathrm{SG}_\varepsilon^+$ defined in Section 2.5.

6.4 Model parameters, stochastic solver and convergence analysis

(i) In Section 2.5, we have seen that the dispersion parameter δ allowed the dispersion of the random matrix $[\mathbf{G}]$ to be controlled. The dispersion parameters of random matrices $[\mathbf{G}_H]$, $[\mathbf{G}_{M^s}]$, $[\mathbf{G}_{D^s}]$, $[\mathbf{G}_{K^s}]$, $[\mathbf{G}_{M^a}]$, $[\mathbf{G}_{D^a}]$ and $[\mathbf{G}_{K^a}]$ will then be denoted by

$$\delta_H\, ,\, \delta_{M^s}\, ,\, \delta_{D^s}\, ,\, \delta_{K^s}\, ,\, \delta_{M^a}\, ,\, \delta_{D^a}\, ,\, \delta_{K^a} \quad . \tag{75}$$

and allow the level of uncertainties to be controlled.

(ii) For n and m fixed to the values n_0 and m_0, and for all ω fixed in $\mathcal{F}$, the random equation defined by Eq. (57) is solved by the Monte Carlo method with ν independent realizations. Using the usual statistical estimator of the mathematical expectation E, the mean-square convergence of the stochastic solution with respect to ν can be analyzed in studying the function $\nu \mapsto \mathcal{E}(\nu)$ such that

$$\mathcal{E}(\nu) = \nu^{-1/2}\{\sum_{\ell=1}^{\nu} \int_{\mathcal{F}} (\|\mathbf{U}^n(\omega,\theta_\ell)\|^2 + \|\mathbf{P}^m(\omega,\theta_\ell)\|^2)\, d\omega\}^{1/2} . \tag{76}$$

6.5 Estimation of the parameters of the prior probability model of uncertainties

There are several possible methodologies to identify the optimal values of the dispersion parameters δ_{M^s}, δ_{D^s}, δ_{K^s} for the structure, δ_{M^a}, δ_{D^a}, δ_{K^a} for the internal acoustic cavity and δ_H for the structural-acoustic coupling interface, using experimental data. The methodology described in Section 4.3 can be used. For the identification using experimental data, many details can be found in (Durand et al., 2008; Soize et al., 2008a; Fernandez et al., 2009, 2010).

6.6 Comments about the applications and the validations

Concerning the applications to structural-acoustic problems, we refer the reader to (Durand et al., 2008; Fernandez et al., 2009; Kassem et al., 2009; Fernandez et al., 2010; Kassem et al., 2011) for structural acoustics of automotive vehicles with experimental comparisons, and to (Capiez-Lernout and Soize, 2008b) for an example of the design optimization with an uncertain structural-acoustic model.

7 Conclusion

We have presented an ensemble of main concepts, methodologies and results allowing important problems to be addressed in the domain of the stochastic modeling of uncertainties, in particular in presence of model uncertainties induced by modeling errors in computational models. The approach proposed is based on the use of the random matrix theory allowing modeling errors to be directly taken into account at the operator levels of the boundary value problems. The fundamental aspects related to the identification of the prior and the posterior probability models using experimental data are presented. We have limited the presentation to the context of representative dynamical and structural-acoustical problems. Such a probability methodology can be used for elliptic (static) problems and parabolic (diffusion) problems. All the presented methodologies and presented formulations have been validated with simulate experiments and/or with experimental data and we have referred the readers to papers in which the details can be found.

Bibliography

T. W. Anderson. *Introduction to Multivariate Statistical Analysis.* John Wiley & Sons, New York, 1958.

M. Arnst and R. Ghanem. Probabilistic equivalence and stochastic model reduction in multiscale analysis. *Computer Methods in Applied Mechanics and Engineering*, 197:3584–3592, 2008.

M. Arnst, D. Clouteau, H. Chebli, R. Othman, and G. Degrande. A nonparametric probabilistic model for ground-borne vibrations in buildings. *Probabilistic Engineering Mechanics*, 21(1):18–34, 2006.

M. Arnst, D. Clouteau, and M. Bonnet. Inversion of probabilistic structural models using measured transfer functions. *Computer Methods in Applied Mechanics and Engineering*, 197(6-8):589–608, 2008.

M. Arnst, R. Ghanem, and C. Soize. Identification of bayesian posteriors for coefficients of chaos expansions. *Journal of Computational Physics*, 229(9):3134–3154, 2010.

S.K. Au and J.L. Beck. Subset simulation and its application to seismic risk based on dynamic analysis. *Journal of Engineering Mechanics - ASCE*, 129(8):901–917, 2003a.

S.K. Au and J.L. Beck. Important sampling in high dimensions. *Structural Safety*, 25(2):139–163, 2003b.

I. Babuska, R. Tempone, and G. E. Zouraris. Solving elliptic boundary value problems with uncertain coefficients by the finite element method: the stochastic formulation. *Computer Methods in Applied Mechanics and Engineering*, 194(12-16):1251–1294, 2005.

I. Babuska, F. Nobile, and R. Tempone. A stochastic collocation method for elliptic partial differential equations with random input data. *SIAM Journal on Numerical Analysis*, 45(3):1005–1034, 2007.

A. Batou and C. Soize. Identification of stochastic loads applied to a nonlinear dynamical system using an uncertain computational model and experimental responses. *Computational Mechanics*, 43(4):559–571, 2009a.

A. Batou and C. Soize. Experimental identification of turbulent fluid forces applied to fuel assemblies using an uncertain model and fretting-wear estimation. *Mechanical Systems and Signal Processing*, 23(7):2141–2153, 2009b.

A. Batou, C. Soize, and M. Corus. Experimental identification of an uncertain computational dynamical model representing a family of structures. *Computer and Structures*, pages In press,doi:10.1016/j.compstruc.2011.03.004, 2011.

J. L. Beck and L. S. Katafygiotis. Updating models and their uncertainties. i: Bayesian statistical framework. *Journal of Engineering Mechanics*, 124(4):455–461, 1998.

J.L. Beck. Bayesian system identification based on probability logic. *Structural Control and Health Monitoring*, 17(7):825–847, 2010.

J.L. Beck and S.K. Au. Bayesian updating of structural models and reliability using markov chain monte carlo simulation. *Journal of Engineering Mechanics - ASCE*, 128(4):380–391, 2002.

J.L. Beck, E. Chan, A.Irfanoglu, and et al. Multi-criteria optimal structural design under uncertainty. *Earthquake Engineering and Structural Dynamics*, 28(7):741–761, 1999.

J. M. Bernardo and A. F. M. Smith. *Bayesian Theory*. John Wiley & Sons, Chichester, 2000.

M. Berveiller, B. Sudret, and M. Lemaire. Stochastic finite element: a non-intrusive approach by regression. *European Journal of Computational Mechanics*, 15:81–92, 2006.

G. Blatman and B. Sudret. Sparse polynomial chaos expansions and adaptive stochastic finite elements using a regression approach. *Comptes Rendus Mcanique*, 336(6):518–523, 2007.

A. W. Bowman and A. Azzalini. *Applied Smoothing Techniques for Data Analysis*. Oxford University Press, Oxford, 1997.

E. Capiez-Lernout and C. Soize. Nonparametric modeling of random uncertainties for dynamic response of mistuned bladed disks. *Journal of Engineering for Gas Turbines and Power*, 126(3):600–618, 2004.

E. Capiez-Lernout and C. Soize. Robust design optimization in computational mechanics. *Journal of Applied Mechanics - Transactions of the ASME*, 75(2):021001–1 – 021001–11, 2008a.

E. Capiez-Lernout and C. Soize. Design optimization with an uncertain vibroacoustic model. *Journal of Vibration and Acoustics*, 130(2):021001–1 – 021001–8, 2008b.

E. Capiez-Lernout and C. Soize. Robust updating of uncertain damping models in structural dynamics for low- and medium-frequency ranges. *Mechanical Systems and Signal Processing*, 22(8):1774–1792, 2008c.

E. Capiez-Lernout, C. Soize, J.-P. Lombard, C. Dupont, and E. Seinturier. Blade manufacturing tolerances definition for a mistuned industrial bladed disk. *Journal of Engineering for Gas Turbines and Power*, 127(3):621–628, 2005.

E. Capiez-Lernout, M. Pellissetti, H. Pradlwarter, G. I. Schueller, and C. Soize. Data and model uncertainties in complex aerospace engineering systems. *Journal of Sound and Vibration*, 295(3-5):923–938, 2006.

B. P. Carlin and T. A. Louis. *Bayesian Methods for Data Analysis*. Third Edition, Chapman & Hall / CRC Press, Boca Raton, 2009.

G. Casella and E.I. George. Explaining the gibbs sampler. *The American Statistician*, 46(3):167–174, 1992.

E. Cataldo, C. Soize, R. Sampaio, and C. Desceliers. Probabilistic modeling of a nonlinear dynamical system used for producing voice. *Computational Mechanics*, 43(2):265–275, 2009.

H. Chebli and C. Soize. Experimental validation of a nonparametric probabilistic model of non homogeneous uncertainties for dynamical systems. *Journal of the Acoustical Society of America*, 115(2):697–705, 2004.

C. Chen, D. Duhamel, and C. Soize. Probabilistic approach for model and data uncertainties and its experimental identification in structural dynamics: Case of composite sandwich panels. *Journal of Sound and Vibration*, 294(1-2):64–81, 2006.

S.H. Cheung and J.L. Beck. Bayesian model updating using hybrid monte carlo simulation with application to structural dynamic models with many uncertain parameters. *Journal of Engineering Mechanics - ASCE*, 135(4):243–255, 2009.

S.H. Cheung and J.L. Beck. Calculation of posterior probabilities for bayesian model class assessment and averaging from posterior samples based on dynamic system data. *Computer-Aided Civil and Infrastructure Engineering*, 25(5):304–321, 2010.

J.Y. Ching, J.L. Beck, and K.A. Porter. Bayesian state and parameter estimation of uncertain dynamical systems. *Probabilistic Engineering Mechanics*, 21(1):81–96, 2006.

P. Congdon. *Bayesian Statistical Modelling*. Second Edition, John Wiley & Sons, Chichester, 2007.

R. Cottereau, D. Clouteau, and C. Soize. Construction of a probabilistic model for impedance matrices. *Computer Methods in Applied Mechanics and Engineering*, 196(17-20):2252–2268, 2007.

R. Cottereau, D. Clouteau, and C. Soize. Probabilistic impedance of foundation, impact of the seismic design on uncertain soils. *Earthquake Engineering and Structural Dynamics*, 37(6):899–918, 2008.

S. Das and R. Ghanem. A bounded random matrix approach for stochastic upscaling. *Multiscale Model. Simul.*, 8(1):296325, 2009.

S. Das, R. Ghanem, and J. C. Spall. Asymptotic sampling distribution for polynomial chaos representation from data: a maximum entropy and fisher information approach. *SIAM Journal on Scientific Computing*, 30 (5):2207–2234, 2008.

S. Das, R. Ghanem, and S. Finette. Polynomial chaos representation of spatio-temporal random field from experimental measurements. *Journal of Computational Physics*, 228:8726–8751, 2009.

M. Deb, I. Babuska, and J.T. Oden. Solution of stochastic partial differential equations using galerkin finite element techniques. *Computer Methods in Applied Mechanics and Engineering*, 190:6359–6372, 2001.

B.J. Debusschere, H.N. Najm, P.P. Pebay, and *et al.* Numerical challenges in the use of polynomial chaos representations for stochastic processes. *SIAM Journal on Scientific Computing*, 26(2):698–719, 2004.

G. Deodatis and P. D. Spanos. 5th international conference on computational stochastic mechanics. *Special issue of the Probabilistic Engineering Mechanics*, 23(2-3):103–346, 2008.

C. Desceliers, C. Soize, and S. Cambier. Non-parametric - parametric model for random uncertainties in nonlinear structural dynamics - application to earthquake engineering. *Earthquake Engineering and Structural Dynamics*, 33(3):315–327, 2004.

C. Desceliers, R. Ghanem, and C. Soize. Maximum likelihood estimation of stochastic chaos representations from experimental data. *International Journal for Numerical Methods in Engineering*, 66(6):978–1001, 2006.

C. Desceliers, C. Soize, and R. Ghanem. Identification of chaos representations of elastic properties of random media using experimental vibration tests. *Computational Mechanics*, 39(6):831–838, 2007.

C. Desceliers, C. Soize, Q. Grimal, M. Talmant, and S. Naili. Determination of the random anisotropic elasticity layer using transient wave propagation in a fluid-solid multilayer: Model and experiments. *Journal of the Acoustical Society of America*, 125(4):2027–2034, 2009.

A. Doostan and G. Iaccarino. A least-squares approximation of partial differential equations with highdimensional random inputs. *Journal of Computational Physics*, 228(12):4332–4345, 2009.

A. Doostan, R. Ghanem, and J. Red-Horse. Stochastic model reductions for chaos representations. *Computer Methods in Applied Mechanics and Engineering*, 196(37-40):3951–3966, 2007.

J. Duchereau and C. Soize. Transient dynamics in structures with nonhomogeneous uncertainties induced by complex joints. *Mechanical Systems and Signal Processing*, 20(4):854–867, 2006.

J.-F. Durand, C. Soize, and L. Gagliardini. Structural-acoustic modeling of automotive vehicles in presence of uncertainties and experimental identification and validation. *Journal of the Acoustical Society of America*, 124(3):1513–1525, 2008.

B. Faverjon and R. Ghanem. Stochastic inversion in acoustic scattering. *Journal of the Acoustical Society of America*, 119(6):3577–3588, 2006.

C. Fernandez, C. Soize, and L. Gagliardini. Fuzzy structure theory modeling of sound-insulation layers in complex vibroacoustic uncertain sytems - theory and experimental validation. *Journal of the Acoustical Society of America*, 125(1):138–153, 2009.

C. Fernandez, C. Soize, and L. Gagliardini. Sound-insulation layer modelling in car computational vibroacoustics in the medium-frequency range. *Acta Acustica united with Acustica (AAUWA)*, 96(3):437–444, 2010.

G.S. Fishman. *Monte Carlo: Concepts, algorithms, and applications.* Springer Verlag, New York, 1996.

C. Fougeaud and A. Fuchs. *Statistique.* Dunod, Paris, 1967.

P. Frauenfelder, C. Schwab, and R.A. Todor. Finite elements for elliptic problems with stochastic coefficients. *Computer Methods in Applied Mechanics and Engineering*, 194(2-5):205–228, 2005.

B. Ganapathysubramanian and N. Zabaras. Sparse grid collocation schemes for stochastic natural convection problems. *Journal of Computational Physics*, 225(1):652–685, 2007.

S. Geman and D. Geman. Stochastic relaxation, gibbs distribution and the bayesian distribution of images. *IEEE Transactions on Pattern Analysis and Machine Intelligence*, Vol PAM I-6(6):721–741, 1984.

R. Ghanem. Ingredients for a general purpose stochastic finite elements formulation. *Computer Methods in Applied Mechanics and Engineering*, 168(1-4):19–34, 1999.

R. Ghanem and S. Dham. Stochastic finite element analysis for multiphase flow in heterogeneous porous media. *Transp. Porous Media*, 32:239–262, 1998.

R. Ghanem and R. Doostan. Characterization of stochastic system parameters from experimental data: A bayesian inference approach. *Journal of Computational Physics*, 217(1):63–81, 2006.

R. Ghanem and D. Ghosh. Efficient characterization of the random eigenvalue problem in a polynomial chaos decomposition. *International Journal for Numerical Methods in Engineering*, 72(4):486–504, 2007.

R. Ghanem and R. M. Kruger. Numerical solution of spectral stochastic finite element systems. *Computer Methods in Applied Mechanics and Engineering*, 129:289–303, 1996.

R. Ghanem and M. Pellissetti. Adaptive data refinement in the spectral stochastic finite element method. *Comm. Numer. Methods Engrg.*, 18: 141–151, 2002.

R. Ghanem and J. Red-Horse. Propagation of probabilistic uncertainty in complex physical systems using a stochastic finite element approach. *Physica D*, 133(1-4):137–144, 1999.

R. Ghanem and A. Sarkar. Reduced models for the medium-frequency dynamics of stochastic systems. *Journal of the Acoustical Society of America*, 113(2):834–846, 2003.

R. Ghanem and P. D. Spanos. *Stochastic finite elements: a spectral approach.* Springer-Verlag, New York, 1991.

R. Ghanem and P.D. Spanos. Polynomial chaos in stochastic finite elements. *Journal of Applied Mechanics - Transactions of the ASME*, 57(1):197–202, 1990.

R. Ghanem and P.D. Spanos. *Stochastic Finite Elements: A spectral Approach.* (revised edition) Dover Publications, New York, 2003.

R. Ghanem, S. Masri, M. Pellissetti, and R. Wolfe. Identification and prediction of stochastic dynamical systems in a polynomial chaos basis. *Computer Methods in Applied Mechanics and Engineering*, 194(12-16): 1641–1654, 2005.

R. Ghanem, R. Doostan, and J. Red-Horse. A probability construction of model validation. *Computer Methods in Applied Mechanics and Engineering*, 197(29-32):2585–2595, 2008.

D. Ghosh and R. Ghanem. Stochastic convergence acceleration through basis enrichment of polynomial chaos expansions. *International Journal for Numerical Methods in Engineering*, 73(2):162–184, 2008.

B. Goller, H.J. Pradlwarter, and G.I. Schueller. Robust model updating with insufficient data. *Computer Methods in Applied Mechanics and Engineering*, 198(37-40):3096–3104, 2009.

J. Guilleminot and C. Soize. Non-gaussian positive-definite matrix-valued random fields with constrained eigenvalues: application to random elasticity tensors with uncertain material symmetries. *International Journal for Numerical Methods in Engineering*, page (To appear), 2011.

J. Guilleminot, C. Soize, D. Kondo, and C. Benetruy. Theoretical framework and experimental procedure for modelling volume fraction stochastic fluctuations in fiber reinforced composites. *International Journal of Solid and Structures*, 45(21):5567–5583, 2008.

J. Guilleminot, C. Soize, and D. Kondo. Mesoscale probabilistic models for the elasticity tensor of fiber reinforced composites: experimental identification and numerical aspects. *Mechanics of Materials*, 41(12):1309–1322, 2009.

J. Guilleminot, A. Noshadravanb, C. Soize, and R. Ghanem. A probabilistic model for bounded elasticity tensor random fields with application to polycrystalline microstructures. *Computer Methods in Applied Mechanics and Engineering*, 200(17-20):1637–1648, 2011.

W. K. Hastings. Monte carlo sampling methods using markov chains and their applications. *Biometrika*, 109:57–97, 1970.

E. T. Jaynes. Information theory and statistical mechanics. *Physical Review*, 108(2):171–190, 1957.

J. Kaipio and E. Somersalo. *Statistical and Computational Inverse Problems*. Springer-Verlag, New York, 2005.

M. Kassem, C. Soize, and L. Gagliardini. Energy density field approach for low- and medium-frequency vibroacoustic analysis of complex structures using a stochastic computational model. *Journal of Sound and Vibration*, 323(3-5):849–863, 2009.

M. Kassem, C. Soize, and L. Gagliardini. Structural partitioning of complex structures in the medium-frequency range. an application to an automotive vehicle. *Journal of Sound and Vibration*, 330(5):937–946, 2011.

L.S. Katafygiotis and J.L. Beck. Updating models and their uncertainties. ii: Model identifiability. *Journal of Engineering Mechanics - ASCE*, 124 (4):463–467, 1998.

O.M. Knio and O.P. Le Maitre. Uncertainty propagation in cfd using polynomial chaos decomposition. *Fluid Dynamics Research*, 38(9):616–640, 2006.

T. Leissing, C. Soize, P. Jean, and J. Defrance. Computational model for long-range non-linear propagation over urban cities. *Acta Acustica united with Acustica (AAUWA)*, 96(5):884–898, 2010.

O. P. LeMaitre, O. M. Knio, H. N. Najm, and R. Ghanem. A stochastic projection method for fluid flow. ii. random process. *Journal of Computational Physics*, 181:9–44, 2002.

O. P. LeMaitre, O. M. Knio, H. N. Najm, and R. Ghanem. Uncertainty propagation using wiener-haar expansions. *Journal of Computational Physics*, 197(1):28–57, 2004a.

O. P. LeMaitre, H. N. Najm, R. Ghanem, and O. Knio. Multi-resolution analysis of wiener-type uncertainty propagation schemes. *Journal of Computational Physics*, 197(2):502–531, 2004b.

O. P. LeMaitre, H. N. Najm, P. P. Pebay, R. Ghanem, and O. Knio. Multi-resolution-analysis scheme for uncertainty quantification in chemical systems. *SIAM Journal on Scientific Computing*, 29(2):864–889, 2007.

O.P. LeMaitre and O.M. Knio. *Spectral Methods for Uncerainty Quantification with Applications to Computational Fluid Dynamics.* Springer, Heidelberg, 2010.

D. Lucor, C.H. Su, and G.E. Karniadakis. Generalized polynomial chaos and random oscillators. *International Journal for Numerical Methods in Engineering*, 60(3):571–596, 2004.

D. Lucor, J. Meyers, and P. Sagaut. Sensitivity analysis of large-eddy simulations to subgrid-scale-model parametric uncertainty using polynomial chaos. *Journal of Fluid Mechanics*, 585:255–279, 2007.

X. Ma and N. Zabaras. An efficient bayesian inference approach to inverse problems based on an adaptive sparse grid collocation method. *Inverse Problems*, 25(3):Article Number: 035013, 2009.

R. Mace, W. Worden, and G. Manson. Uncertainty in structural dynamics. *Special issue of the Journal of Sound and Vibration*, 288(3):431–790, 2005.

Y.M. Marzouk and H.N. Najm. Dimensionality reduction and polynomial chaos acceleration of bayesian inference in inverse problems. *Journal of Computational Physics*, 228(6):1862–1902, 2009.

Y.M. Marzouk, H.N. Najm, and L.A. Rahn. Stochastic spectral methods for efficient bayesian solution of inverse problems. *Journal of Computational Physics*, 224(2):560–586, 2007.

L. Mathelin and O. LeMaitre. Dual based a posteriori estimation for stochastic finite element method. *Comm. App. Math. Comp. Sci.*, 2(1):83–115, 2007.

H. G. Matthies. Stochastic finite elements: Computational approaches to stochastic partial differential equations. *Zamm-Zeitschrift Fur Angewandte Mathematik Und Mechanik*, 88(11):849–873, 2008.

H.G. Matthies and A. Keese. Galerkin methods for linear and nonlinear elliptic stochastic partial differential equations. *Computer Methods in Applied Mechanics and Engineering*, 194(12-16):1295–1331, 2005.

M. L. Mehta. *Random Matrices, Revised and Enlarged Second Edition.* Academic Press, New York, 1991.

N. Metropolis and S. Ulam. The monte carlo method. *Journal of American Statistical Association*, 49:335–341, 1949.

M. P. Mignolet and C. Soize. Nonparametric stochastic modeling of linear systems with prescribed variance of several natural frequencies. *Probabilistic Engineering Mechanics*, 23(2-3):267–278, 2008a.

M. P. Mignolet and C. Soize. Stochastic reduced order models for uncertain nonlinear dynamical systems. *Computer Methods in Applied Mechanics and Engineering*, 197(45-48):3951–3963, 2008b.

H.N. Najm. Uncertainty quantification and polynomial chaos techniques in computational fluid dynamics. *Journal Review of Fluid Mechanics*, pages 35–52, 2009.

A. Nouy. A generalized spectral decomposition technique to solve a class of linear stochastic partial differential equations. *Computer Methods in Applied Mechanics and Engineering*, 196(45-48):4521–4537, 2007.

A. Nouy. Generalized spectral decomposition method for solving stochastic finite element equations: Invariant subspace problem and dedicated algorithms. *Computer Methods in Applied Mechanics and Engineering*, 197(51-52):4718–4736, 2008.

A. Nouy. Recent developments in spectral stochastic methods for the numerical solution of stochastic partial differential equations. *Archives of Computational Methods in Engineering*, 16(3):251–285, 2009.

A. Nouy. Proper generalized decomposition and separated representations for the numerical solution of high dimensional stochastic problems. *Archives of Computational Methods in Engineering*, 17(4):403–434, 2010.

A. Nouy and O. P. Le Maitre. Generalized spectral decomposition for stochastic nonlinear problems. *Journal of Computational Physics*, 228 (1):202–235, 2009.

R. Ohayon and C. Soize. *Structural Acoustics and Vibration*. Academic Press, San Diego, London, 1998.

C. Papadimitriou, J.L. Beck, and S.K. Au. Entropy-based optimal sensor location for structural model updating. *Journal of Vibration and Control*, 6(5):781–800, 2000.

C. Papadimitriou, J.L. Beck, and L.S. Katafygiotis. Updating robust reliability using structural test data. *Probabilistic Engineering Mechanics*, 16(2):103–113, 2001.

M. Papadrakakis and A. Kotsopulos. Parallel solution methods for stochastic finite element analysis using monte carlo simulation. *Computer Methods in Applied Mechanics and Engineering*, 168(1-4):305–320, 1999.

M. Papadrakakis and N.D. Lagaros. Reliability-based structural optimization using neural networks and monte carlo simulation. *Computer Methods in Applied Mechanics and Engineering*, 191(32):3491–3507, 2002.

M. Papadrakakis and V. Papadopoulos. Robust and efficient methods for stochastic finite element analysis using monte carlo simulation. *Computer Methods in Applied Mechanics and Engineering*, 134(134):325–340, 1996.

M. Pellissetti, E. Capiez-Lernout, H. Pradlwarter, C. Soize, and G. I. Schueller. Reliability analysis of a satellite structure with a parametric and a non-parametric probabilistic model. *Computer Methods in Applied Mechanics and Engineering*, 198(2):344–357, 2008.

B. Peters and G. De Roeck. Stochastic system identification for operational modal analysis: A review. *Journal of Dynamic Systems Measurement and Control-Transactions of The Asme*, 123(4):659–667, 2001.

H. J. Pradlwarter, G. I. Schueller, and G. S. Szekely. Random eigenvalue problems for large systems. *Computer and Structures*, 80:2415–2424, 2002.

H.J. Pradlwarter and G.I. Schueller. On advanced monte carlo simulation procedures in stochastic structural dynamics. *International Journal of Non-Linear Mechanics*, 32(4):735–744, 1997.

H.J. Pradlwarter and G.I. Schueller. Local domain monte carlo simulation. *Structural Safety*, 32(5):275–280, 2010.

T.G. Ritto, C. Soize, and R. Sampaio. Nonlinear dynamics of a drill-string with uncertainty model of the bit-rock interaction. *International Journal of Non-Linear Mechanics*, 44(8):865–876, 2009.

T.G. Ritto, C. Soize, and R. Sampaio. Robust optimization of the rate of penetration of a drill-string using a stochastic nonlinear dynamical model. *Computational Mechanics*, 45(5):415–427, 2010.

R. Y. Rubinstein and D. P. Kroese. *Simulation and the Monte Carlo Method.* Second Edition, John Wiley & Sons, New York, 2008.

C.P. Rupert and C.T. Miller. An analysis of polynomial chaos approximations for modeling single-fluid-phase flow in porous medium systems. *Journal of Computational Physics*, 226(2):2175–2205, 2007.

S. Sakamoto and R. Ghanem. Polynomial chaos decomposition for the simulation of non-gaussian nonstationary stochastic processes. *Journal of Engineering Mechanics-ASCE*, 128(2):190–201, 2002.

R. Sampaio and C. Soize. On measures of non-linearity effects for uncertain dynamical systems - application to a vibro-impact system. *Journal of Sound and Vibration*, 303(3-5):659–674, 2007a.

R. Sampaio and C. Soize. Remarks on the efficiency of pod for model reduction in nonlinear dynamics of continuous elastic systems. *International Journal for Numerical Methods in Engineering*, 72(1):22–45, 2007b.

G. I. Schueller. Uncertainties in structural mechanics and analysis-computational methods. *Special issue of Computer and Structures*, 83(14):1031–1150, 2005.

G. I. Schueller. On the treatment of uncertainties in structural mechanics and analysis. *Computer and Structures*, 85(5-6):235–243, 2007.

G. I. Schueller and H. A. Jensen. Computational methods in optimization considering uncertainties - an overview. *Computer Methods in Applied Mechanics and Engineering*, 198(1):2–13, 2008.

G.I. Schueller. Efficient monte carlo simulation procedures in structural uncertainty and reliability analysis - recent advances. *Structural Engineering and Mechanics*, 32(1):1–20, 2009.

G.I. Schueller and H.J. Pradlwarter. Uncertain linear systems in dynamics: Retrospective and recent developments by stochastic approaches. *Engineering Structures*, 31(11):2507–2517, 2009.

R. J. Serfling. *Approximation Theorems of Mathematical Statistics.* John Wiley & Sons, 1980.

C. E. Shannon. A mathematical theory of communication. *Bell System Technology Journal*, 27(14):379–423 & 623–659, 1948.

C. Soize. *The Fokker-Planck Equation for Stochastic Dynamical Systems and its Explicit Steady State Solutions.* World Scientific Publishing Co Pte Ltd, Singapore, 1994.

C. Soize. A nonparametric model of random uncertainties on reduced matrix model in structural dynamics. *Probabilistic Engineering Mechanics*, 15 (3):277–294, 2000.

C. Soize. Maximum entropy approach for modeling random uncertainties in transient elastodynamics. *Journal of the Acoustical Society of America*, 109(5):1979–1996, 2001.

C. Soize. Random matrix theory and non-parametric model of random uncertainties. *Journal of Sound and Vibration*, 263(4):893–916, 2003a.

C. Soize. Uncertain dynamical systems in the medium-frequency range. *Journal of Engineering Mechanics*, 129(9):1017–1027, 2003b.

C. Soize. A comprehensive overview of a non-parametric probabilistic approach of model uncertainties for predictive models in structural dynamics. *Journal of Sound and Vibration*, 288(3):623–652, 2005a.

C. Soize. Random matrix theory for modeling uncertainties in computational mechanics. *Computer Methods in Applied Mechanics and Engineering*, 194(12-16):1333–1366, 2005b.

C. Soize. Non gaussian positive-definite matrix-valued random fields for elliptic stochastic partial differential operators. *Computer Methods in Applied Mechanics and Engineering*, 195(1-3):26–64, 2006.

C. Soize. Tensor-valued random fields for meso-scale stochastic model of anisotropic elastic microstructure and probabilistic analysis of representative volume element size. *Probabilistic Engineering Mechanics*, 23(2-3): 307–323, 2008a.

C. Soize. Construction of probability distributions in high dimension using the maximum entropy principle. applications to stochastic processes, random fields and random matrices. *International Journal for Numerical Methods in Engineering*, 76(10):1583–1611, 2008b.

C. Soize. Generalized probabilistic approach of uncertainties in computational dynamics using random matrices and polynomial chaos decompositions. *International Journal for Numerical Methods in Engineering*, 81 (8):939–970, 2010a.

C. Soize. Random matrices in structural acoustics. In R. Weaver and M. Wright, editors, *New Directions in Linear Acoustics: Random Matrix Theory, Quantum Chaos and Complexity*, pages 206–230. Cambridge University Press, Cambridge, 2010b.

C. Soize. Identification of high-dimension polynomial chaos expansions with random coefficients for non-gaussian tensor-valued random fields using partial and limited experimental data. *Computer Methods in Applied Mechanics and Engineering*, 199(33-36):2150–2164, 2010c.

C. Soize. A computational inverse method for identification of non-gaussian random fields using the bayesian approach in very high dimension. *Computer Methods in Applied Mechanics and Engineering*, 200(45-46):3083–3099, 2011.

C. Soize and H. Chebli. Random uncertainties model in dynamic substructuring using a nonparametric probabilistic model. *Journal of Engineering Mechanics*, 129(4):449–457, 2003.

C. Soize and C. Desceliers. Computational aspects for constructing realizations of polynomial chaos in high dimension. *SIAM Journal On Scientific Computing*, 32(5):2820–2831, 2010.

C. Soize and R. Ghanem. Physical systems with random uncertainties : Chaos representation with arbitrary probability measure. *SIAM Journal On Scientific Computing*, 26(2):395–410, 2004.

C. Soize and R. Ghanem. Reduced chaos decomposition with random coefficients of vector-valued random variables and random fields. *Computer Methods in Applied Mechanics and Engineering*, 198(21-26):1926–1934, 2009.

C. Soize, E. Capiez-Lernout, J.-F. Durand, C. Fernandez, and L. Gagliardini. Probabilistic model identification of uncertainties in computational models for dynamical systems and experimental validation. *Computer Methods in Applied Mechanics and Engineering*, 198(1):150–163, 2008a.

C. Soize, E. Capiez-Lernout, and R. Ohayon. Robust updating of uncertain computational models using experimental modal analysis. *AIAA Journal*, 46(11):2955–2965, 2008b.

J. C. Spall. *Introduction to Stochastic Search and Optimization*. John Wiley, 2003.

G. Stefanou, A. Nouy, and A. Clément. Identification of random shapes from images through polynomial chaos expansion of random level set functions. *International Journal for Numerical Methods in Engineering*, 79(2):127–155, 2009.

G.S. Szekely and G.I. Schuller. Computational procedure for a fast calculation of eigenvectors and eigenvalues of structures with random properties. *Computer Methods in Applied Mechanics and Engineering*, 191 (8-10):799–816, 2001.

A.A. Taflanidis and J.L. Beck. An efficient framework for optimal robust stochastic system design using stochastic simulation. *Computer Methods in Applied Mechanics and Engineering*, 198(1):88–101, 2008.

M. T. Tan, G.-L. Tian, and K. W. Ng. *Bayesian Missing Data Problems, EM, Data Augmentation and Noniterative Computation.* Chapman & Hall / CRC Press, Boca Raton, 2010.

E. Walter and L. Pronzato. *Identification of Parametric Models from Experimental Data.* Springer-Verlag, Berlin, 1997.

X.L. Wan and G.E. Karniadakis. An adaptive multi-element generalized polynomial chaos method for stochastic differential equations. *Journal of Computational Physics*, 209(2):617–642, 2005.

X.L. Wan and G.E. Karniadakis. Multi-element generalized polynomial chaos for arbitrary probability measures. *SIAM Journal on Scientific Computing*, 28(3):901–928, 2006.

X.L. Wan and G.E. Karniadakis. Error control in multielement generalized polynomial chaos method for elliptic problems with random coefficients. *Comm. Comput. Phys.*, 5(2-4):793–820, 2009.

X.Q. Wang, M.P Mignolet, C. Soize, and V. Khannav. Stochastic reduced order models for uncertain infinite-dimensional geometrically nonlinear dynamical system - stochastic excitation cases. In *IUTAM Symposium on Nonlinear Stochastic Dynamics and Control*, Hangzhou, China, May 10-14 2010.

C.G. Webster, F. Nobile, and R. Tempone. A sparse grid stochastic collocation method for partial differential equations with random input data. *SIAM Journal on Numerical Analysis*, 46(5):2309–2345, 2007.

N. Wiener. The homogeneous chaos. *American Journal of Mathematics*, 60 (1):897–936, 1938.

M. Wright and R. Weaver. *New Directions in Linear Acoustics: Random Matrix Theory, Quantum Chaos and Complexity.* Cambridge University Press, Cambridge, 2010.

D.B. Xiu and G.E. Karniadakis. Wiener-askey polynomial chaos for stochastic differential equations. *SIAM Journal on Scientific Computing*, 24(2): 619–644, 2002a.

D.B. Xiu and G.E. Karniadakis. Modeling uncertainty in steady state diffusion problems via generalized polynomial chaos. *Computer Methods in Applied Mechanics and Engineering*, 191(43):4927–4948, 2002b.

D.B. Xiu and G.E. Karniadakis. Modeling uncertainty in flow simulations via generalized polynomial chaos. *Journal of Computational Physics*, 187 (1):137–167, 2003.

N. Zabaras and B. Ganapathysubramanian. A scalable framework for the solution of stochastic inverse problems using a sparse grid collocation approach. *Journal of Computational Physics*, 227(9):4697–4735, 2008.

The Tale of Stochastic Linearization Technique: Over Half a Century of Progress

Isaac Elishakoff[1] and Lova Andriamasy[2]

[1]*Department of Ocean and Mechanical Engineering,*
Florida Atlantic University
Boca Raton, FL 33431-0991, U.S.A.
e-mail: elishako@fau.edu
Fax: 561-297-3885

[2]*Electricite de France,*
79/81 rue Pierre Sémard
92320 Châtillon, FRANCE

Abstract. Developments in stochastic linearization since its inception in 1953 are summarized in this review paper along with the new, statistical orthogonality based derivation of the method. The developments that are described in this review mostly took place after extensive accounts on the classical version of the stochastic linearization technique, such as the monographs by Roberts and Spanos (1990), and by Socha (2008), and the review articles by Socha (2005) and Crandall (2006), have been published the recent decade. This essay is an updated version of our previous reviews (1995, 2000) along with the new derivation of the formulas of the classical stochastic linearization technique.

Introduction

Stochastic linearization is the most versatile method for analysis of general non-linear systems and structures under random excitation. As Roberts and Spanos(1990 pp. 12 and 7) write, "Most of the approximate methods of analysis...are best suited to single degree of freedom systems, with stationary random excitation. For multi-degree of freedom systems, which are prevalent in most engineering applications, these methods are very difficult to apply; they tend to involve severe analytical complexity, often combined with excessive, computational requirements, in terms of core storage or execution time. The single exception is the method of statistical linearization, which enables the results to be obtained with relative ease, even in situations where multi-degree of freedom systems subjected to non-stationary random excitations are of concern... From the perspective of treatable physical problems, the statistical linearization method has proved a useful analytical tool across a wide spread of engineering applications." In almost thirty years since its first presentation by Booton (1953) and Kazakov (1954) it has been widely applied for study of various non-linear systems not amenable to exact solution. For example, the monograph by Roberts and Spanos (1990), and the review paper by Socha and Soong (1991) give comprehensive accounts of some of these developments. The fundamental idea of the method consists in replacing the original non-linear system by a linear one such that the difference between two systems is minimal in some probabilistic sense. Following the classical approach, the linearized system parameters are determined so that the difference between the nonlinear force and that in the linearized system is minimal in the mean square value. Falsone (1992.a, 1992.b) demonstrated that when parametric excitations are present, it is preferable to measure the difference on the coefficients of the Itô differential rule. Remarkably when only purely external excitations are present, Falsone's approach coincides with classical stochastic linearization. Recently, for the last named case, new stochastic linearization techniques have been suggested in which the differences between the nonlinear original system and the linearized one are considered in terms of potential energy (Zhang, 1989). In particular, Elishakoff (1991) investigated a Duffing oscillator subjected to white-noise excitation; it was shown that the best results are obtained when the linearized system parameters are obtained through minimizing the mean square error between the potential energies of the two systems. The exact solution for the stationary probability density of the oscillator is readily available and approximate solutions are not needed for this problem; therefore the validity of the modified stochastic linearization technique is easily checked for this case. Moreover, the results obtained by the stochastic linearization technique are compared with those yielded by

Monte Carlo simulation where exact solutions are unattainable.

Elishakof (1991) has demonstrated, that for some combinations of the parameters of the Duffing oscillator the proposed linearization criteria may yield results in perfect agreement with the exact solutions. For other sets of parameters, the proposed linearization yields results that are slightly higher than the exact probabilistic responses, whereas conventional linearization yields responses that are below the exact values. Since engineers generally resort to the notion of safety factors, structures realized through conventional stochastic linearization, may turn out to be "over-designed." Where the Duffing oscillator is extremely weakly nonlinear, the conventional approach may result in less error, than its energy-wise counterpart. However, for these almost linear systems the percentage-wise error is under two percent for either criteria (criterion). Therefore, for the Duffing oscillator, the energy-wise approach is almost always preferable.

The aim of this paper is to give a brief account of these developments, as well as to investigate the accuracy of energy linearization technique in the case of colored noise excitation. The results obtained in this way are compared with those obtained by Monte Carlo simulation.

Combined use of the concept of the potential energy and energy dissipation function is discussed for nonlinearly damped structures with nonlinear restoring force. Finally, combined use of stochastic linearization and Monte Carlo method is discussed.

Present work is motivated by following observations. Socha and Pawleta (1994) and independently Elishakoff and Colajanni (1997, 1998), Colajanni and Elishakoff (1998a, 1998b) criticized the classical derivation of the stochastic linearization technique including several previous works by this writer, and suggested another derivation of it. It turned out that the new technique yielded results that were *farther* from the exact solution, when the latter was available, than the classical scheme. An exception was the work by Elishakoff and Colajanni (1998). Here the authors constructed an example in which the criticized classical technique and the alternative approach yielded the same result. At first glance such a situation may appear to be disturbing, since it may appear natural (although not absolutely necessary) for a consistent derivation to produce "better" results than an inconsistent formulation.

Yet, the following ideas were entertained by this writer: Since the classical results are closer to the exact solution than the results based on the consistent derivation, perhaps it is possible to obtain the classical results via "pure" means, without resorting to unnecessary assumptions. The result of this query turned out to be positive. Hence a new interpretation of the stochastic linearization will be given, and in its light, the entire technique will be reviewed. As far as the detailed account of papers by Socha and Pawleta (1994, 1999), Elishakoff and Colajanni (1997, 1998), Colajanni and Elishakoff (1998a, 1998b) is concerned, we will refrain from it in this review, with their subtle points awaiting a more detailed exposition, once the responses to it will become available.

I. New Derivation of the Classical Linearization Technique

Let us consider the single-degree-of-freedom system governed by the following equation of motion

$$mX + cX + f(X) = W(t) \tag{1.1}$$

where m is mass, c is damping, $f(\bullet)$ is the restoring force (considered here as a non-linear function of the displacement X), $W(t)$ is a random, stationary, Gaussian excitation and the dots denote time derivatives. We are interested in calculating the mean square value of the displacement $E\left[X^2\right]$. The basic idea of any stochastic linearization technique consists of replacing the original non-linear equation (1) by a linear one such that the difference between the two systems is minimal in some probabilistic sense. According to the classical linearization technique equation (1.1) is replaced by the following linear one:

$$mX + cX + k_{eq}^{(1)}X = W(t) \tag{1.2}$$

in which $k_{eq}^{(1)}$ is the spring coefficient. Here, in contrast to classical derivations by Booton (1953), Kazakov (1954) and Caughey (1963), we postulate that $k_{eq}^{(1)}$ must be chosen so that the difference between equations (1.2) and (1.1) is orthogonal to X itself:

$$E\left[\left\{f(X) - k_{eq}^{(1)}X\right\}X\right] = 0 \tag{1.3}$$

where $E[\cdot]$ denotes the mathematical expectation of $(\bullet)$. This method can be dubbed as an *orthogonality based stochastic linearization.* Following this approach one finds the for $k_{eq}^{(1)}$

$$k_{eq}^{(1)} = \frac{E\left[f(X)X\right]}{E\left[X^2\right]} \tag{1.4}$$

The question is how to evaluate $E\left[f(X)X\right]$ and $E[X^2]$. At the first glance Eq. (1.4) appears to be totally useless. Indeed, by using it we have replaced the problem of evaluating $E[X^2]$ to that of calculating Eq. (1.4), whose numerator involves $E\left[f(X)X\right]$ and the denominator $E[X^2]$ is the sought quantity itself!

The question is how to evaluate $E\left[f(X)X\right]$? Obviously, the "best" policy would be utilizing the exact probability density $p_x(x)$ since

$$E\left[f(X)X\right] = \int_{-\infty}^{+\infty} p_x(x)f(x)x\,dx$$

Yet, if we knew $p_x(x)$ then we would not engage ourselves in replacing the original system by a linear one and consequently evaluating $E\left[f\left(X\right)X\right]$.

This situation is not unlike what happens in the Rayleigh quotient. After its derivation one is in a position to pose a following rightful question:" Is it not totally useless? In fact, for determining the natural frequency one needs to know the mode shape. But, if we knew the exact mode shape, we would substitute it into the governing differential equation, thus deriving the natural frequency!" The reply is, of course, that indeed, at the initial level of observation, Rayleigh's quotient appears to be of no use. Rayleigh had a wonderful idea: By approximating the mode shape, one can get the approximate expression for the natural frequency.

The analogous situation occurs with the stochastic linearization technique. At the initial observation, Eq. (1.4) appears impractical, since exact $p_x(x)$ is unknown. But if we approximate it, we can get an approximation of $k_{eq}^{(1)}$.

Which approximation should on utilize? It appears that one would use the probability density of the *replacing* linear system. If the input to the linear system is Gaussian, so is the output. Thus, one can use the Gaussian assumption for $p_x(x)$. If there is strong evidence or physical considerations for using another probability density, one could do so. This allowed developing the so-called non-Gaussian linearization scheme (Pradlwarter, 1991).

The above derivation also explains the fact that the classical stochastic linearization technique, when using the exact probability density, yields the exact mean square values for $E[X^2]$ (Caughey, 1998). This remarkable fact was observed by Caughey in his 1963 paper, apparently for the first time. The explanation is as follows: If $k_{eq}^{(1)}$ is derived via the exact probability density, then its value is exact. When this exact value is employed for the linear system, the output mean-square response must be exact too.

Whereas derivation of Eq. (1.4) differs from the classical derivation by Booton (1953) and Kazakov (1954) it coincides with their expressions, as well as with those utilized by other investigators. The present derivation is however extremely simple.

An alternative criterion was suggested by Kazakov in his pioneering paper (1954). This criterion requires that the mean square values of the nonlinear restoring force $f(X)$ and the equivalent, linear restoring force $k_{eq}^{(1)}X$ be equal

$$E\left[f^2(X)\right] = E\left[\left(k_{eq}^{(2)}X\right)^2\right] \tag{1.5}$$

which leads to expression

$$k_{eq}^{(2)} = \sqrt{\frac{E\left[f^2(X)\right]}{E\left(X^2\right)}} \tag{1.6}$$

It is interesting to note that the result of the Booton-Kazakov criterion (1.4) is almost universally resorted to in literature. To the best of author's knowledge, the Kazakov criterion (1.5) was referred to only in the book by Popov and Paltov (1960), who observed that for numerous cases, the mean-square values of the response, according to the Booton-Kazakov criterion exceeded the result yielded by the Kazakov criterion. According, they suggested that the arithmetic mean of these two results be used to obtain an approximation which would be closer to the exact mean square, than either of them, Booton (1953) or Kazakov (1954) criteria.

Bolotin (1984) demonstrated, however, on the particular example of a half-degree-of-freedom system (the simple oscillator with negligible mass), for which an exact solution was constructed by Caughey and Dienes, (1961) that the exact mean-square displacement may exceed the results yielded by either of the criteria. Therefore the approximation suggested by Popov and Paltov (1960) does not bracket the exact solution as claimed.

In the recent decade, a new stochastic linearization technique has been suggested based on the concept of potential energy; the reader may consult with papers by Zhang, (1989), Zhang, Elishakoff, Zhang (1991); Zhang, Elishakoff and Shinozuka (1992), Elishakoff (1991), Elishakoff and Colombi (1993), Elishakoff and Zhang (1992a), 1992b), Elishakoff and Falsone (1993), Elishakoff, Fang and Caimi (1995), Zhang and Zhang (1999).

Following this technique, equation (1.1) is replaced by the following linear one:

$$mX + cX + k_{eq}^{(3)} X = W(t) \tag{1.7}$$

in which the new stiffness spring $k_{eq}^{(3)}$ is chosen such that the mean square deviation between the potential energies held by the original non-linear system (1.1) and by its linear counterpart (5), is orthogonal to X^2, i.e.

$$E\left[\left\{U(X) - k_{eq}^{(3)} X^2/2\right\} X^2\right] = 0 \tag{1.8}$$

Where

$$k_{eq}^{(3)} = \frac{2\,E\left[U(X)X^2\right]}{E\left[X^4\right]} \tag{1.9}$$

$U(X)$ denotes the potential energy of the original non-linear system.

Elishakoff (1991) applied this new technique to a Duffing oscillator subjected to a white noise excitation, and performed a systematic comparison with an exact solution. It was shown that the new technique yields a "better" result than classical linearization; namely, for particular values of the parameters, it may yield the *exact* response. In the next section the classical and new linearization techniques are applied to a Duffing oscillator to illustrate the superiority of the new technique.

In analogy with Kazakov's criterion, in Eq. (1.5), Elishakoff and Zhang (1992a, 1992b) suggested using a criterion, based on the requirement that the mean-square values of the potential energy of the original system and its linear counterpart be equal:

$$E\left[U(X)^2\right] = E\left\{\left[k_{eq}^{(4)}X^2/2\right]^2\right\} \tag{1.10}$$

where

$$\mathrm{k}_{eq}^{(4)} = \sqrt{\frac{E\left\{\left[U(X)\right]^2\right\}}{E\left(X^4\right)}} \tag{1.11}$$

Criteria (1.3), (1.4), (1.8) and (1.10) refer to systems with linear damping characteristics. For *nonlinearly* damped systems these four criteria can be directly generalized. The conventional stochastic linearization technique stipulates that the mean square difference between the nonlinear damping force $\phi(X)$ and its linear counterpart $c_{eq}^{(1)}X$ is orthogonal to X

$$E\left[\left\{\varphi(X) - c_{eq}^{(1)}X\right\}X\right] = 0 \tag{1.12}$$

hence

$$c_{eq}^{(1)} = \frac{E\left[\varphi(X)X\right]}{E\left(X^2\right)} \tag{1.13}$$

In full analogy with Kazakov's criterion (1.5), applicable for the system with nonlinear restoring force, we can also require that the mean-square values of the dissipation force and its linear equivalent be equal, i.e.

$$\mathrm{E}\left[\varphi^2(\mathrm{X})\right] = \mathrm{E}\left[\mathrm{c}_{\mathrm{eq}}^{(2)}\mathrm{X}^2\right] \tag{1.14}$$

hence

$$c_{eq}^{(2)} = \sqrt{\frac{E\left[\varphi^2(X)\right]}{E\left(X^2\right)}} \tag{1.15}$$

To the best of our knowledge, this criterion is formulated here for the first time.

The third possible criterion of equivalence between the nonlinearly damped system and the one with linear damping was formulated by Wang and Zhang (1985) and Elishakoff and Zhang (1992.b). They required that the mean-square difference between the energy dissipation function $\Phi(X)$ of the original nonlinear system and that of the equivalent linear system $c_{eq}^{(3)}X^2/2$ is orthogonal to X^2

$$\mathrm{E}\left\{\left[\left\{\Phi(\mathrm{X}) - \mathrm{c}_{\mathrm{eq}}^{(3)}\mathrm{X}^2/2\right\}\mathrm{X}^2\right]\right\} = 0 \tag{1.16}$$

resulting in the equivalent damping coefficient

$$c_{eq}^{(3)} = 2\frac{E\left[\Phi(X)X^2\right]}{E\left(X^4\right)} \tag{1.17}$$

The fourth criterion was also suggested by Elishakoff and Zhang (1992.b). It is based on requirement of equality of mean squares of and $c_{eq}^{(4)}X^2/2$, namely

$$\mathrm{E}\left[\Phi^2(\mathrm{X})\right] = \mathrm{E}\left[\left(\mathrm{c}_{\mathrm{eq}}^{(4)}\mathrm{X}^2/2\right)^2\right] \tag{1.18}$$

with the attendant value of $c_{eq}^{(4)}$

$$c_{eq}^{(4)} = 2\sqrt{\frac{E\left[\Phi^2(X)\right]}{E\left(X^4\right)}} \tag{1.19}$$

We will evaluate several examples to illustrate the performance of the proposed stochastic linearization criteria.

Consider a Duffing oscillator subjected to ideal white noise excitation

$$X + \beta X + \alpha X + \varepsilon X^3 = W(t) \tag{1.20}$$

The mean-square displacement for the system with ε≡ 0 reads

$$E\left(X^2\right)_{\varepsilon=0} = \pi S / \alpha\beta \equiv e_0^2 \tag{1.21}$$

where S is the value of the spectral density of excitation $W(t)$. The exact probability density of X is readily available in the literature and allows comparison between the conventional and new stochastic linearization methods:

$$p_X(x) = C_2 \exp\left[-\frac{\pi S}{\alpha\beta}\left(\frac{1}{2}x^2 + \frac{1}{4}\frac{\varepsilon}{\alpha}x^4\right)\right] \tag{1.22}$$

where C_2 is a normalization constant. In view of Eq.(1.21), Eq.(1.22) can be rewritten as

$$p_X(x) = C_2 \exp\left[-\frac{1}{e_0^2}\left(\frac{1}{2}x^2 + \frac{1}{4}\frac{\varepsilon}{\alpha}x^4\right)\right] \tag{1.23}$$

We introduce new variables

$$p = \left(\frac{4\alpha e_0^{\ 2}}{\varepsilon}\right)\frac{1}{4} \quad y = \frac{1}{4e_0}\frac{\alpha}{\varepsilon}\frac{1}{2} \tag{1.24}$$

The normalization condition

$$C_2 \int_{+\infty}^{-\infty} \exp\left[-\frac{1}{e_0^{\ 2}}\left(\frac{1}{2}x^2 + \frac{1}{4}\frac{\varepsilon}{\alpha}x^4\right)\right]dx = 1 \tag{1.25}$$

yields

$$C_2 = \left[pZ_1(y)\right]^{-1} \tag{1.26}$$

where

$$Z_1(y) = 2\int_{-\infty}^{+\infty} \exp(-\tau^4 - 4y\tau^4)dy \tag{1.27}$$

Mean square displacement

$$E\left[X^2\right] = \int_{-\infty}^{+\infty} x^2 p_X(x)dx \tag{1.28}$$

reads

$$E\left[X^2\right] = \frac{p^2 Z_2(y)}{Z_1(y)} = 2e_0\frac{Z_2(y)}{Z_1(y)}\left(\frac{\alpha}{\varepsilon}\right)^{1/2} \tag{1.29}$$

where

$$\mathrm{Z_2(y)} = 2\int_{-\infty}^{+\infty} \tau^2 \mathrm{exp}(-\tau^4 - 4\mathrm{y})\mathrm{dx} \tag{1.30}$$

The functions $Z_1(y)$ and $Z_2(y)$ are defined and tabulated for certain values of y by Stratonovich (1961). Note that these functions can also be reduced to cylindrical functions of a fractional order (Bolotin; 1984, Piszczek and Niziol, 1984; Constantinou, 1985). Consider a particular case

$$e_0^2 = 0.54, \quad \beta = \frac{\alpha}{\varepsilon} = 1 \tag{1.31}$$

In these circumstances,

$$y = 0.34021, \; Z_1(y) = 1.26368, \; Z_2(y) = 0.26310 \tag{1.32}$$

The exact mean square value becomes

$$E\left[X^2\right] = 0.306 \tag{1.33}$$

Let us now contrast the performance of the conventional and new stochastic linearization (equation4) techniques. Conventional stochastic linearization yields

$$k_{eq}^{(1)} = \alpha + 3\varepsilon E\left[X^2\right] \tag{1.34}$$

Substitution the expression for the mean-square value

$$E\left[X^2\right] = \pi S / k_{eq}^{(1)} \beta \tag{1.35}$$

results in

$$E\left[X^2\right] = \pi S / \beta\left[\alpha + 3\varepsilon E\left[X^2\right]\right] \tag{1.36}$$

or, in view of Eq.(1.34)

$$E\left[X^2\right] = e_0^{\,2} / \left[1 + 3(\varepsilon/\alpha)E\left[X^2\right]\right] \tag{1.37}$$

which yields the quadratic

$$3(\varepsilon/\alpha)E\left[X^2\right]^2 + E\left[X^2\right] - e_0^{\,2} = 0 \tag{1.38}$$

For numerical values adopted in Eq.(1.44), we have, instead of Eq.(1.38)

$$3\left[E\left(X^2\right)\right]^2 + E\left(X^2\right) - 0.54 = 0 \tag{1.39}$$

with the attendant mean-square value

$$E\left(X^2\right) = 0.289 \tag{1.40}$$

which constitutes a difference of 6.47% compared to the exact solution.

Consider now the energy based stochastic linearization method. Eq.(6) yields

$$k_{eq}^{(3)} = \alpha + 2.5\varepsilon E\left(X^2\right) \tag{1.41}$$

Substitution $k_{eq}^{(2)}$ instead of $k_{eq}^{(1)}$ in Eq.(1.35) yields

$$E\left(X^2\right) = \pi S / \beta\left[\alpha + 2.5\varepsilon E\left(X^2\right)\right] \tag{1.42}$$

or in view of Eq.(1.21)

$$E\left(X^2\right) = e_0^{\,2} / \left[1 + 2.5(\varepsilon/\alpha)E\left(X^2\right)\right] \tag{1.43}$$

which results in a quadratic

$$2.5(\varepsilon/\alpha)\left[E\left(X^2\right)\right]^2 + E\left(X^2\right) - e_0^{\ 2} = 0 \qquad (1.44)$$

For the values in Eq.(1.44), we have

$$2.5\left[E\left(X^2\right)\right]^2 + E\left(X^2\right) - 0.54 = 0 \qquad (1.45)$$

with the attendant mean-square value $E\left(X^2\right) = 0.306$, which *coincides* with the exact value given in Eq.(1.33). For $e_0^{\ 2}$ in the vicinity of 0.54 and ratio ε/α in the vicinity of unity, the relative error associated with the new stochastic linearization technique may constitute about one percent, much smaller than under the conventional approach.

The above coincidence of the stochastic linearization result with exact solution suggests that for a specific set of parameters (namely $e_0^{\ 2} \approx 0.54$) the new version of stochastic linearization constitutes "*true*" linearization, in terminology of Kozin (1987).

For the Duffing oscillator subjected to Gaussian white noise the problem is amenable to exact solution. Consider now the case where the exact solution is unattainable. Indeed, the power of stochastic linearization lies in its applicability where all other analytical methods may fail. For a Duffing oscillator subjected to a colored noise (Falsone and Elishakoff, 1992), the motion of the system is governed by the differential Eq.(1.20), where, under new circumstances, $Q(t)$ is a filtered noise; in particular, we assume that $Q(t)$ is the response of the following first equation.

$$Q = -\gamma Q + \gamma W(t) \qquad (1.46)$$

$W(t)$ being a white noise with constant spectral density S. In equation (1.46) the parameter γ is the measure of the filtering. In fact, it is easy to verify that the spectral density function $\mathrm{S}_Q(\omega)$ of $Q(t)$ is given by

$$S_Q(\omega) = \frac{\gamma^2}{\gamma^2 + \omega^2} S \qquad (1.47)$$

and, for large values of γ, tends to S.

According to the classical linearization technique, Eq. (1.46) is replaced by one in which the linearized spring $k_{eq}^{(1)}$ is given by Eq. (1.34). In this way, the stationary mean square value response of the linearized system reads

$$E\left[X^2\right] = \frac{\gamma(\gamma + \beta)}{k_{eq}^{(1)}\beta\left(k_{eq}^{(1)} + \beta\gamma + \gamma^2\right)}\pi S \qquad (1.48)$$

which for large values of γ, tends to the mean square response of the linear system subjected to the white noise $W(t)$.

According to the new technique, specified in Eq.(1.11), the linearized spring $k_{eq}^{(3)}$ is given by Eq. (1.41). The stationary mean-square response is given in Eq.(1.48) where $k_{eq}^{(1)}$ is replaced by $k_{eq}^{(3)}$.

The two linearization techniques were first applied to the Duffing oscillator by varying the filter parameter γ and fixing the system parameters at the values used by (Elishakoff, 1991):

$$\pi \mathrm{S}/\alpha\beta = 0.54, \ \varepsilon/\alpha = 1 \qquad (1.49)$$

These values are the same as those for which the new technique yielded the exact stationary ready-wave solution, when the input is a white noise (Elishakoff, 1991). Falsone and Elishakoff (1992) evaluated the percentage-wise error between the results obtained by means of each of the two schemes, and the results obtained via the Monte Carlo simulation. The superior accuracy of the new approach was evident for every value of the filter parameter γ examined. Moreover, it is worth noting that, for $\gamma = 10$, the percentage error under the new approach is practically zero, confirming the previous result obtained by Elishakoff (1991). In the same occasion, the percentage error was studied for $\gamma = 1$, $\pi S / \alpha\beta = 0.54$ and varying the ε / α. The superior accuracy of the new approach was yet established even for a high level of non-linearity; indeed, for $\varepsilon / \alpha = 2$ the classical linearization yielded an error of 12% whereas the error for the new method was under two percent.

These results confirm that, in the case of a Duffing oscillator, linearization with respect to the potential energy yields much more accurate results than the classical approach, even when the input is a colored noise and the level of non-linearity is high.

Let us now test the proposed criteria on an example studied earlier via the perturbation method (Khabbaz, 1964), and involving a system governed by the differential equation

$$mX + c_0 X \left(1 + b_2 X^2 + b_4 X^4\right) + k_0 X \left(1 + a_2 X^2 + a_4 X^4\right) = W(t) \quad (1.50)$$

where m is the mass, c_0 and k_0 are coefficients modeling the damping and spring stiffness of the linear system obtained by the formal substitution: $a_2 = b_2 = a_4 = b_4 = 0$. In the original system the four coefficients are assumed to be positive and specified, $W(t)$ is a *stationary* Gaussian white noise. In the absence of an exact solution to Eq. (1.50), we derive approximate solutions through the various stochastic linearization criteria and the results are compared with those yielded by the perturbation method and the Monte Carlo simulation.

We first consider the analysis through conventional stochastic linearization. Conditions (1.4) and (1.13) yield respectively,

$$k_{eq}^{(1)} = k_0 \left[E\left(X^2\right) + a_2 E\left(X^4\right) + a_4 E\left(X^6\right) \right] / E\left(X^2\right) \quad (1.51)$$

$$c_{eq}^{(1)} = c_0 \left[E\left(X^2\right) + b_2 E\left(X^4\right) + b_4 E\left(X^6\right) \right] / E\left(X^2\right) \quad (1.52)$$

The mean square displacement of the linear system

$$\sigma_{X_0}{}^2 = \pi S / c_0 k_0 \quad (1.53)$$

where S is the spectral density of $W(t)$ .For the equivalent linear system the mean-square displacement is

$$\sigma_X{}^2 = \pi S / c_{eq} k_{eq} \quad (1.54)$$

Moreover, the mean-square velocity equals

$$\sigma_{\dot{X}}{}^2 = k_{eq} \sigma^2{}_X / m = \pi S / c_{eq} m \quad (1.55)$$

Bearing in mind the postulated jointly normal probability density of the displacement and velocity for

$$\begin{aligned} c_{eq}^{(1)} &= c_0 \left[1 + 3b_2 \left(k_0 / m\right)\left(1 + 3a_2\sigma^2{}_X + 15a_4\sigma^4{}_X\right)\sigma^2{}_X \right. \\ &\left. +15b_4\left(k_0^2 / m^2\right)\left(1 + 3a_2\sigma^2{}_X + 15a_4\sigma^4{}_X\right) 2\sigma^4{}_X\right] \end{aligned} \quad (1.56)$$

evaluation of Eqs. (1.51) and (1.52), we find

$$k_{eq}^{(1)} = k_0\left(1 + 3a_2\sigma_X^2 + 15a_4\sigma_X^4\right) \quad (1.57)$$

Substitution of Eqs. (1.56) and (1.57) in (1.54) yields a polynomial equation

$$\sum_{i=0}^{9} A_i \left(\sigma_X^2\right)^i = 0 \quad (1.58)$$

where the coefficients A_i read

$$\begin{aligned}
A_0 &= -\sigma_{X_0}, \quad A_1 = 1, \quad A_2 = 3(a_2 m + k_0 b_2)/m \\
A_3 &= 3(5a_4 m^2 + 6mk_0 a_2 b_2 + 5k_0^2 b_4)\ /m^2 \\
A_4 &= 9k_0(3ma_2^2 b_2 + 10ma_4 b_2 + 15k_0 a_2 b_4)/m^2 \\
A_5 &= 135k_0(2ma_2 a_4 b_2 + 3k_0 a_2^2 b_4 + 5k_0 a_4 b_4)/m^2 \\
A_6 &= 135k_0(5ma_2^4 b_2 + 3k_0 a_2^3 b_4 + 30k_0 a_2 a_4 b_4)/m^2 \\
A_7 &= 2025k_0^2 a_4 b_4(3a_2^2 + 5a_4)/m^2 \\
A_8 &= 30375\mathrm{k}_0^2 \mathrm{a}_2 \mathrm{a2}_4 \mathrm{b}_4\ /\mathrm{m}^2 \\
A_9 &= 50625k_0^2 a_4^3 b_4\ /m^2
\end{aligned} \tag{1.59}$$

Here $\sigma_{X_0}{}^2$ is the mean-square response of the corresponding linear system. According to the Descartes' rule of signs, the polynomial Eq.(1.58) has a single positive root for $\sigma_X{}^2$, since the sequence of coefficients in Eq.(18) has only one change in sign. Evaluation of the mean-square displacement should be performed numerically.

Let us now reason to the criteria (1.8) and (1.16). The potential energy of the system reads

$$U(X) = k_0\ \{(1/2)X^2 + (1/4)a_2 X^4 + (1/6)a_4 X^6\} \tag{1.60}$$

For the energy dissipation function we obtain

$$\begin{aligned}
\Phi(X) &= \int_0^x c_0\ u(1 + b_2 u_2 + b_4 u_4)du \\
&= c_0(1/2)X^2 + (1/4)b_2 X^4 + (1/6)b_4 X^6
\end{aligned} \tag{1.61}$$

The expression for $k_{eq}^{(3)}$ reads

$$k_{eq}^{(3)} = 2k_0 \, K_{eq}^{(2)} \tag{1.62}$$

where

$$K_{eq}^{(2)} = (1/2) + (5/4)a_2\sigma_x^2 + (35/6)a_4\sigma_x^4 \tag{1.63}$$

The expression for the equivalent linear damping reads

$$c_{eq}^{(2)} = 2c_0 \left[(1/2) + (5/4)b_2\sigma_X^2 + (35/6)b_4\sigma_X^4\right] = 2c_0 C_{eq}^{(2)} \tag{1.64}$$

where

$$\begin{aligned} C_{eq}^{(2)} &= \frac{1}{2} + \frac{5}{4}b_2 2\frac{k_0}{m}\left(\frac{1}{2} + \frac{5}{4}a_2\sigma_X^2 + \frac{35}{6}a_4\sigma_X^4\right)^2 \sigma_X^4 \\ &+ \frac{35}{6}b_4\frac{4k_0^2}{m^2}\left(\frac{1}{2} + \frac{5}{4}a_2\sigma_X^2 + \frac{35}{6}a_4\sigma_X^4\right)\sigma_X^2 \end{aligned} \tag{1.65}$$

The mean-square displacement equals

$$\sigma_X^2 = \frac{\pi S}{c_{eq}^{(2)}k_{eq}^{(2)}} = \frac{\sigma_{X_0}^2}{4C_{eq}^{(2)}K_{eq}^{(2)}} \tag{1.66}$$

Substitution of Eqs.(1.63) and (1.64) in Eq.(1.66) leaves us with a polynomial equation$

$$\sum_{i=0}^{9} B_i \left(\sigma_X^2\right)^i = 0 \tag{1.67}$$

where the coefficients B_i read

$$
\begin{aligned}
B_0 &= -\sigma_{X_0}^2, \quad B_1 = 1, \quad B_2 = 5\left(a_2 m + k_0 b_2\right)/2m \\
B_3 &= 5\left(14a_4 m^2 + 15mk_0 a_2 b_2 + 14k_0^2 b_4\right)6m^2 \\
B_4 &= 25\,k_0\left(15a_2^2 b_2 m + 56a_4 b_2 m + 84k_0 a_2 b_4\right)/24m^2 \\
B_5 &= 175\,k_0\left(10a_2 a_4 b_2 m + 15k_0 a_2^2 b_4 + 28k_0 a_4 b_4\right)/12m^2 \\
B_6 &= 875\,k_0\left(28a_4^2 b_2 m + 15k_0 a_2^3 b_4 + 168k_0 a_2 a_4 b_4\right)/72m^2 \\
B_7 &= 6125\,k_0^2 a_4 b_4\left(15a_2^2 + 28a_4\right)/36m^2 \\
B_8 &= 214375k_0^2 a_2 a_4^2 b_4/18m^2 \\
B_9 &= 1500625k_0^2 a_4^3 b_4/81m^2
\end{aligned}
\tag{1.68}
$$

Again, Eq.(1.67), like Eq.(1.58) has a single positive root. Combined application of criteria (10) and (18) yields

$$k_{eq}^{(3)} = 2k_0 K_{eq}^{(3)} \tag{1.69}$$

Where

$$K_{eq}^{r(3)} = \left[\frac{1}{4} + \frac{5}{4}a_2\sigma_X^? + 35\left(\frac{1}{16}a_2 + \frac{1}{6}a_4\right)\sigma_X^4 + \frac{105}{4}a_2 a_4\sigma_X^6 + 105a_4\sigma_X^8\right]^{1/2} \tag{1.70}$$

Also

$$C_{eq}^{(3)} = 2c_0 C_{eq}^{(3)} \tag{1.71}$$

Where

$$
\begin{aligned}
\mathrm{C}_{eq}^{(3)} = &\left\{\frac{1}{4} + \frac{5}{4}\frac{b_2 k_{eq}^{(3)}}{m}\sigma_X^2 + 35\left(\frac{1}{16}b_2 + 16b_4\right)\left[\frac{k_{eq}^{(3)}}{m}\right]^2\sigma_X^4 + \right. \\
&\left.\frac{105}{4}b_2 b_4\left[\frac{k_{eq}^{(3)}}{m}\right]3\sigma_X^6 + 105b_4\left[\frac{k_{eq}^{(3)}}{m}\right]4\sigma_X^8\right\}^{1/2}
\end{aligned}
\tag{1.72}
$$

Substitution of expressions for equivalent spring stiffness and equivalent damping coefficient in Eq.(54) yields an equation for ${\sigma_X}^2$:

$$\sigma_X^2 - \frac{\pi S}{c_{eq}^{(3)} k_{eq}^{(3)}} \equiv \sigma_X^2 - \frac{\sigma 2X}{4K_{eq}^{(3)} C_{eq}^{(3)}} = 0 \tag{1.73}$$

The explicit form of the resulting polynomial equation in terms of ${\sigma_X}^2$ is cumbersome and is not reproduced here. One can show, however, that the equation also possesses a single root, as in the previous cases. Numerical results for Eqs.(1.58)~(1.67) have been obtained by Elishakoff and Zhang (1992a). Comparison with those results of Monte Carlo method (see Fig. 1 in the above paper) demonstrated that the conventional technique results in the largest error. For small values of the parameter ε criteria 1.3 and 1.4 yielded values extremely close to the simulated mean-square responses. For intermediate values of ε, namely $\varepsilon \approx 1$, the fourth criterion performed the best, whereas for larger values, namely $\varepsilon \approx 2$, the third criterion yielded results in close proximity to those of the Monte Carlo method.

Elishakoff and Zhang (1992.b) considered also a softening system:

$$mX + cX + (k_0 / \varepsilon)\operatorname{sgn}(X)(1 - e^{-\varepsilon|X|}) = W(t) \tag{1.74}$$

For values of ε up to unity the criterion of equal energy variances yielded results in best agreement with the simulation. In the range $1 \le \varepsilon \le 2$ the energy wise minimum mean square criterion turned out to be superior to the other criteria.

2. Hybrid Stochastic Linearization-Monte Carlo Method

The motivating considerations for developing a hybrid stochastic linearization-Monte Carlo Method are as follows.

The fraction of problems amenable to exact solution is very small. In these circumstances the answer lies either in purely numerical approximate techniques or in the Monte Carlo method. In the latter case, the volume of computation may be enormous for a large system hence the desirability of an analytical method combined with small scale simulation. To do this we choose weighing functions in the following form

$$w(X) = \left(1 + \sum_{i=1}^{n} \alpha i U i\right)^{1/2} \tag{2.1}$$

where coefficients α_i and β_i should be determined from the numerical experiments by the Monte Carlo method; n signifies the number of series of Monte Carlo simulations. Note that the weighing functions have also been considered, although in totally different contexts by Wang and Zhang (1985), Yu and Cao (1988), Izumi et al (1989), Fang and Fang (1991), Elishakoff and Zhang (1991a) and Zhang (1992).

In other respects, the proposed method is one of stochastic linearization and the four alternative criteria replaced by their equivalent versions including the weighing functions. For example, the third, criterion is replaced by

$$E\left[w(X)\left\{U(X) - k_{eq}^{(3)} X^2/2\right\}\right]^2 = \min \tag{2.2}$$

yielding

$$k_{eq}^{(3)} = \frac{2E\left\{w^2(X)U(X)X^2\right\}}{Ew^2(X)X^4} \tag{2.3}$$

Each simulation series should be conducted for specified sets of parameters of the system. Thus for n such sets of the results of Monte Carlo simulations will numerically coincide with these of the stochastic linearization technique. It is expected that for other sets of parameters the accuracy of the results yielded by the stochastic linearization technique with the weighing function will be quite satisfactory than without the latter.

Consider, for illustration purposes, a system governed by the following differential equation

$$mX + cX + k_1 X + k_2 X^5 = W(t) \tag{2.4}$$

where m, c, k_1 and k_2 are positive constants, $W(t)$ is an ideal white noise with zero mean. The system (1.82) is amenable to exact solution. It was originally chosen in the context of the possibility of quantifying the error associated with the approximate techniques. It was shown by Zhang, Elishakoff and Zhang (1991) that energy based linearization technique, even without recourse to weighing functions reduces the errors in determining the mean-square displacement of the system by about 50%.

We will illustrate the application of the energy based technique with weighing function. The potential energy of the nonlinear system in Eq. (2.4) reads

$$U(X) = (1/2)k_1 X^2 + (1/6)k_2 X^6 \tag{2.5}$$

We perform a single series of Monte Carlo simulations. Accordingly, in Eq.(2.1), $n = 1$ with

$$w(X) = (1 + \alpha_1 U)^{1/2} \tag{2.6}$$

We obtain following expressions for k_{eq}:

$$k_{eq} = k_1 + 2k_2 \frac{E\left[6X^8 + 3\alpha k_1 X^{10} + \alpha k_2 X^{14}\right]}{E\left[36X^4 + 18\alpha k_1 X^6 + 6\alpha k_2 X^{10}\right]} \tag{2.7}$$

The spectral analysis, yields the following mean-square displacement

$$\sigma_X^2 = \frac{\pi S}{c_{eq}^k} = \frac{\sigma_{X_0}^2}{1 + A} \tag{2.8}$$

where $\sigma_{X_0}{}^2$ is defined as

$$\sigma_{X_0}^2 = \pi S / ck_1 \tag{2.9}$$

and the "corrective" term A reads

$$A = 2\frac{k_2 E\left[6X^8 + 3\alpha k_1 X^{10} + \alpha k_2 X^{14}\right]}{k_1 E\left[36X^4 + 12\alpha k_1 X^6 + 6\alpha X^{10}\right]} \tag{2.10}$$

To calculate $E\left[X^{2J}\right]$ we use the usual approximation whereby that X has a normal probability density with zero mean Eq. (2.14) becomes:

$$\begin{aligned} &15015\alpha k_2^2\sigma_X^{12} + 630\alpha k_2\sigma_X^8 + \left(70k_2 - 315\alpha k_2 k_1 \sigma_{X_0}\right)\sigma_X^6 \\ &+ 15\alpha k_1^2\sigma_X^4 + \left(6k_1 - 15\alpha k_1^2\sigma_{X_0}\sigma_X^2\right)\sigma_X^2 - 6k_1\sigma_{X_0} = 0 \end{aligned} \tag{2.11}$$

In the latter equation the parameter α is not known. In order to determine it, we first solve an auxiliary problem of evaluation the mean square displacement $\sigma_X{}^2$ for specified set of parameters, by the Monte Carlo method, say for

$$m = m^{(1)}, k_1 = k_1^{(1)}, c = c^{(1)}, k2 = k_2^{(1)} \tag{2.12}$$

The result of simulation is denoted by $\sigma_X{}^2$. This allows one to determine the value of α corresponding the simulation results

$$\alpha = -2\frac{35k_2\sigma_X^6 - 3k_1\sigma_X^2 + 3k_1\sigma_0^X}{15\sigma_X\left(1001k_2^2\sigma_X^{12} + 42k_2k_1\sigma_X^8 - 21k_2k_1\sigma_X^6\sigma_{X_0} + k_1^2\sigma_X^4 - k_1^2\sigma_{X_0}\sigma_X^4\right)} \quad (2.13)$$

We set α at the value α1 determined from this equation, and use it in (2.11) for values of parameters, other than those listed in (2.12). Once α is substituted, (2.11) becomes a polynomial equation in ${\sigma_X}^2$.

$$\begin{aligned}
&35035k_2^3\sigma_X^{15} - 35035k_2^3 3\sigma_X^{14} + 4473k_1k_2^2\sigma_X^{11} + 3003k_1k_2^2\sigma_X^{10} \\
&- 1470k_2^2\sigma_X^{10} - 3738k_1k_2^2\sigma_{X_0}\sigma_X^9 - 2268k_1k_2^2\sigma_{X_0}\sigma_X^8 - 35k_1^2k_2\sigma_{X_0}\sigma_X^7 \\
&+ 161k_1^2k_2\sigma_X^7 + 35k_1^2k_2\sigma_{X_0}\sigma_X^6 - 35k_1^2k_2\sigma_X^6 + 126k_1k_2\sigma_X^6 - 189k_1^2k_2\sigma_{X_0}\sigma_X^5 \\
&- 63k_1^2k_2\sigma_{X_0}\sigma_X^4 - 126k_1k_2\sigma_{X_0}\sigma_X^4 + 63k_1^2k_2\sigma_{X_0}^2\sigma_X^3 - 3k_1^3\sigma_{X_0}\sigma_X^3 \\
&+ 3k_1^3\sigma_X^3 + 63k_1^2k_2\sigma_{X_0}^2\sigma_X^2 - 3k_1^3\sigma_{X_0}\sigma_X^2 + 3k_1^3\sigma_X^2 + 3k_1^3\sigma_{X_0}^2\sigma_X \\
&- 3k_1^3\sigma_{X_0}\sigma_X + 3k_1^3\sigma_{X_0}^2 - 3k_1^3\sigma_{X_0} = 0
\end{aligned} \quad (2.14)$$

For some insight, let us consider some numerical results. Let $m^{(1)} = 1$, $c^{(1)} = 0.1$, $k_1^{(1)} = 10$, $k_2^{(1)} = 15$. Simulation results are in close proximity the exact solution. Calculations yield ${\sigma_X}^2 = 0.6879$. The conventional approach yields for this set of parameters ${\sigma_X}^2 = 0.5043$, or 23.44% off the exact solution, and about 20% off the simulation for a sample of 10^6 simulating systems. The energy based approach without a weighing function results in an estimate ${\sigma_X}^2 = 0.5456$, or 17% off the exact value.

The value of α, matching the results of the Monte Carlo method and stochastic linearization is $\alpha = -0.000607$ As noted above, at ${k_1}^{(1)} = 10$ the stochastic technique yields results coincident with their Monte Carlo counterpart. At ${k_1}^{(2)} = 11$, the conventional is off the solution predicted by the Monte Carlo method by 22.77%The energy based criterion without weighing function results in 17% of error. The proposed combination of energy-wise linearization and the Monte Carlo method yields an error under 5%, namely reduction of error under the conventional approach by a factor of 4.6. Closer to the parameters for which Monte Carlo analysis was performed error may be still smaller - for example, for $k_1^{(3)} = 10.5$, the proposed method yields an error of only 3.85%.

3. Energy Linearization Criteria Based on Energy

The stochastic linearization technique was suggested almost simultaneously by Booton (1954), Kazakov (1954) and Caughey (1953, 1963) over half a century ago. Perhaps in over 99% of the studies on stochastic linearization the requirement of the minimum mean square difference between the nonlinear restoring force and its linear counterparts in the replacing system is utilized. Kazakov [2] also proposed a criterion that demands the equality of the mean-square values of the nonlinear force and its linear counterpart.

For extensive reviews of this method the reader can consult with papers by Falsone and Ricchiardi (2003), Socha (2005) and Bernard (1998). Elishakoff (2000) gave a reinterpretation of the classical method. For the purpose of this study it is pertinent to mention the alternative criterion, based on energy, as suggested by Wang and X.T. Zhang (1985), X.T. Zhang, Elishakoff and R.C. Zhang (1991), Elishakoff, R.C. Zhang and X.T. Zhang (1992), Falsone and Elishakoff (1994). This method was realized for various applications by Fang, Elishakoff and Caimi (1995), Muravyov, Turner, Robinson and Rizzi (1999), X.T. Zhang (1992), R.C. Zhang, Elishakoff and Shinozuka (1992), and X.T. Zhang and R.C. Zhang (1999). Elishakoff and Bert (2000) developed a complementary energy criterion, but no numerical evolution was performed to compare it with the potential energy based procedures.

In this paper we provide a systematic comparison of the energy criteria on the example of the Duffing oscillator. This nonlinear oscillator is chosen since there is an exact solution readily available for it. This allows a straightforward comparison to be conducted of the approximate solutions with the exact one, allowing to discern the relative efficacy of the approximations.

To gain more insight into the appraisal of various energy-based techniques let us consider a specific nonlinear system, namely, the oscillator with polynomial nonlinearity :

$$m\ddot{X}+c\dot{X}+k_0X+k_nX^{2n+1}=f(t) \tag{3.1}$$

where $X(t)$ = displacement, m= mass, c= damping, k_0= linear stiffness coefficient, n is taken as a positive integer, for the sake of the simplicity, k_n= nonlinear stiffness coefficient associated with the nonlinear behavior of the system. When n= 1, we get a particular case of the Duffing oscillator. For general, non-white noise excitation there is no exact solution available for the probability density of $X(t)$, or even for the mean-square displacement $E(X^2)$, where $E(\bullet)$ denotes mathematical expectation. The potential energy of the system reads:

$$P(X)=\frac{1}{2}X\,F(X) \tag{3.2}$$

where $F(X)$ is the restoring force

$$F(X)=k_0X+k_nX^{2n+1} \tag{3.3}$$

Thus,

$$P(X)=\frac{1}{2}k_0X^2+\frac{1}{2(n+1)}k_nX^{2(n+1)} \tag{3.4}$$

The complementary energy $C(X,t)$ defined as

$$C(X)=X\,F(X)-P(X) \tag{3.5}$$

is obtained as follows

$$C(X) = X\left(k_0 X + k_n X^{2n+1}\right) - P(X)$$
$$= X\left(k_0 X + k_n X^{2n+1}\right) - \left(\frac{1}{2}k_0 X^2 + \frac{1}{2(n+1)}k_n X^{2(n+1)}\right) \quad (3.6)$$
$$= \frac{1}{2}k_0 X^2 + \left(\frac{2n+1}{2(n+1)}\right)k_n X^{2n+1}$$

We replace the original system (1.1) by its linear counterpart

$$m\ddot{X} + c\dot{X} + k_{eq}X = f(t) \quad (3.7)$$

with attendant equivalent potential energy designated as $P_{eq}(X)$

$$P_{eq}(X) = \frac{1}{2}k_{eq}X^2 \quad (3.8)$$

The associated equivalent complementary energy

$$C_{eq}(X) = X\,F_{eq}(X) - P_{eq}(X) \quad (3.9)$$

naturally coincides with the potential energy of the replacing system P_{eq}.

Let us now reiterate the orthogonality based stochastic linearizations (Elishakoff, 2000).

To deal with the **potential energy linearization** we demand first that the difference between potential energies in the original and replacing system be orthogonal to X^{2n}, i.e.

$$E\{[P(X) - P_{eq}(X)]X^{2n}\} = 0 \quad (3.10)$$

Or

$$E\{[\frac{1}{2}k_0X^2 + \frac{1}{2(n+1)}k_nX^{2(n+1)} - \frac{1}{2}k_{eq}X^2]X^{2n}\} = 0 \tag{3.11}$$

leading to the following expression of the equivalent stiffness

$$k_{eq} = k_0 + \frac{1}{n+1}k_n\frac{E\left(X^{2(2n+1)}\right)}{E(X^{4n})} \tag{3.12}$$

This formula, although derived differently, coincides with the expression proposed by Wang and Zhang (1995).

In the context of the **complementary energy linearization**, we require that the difference between the complementary energies in the original and the replacing system be orthogonal to X^{2n}, i.e.

$$E\{[C(X) - C_{eq}(X)]X^{2n}\} = 0 \tag{3.13}$$

$$E\{[\frac{1}{2}k_0X^2 + \frac{2n+1}{2(n+1)}k_nX^{2(n+1)} - \frac{1}{2}k_{eq}X^2]X^{2n}\} = 0 \tag{3.14}$$

This leads to the following equivalent stiffness

$$k_{eq} = k_0 + \frac{2n+1}{n+1}k_n\frac{E\left(X^{2(2n+1)}\right)}{E(X^{4n})} \tag{3.15}$$

The particular case for $n = 1$ of this formula was derived by Elishakoff and Bert (200) but no numerical evaluation of its efficacy was studied. We will compare efficacy of the approximations obtained with Eq. (3.10) and (3.13), with the exact solution available in the case that $f(t)$ is a zero-mean Gaussian process with white-noise auto-correlation function.

We now want to compare these two energy criteria. Due to the assumption of normal distribution for $X(t)$ we have

$$E\left(X^{2n}\right)=(2n-1)!![E(X^2)]^n \tag{3.16}$$

Hence, equations (3.12) and (3.15) on be cast in the same form

$$k_{eq}=k_0+\alpha\, k_n E(X^2) \tag{3.17}$$

where

$$\alpha=\begin{cases}\dfrac{4n+1}{n+1}, \text{ for potential energy linearization}\\ (4n+1)\dfrac{2n+1}{n+1}, \text{ for complementary energy linearization}\end{cases} \tag{3.18}$$

The mean-square response of the replacing system reads:

$$E\left(X^2\right)=\frac{\pi S}{c\, k_{eq}} \tag{3.19}$$

Substituting Eq. (3.17) into Eq. (3.19) we arrive at

$$E\left(X^2\right)=\frac{\pi S}{c\, k_0}\frac{1}{1+\alpha\,(k_n/k_0)\,E(X^2)} \tag{3.20}$$

which leads to a quadratic equation

$$\alpha\,(k_n/k_0)\left[E\left(X^2\right)\right]^2+E\left(X^2\right)-E\left(X_0^2\right)=0 \tag{3.21}$$

Where

$$E\left(X_0^2\right)=\frac{\pi S}{c\, k_0} \tag{3.22}$$

is the mean-square response of the system without the nonlinear spring altogether. Eq. (3.21) has a single positive root for $E(X^2)$ reading

$$E\left(X^2\right)=\frac{\sqrt{1+4\alpha\,(k_n/k_0)\,E\left(X_0^2\right)}-1}{2\alpha\,(k_n/k_0)} \tag{3.23}$$

If the nonlinearity is small, i.e. if

$$4\alpha\,(k_n/k_0)\,E\left(X_0^2\right)<<1 \tag{3.24}$$

then the asymptotic solution $E\left(X_0^2\right)$ is obtained, coinciding with linear solution when $\sqrt{1+4\alpha\,(k_n/k_0)\,E\left(X_0^2\right)}$ is approximated as $1+2\alpha\,(k_n/k_0)E\left(X_0^2\right)$. When the value $4\alpha\,(k_n/k_0)\,E\left(X_0^2\right)$ is moderate, we approximate as follows

$$\sqrt{1+4\alpha\,(k_n/k_0)\,E\left(X_0^2\right)}\approx 1+2\alpha\,(k_n/k_0)E\left(X_0^2\right)-\frac{1}{2}\alpha^2\,(k_n/k_0)^2E^2\left(X_0^2\right) \tag{3.25}$$

resulting in

$$E\left(X^2\right)\approx E\left(X_0^2\right)-\frac{1}{4}\alpha\,(k_n/k_0)E^2\left(X_0^2\right) \tag{3.26}$$

Let us now deal with the **mean square equality** criteria.

Alternative criteria may be developed. Elishakoff, Fang and Caimi (1995) suggested that the mean-square energies in the original and the replacing system equal:

$$E[P^2(X)]=E[P_{eq}^2(X)] \tag{3.27}$$

This requirement results in the following value of k_{eq}, for the polynomial oscillator at hand

$$k_{eq} = \left[k_0^2 + \frac{1}{(n+1)^2} k_n^2 \frac{E(X^{4(n+1)})}{E(X^4)} + \frac{2}{(n+1)} k_0\, k_n \frac{E(X^{2(n+2)})}{E(X^4)} \right]^{1/2} \tag{3.28}$$

We can demand that the mean-square complementary potential energies be equal, i.e.

$$E[C^2(X)] = E[C_{eq}^2(X)] \tag{3.29}$$

For the polynomial oscillator this requirement is associated with the following equivalent stiffness:

$$k_{eq} = \left[k_0^2 + \left(\frac{2n+1}{(n+1)} \right)^2 k_n^2 \frac{E(X^{4(n+1)})}{E(X^4)} + 2\left(\frac{(2n+1)}{(n+1)} \right) k_0\, k_n \frac{E(X^{2(n+2)})}{E(X^4)} \right]^{1/2} \tag{3.30}$$

With Eq. (3.16) taken into account, we can re-write Eqs. (3.26) and (3.28) in the similar manner

$$k_{eq} = k_0 \left[1 + \beta \left(\frac{k_n}{k_0} \right)^2 \left\lfloor E(X^2) \right\rfloor^{2n} + \gamma \frac{k_n}{k_0} E(X^2)^n \right]^{1/2} \tag{3.31}$$

where

$$\beta = \left\{ \begin{array}{l} \dfrac{1}{(n+1)^2} \dfrac{(4n+3)!!}{3}, \text{ for potential energy linearization} \\[2ex] \left(\dfrac{2n+1}{(n+1)} \right)^2 \dfrac{(4n+3)!!}{3}, \text{ for complementary energy linearization} \end{array} \right\} \tag{3.32}$$

$$\gamma = \left\{ \begin{array}{l} \dfrac{2}{(n+1)} \dfrac{(2n+1)!!}{3}, \text{ for potential energy linearization} \\[2ex] 2\left(\dfrac{(2n+1)}{(n+1)} \right) \dfrac{(2n+1)!!}{3}, \text{ for complementary energy linearization} \end{array} \right\} \tag{3.33}$$

For the case where n=1, we obtain the Duffing oscillator. The coefficients for the previous linearization method read

$$\beta = \begin{cases} \dfrac{35}{4}, \text{ for potential energy linearization} \\ \dfrac{315}{4}, \text{ for complementary energy linearization} \end{cases} \tag{3.34}$$

$$\gamma = \begin{cases} 1, \text{ for potential energy linearization} \\ 3, \text{ for complementary energy linearization} \end{cases} \tag{3.35}$$

Density of exact solution is given by

$$p_{X(t)}(u) = A\exp\left(\frac{-P(u)}{\pi S_0}\right) \tag{3.36}$$

where $P(u) = k_0u^2/2 + k_1u^4/4$ is the potential energy of the system, and A is a normalization constant evaluated from

$$A^{-1} = \int_{-\infty}^{+\infty} \exp\left(\frac{-P(u)}{\pi S_0}\right) du \tag{3.37}$$

Thus Eq. (3.36) becomes

$$p_{X(t)}(u) = \frac{\exp\left(\dfrac{-1}{\pi S_0}\left[k_0u^2/2 + k_1u^4/4\right]\right)}{\displaystyle\int_{-\infty}^{+\infty} \exp\left(\dfrac{-1}{\pi S_0}\left[k_0v^2/2 + k_1v^4/4\right]\right) dv} \tag{3.38}$$

It is convenient to introduce a normalizing constant $\sigma_0^2 = \pi S_0 / k_0$, which is the variance of $X(t)$ for the limiting linear case $k_1 = 0$. Using a change of variable of $u = w\sigma_0$ then gives the probability density function as

$$p_{X(t)}(u)=\frac{\exp\left(-w^2/2-\alpha w^4/4\right)}{\int_{-\infty}^{+\infty}\exp\left(-v^2/2-\alpha v^4/4\right)dv} \quad (3.39)$$

with $\alpha = k_n\sigma_0 / k_0$.

We introduce notations

$$k_n u^4 / 4\pi S_0 = t^4, y = k_0 / 4\sqrt{\pi S_0 k_n} \quad (3.40)$$

Equation (3.37) is rewritten as

$$A^{-1} = Z_1(y) = 2\int_0^{+\infty} e^{-t^4-4yt^2}\, dt \quad (3.41)$$

where the function $Z_1(y)$ was introduced by Stratonovich (1961).

The mean-square response reads

$$E(X^2)=\int_{-\infty}^{+\infty} x^2 p_{X(t)}(u)\, du = A^{-1}\int_{-\infty}^{+\infty} u^2 \exp\left(-P(u)/\pi S_0\right) du \quad (3.42)$$

with notations in Eq.(3.40), the mean square response becomes

$$E(X^2)=2\sqrt{\pi S_0 / k_n}\; A^{-1}Z_2(y) \quad (3.43)$$

where

$$Z_2(y)=2\int_0^{+\infty} e^{-t^4-4yt^2}t^2\, dt \quad (3.44)$$

is a second Stratonovich (1961) function.

For $y > 0$, the function $Z_1(y)$ and $Z_2(y)$ turn out to be connected with the modified Bessel functions of fractional order. The functions $Z_1(y)$ and $Z_2(y)$ were numerically evaluated by Stratonovich (1961, p.545).

Here we reproduce his results along with present derivations.

$$\begin{aligned} Z_1(y) &= \sqrt{y}\, e^{2y^2} K_{1/4}(2y^2) \\ Z_2(y) &= y\sqrt{y}\, e^{2y^2} [K_{3/4}(2y^2) - K_{1/4}(2y^2)] \end{aligned} \tag{3.45}$$

Analogous expression for A^{-1} was found by Piszczek and Niziol (1984, p.173). For the mean-square response some other authors use the parabolic cylinder function $D_{-3/2}$.

To shed more light on the accuracy of the nonclassical approximations, it is necessary to conduct comparison between exact solution and approximate ones. Figure 1 shows a comparison between exact solution, and various energy criteria.

Fig. 1 represents the comparison of the exact solution with the potential energy linearization as performed in Eq. (3.10). The curve 1 is an exact solution. The curve 2 is based on Eq. (3.10) whereas the curve 3 is based on classical linearization. Energy criterion is observed as superior to the force linearization.

Fig. 2 contrasts various energy criteria. Curve (a) depicts the exact solution, curve (b) is associated with the criterion of equal mean-square potential energies in Eq. (3.27), curve (c) depicts the results based on orthogonality of the mean-square difference of complementary energies, whereas the curve (d) is based on equality of mean-squares of complementary energies. As is seen, the criterion based on equality of mean-square potential energies is closer to the exact solution than that based on equality of mean-square complementary energies. Likewise, comparison of Fig. 1, curve 2 and Fig. 2, curve (b) suggests that the potential energy orthogonalization criterion (Eq. 2.10) is superior to the complementary energy orthogonalization criterion (Eq. (3.13)).

Some quarter century ago, Roberts (1981) noted: "Because linear systems are so much easier to analyze than nonlinear ones, a natural method of attacking nonlinear problems is to replace a given set of nonlinear equations by an equivalent set of linear ones; the difference between the sets of equations is minimized in some appropriate sense." Likewise, one of the three pioneers of the stochastic linearization technique, Kazakov (1954, p.51) mentioned over half a century ago: "the sufficiently accurate approximation of essentially nonlinear characteristics is extremely difficult." Hence it appears very important to search for the *nonclassical* criteria that may lead to better approximations.

The effectiveness of the energy concepts in the nonlinear stochastic dynamics is not accidental. Indeed, the expression of the exact probability density contains the expression of the potential energy.

In this study, we investigated a system with nonlinear restoring force. Extension of energy concepts can be performed for the systems with nonlinear damping. Such approaches have been initiated in papers by Li, Chang and Tseng (2006) and Liang and Feeny (2006). It appears interesting to combine the present approach with that developed by Casciati, Faravelli and Hasofer (2006).

4. Generalization of Anh and Di Paola Version

In terminology of Kozin (1987), "the method of statical linearization has remained a surprisingly popular tool over the many years since it was first formulated." The method is based on replacing the original nonlinear system by a linear one, that is equivalent to the original one in some probabilistic sense. Several criteria have been suggested to arrive at the expressions of the equivalent stiffness and equivalent damping.

Anh and Di Paola (1996) suggested new realization of the stochastic linearization, that appears to be extremely unusual at the first glance. Instead of simplifying a nonlinear expression appearing in the differential equation, they, in essence, suggested to seemingly first complicate it by replacing it by higher order terms. These higher order terms then were replaced by the linear approximation, in several steps. This indirect linearization certainly prolongs, as it were, the linearization process. Yet it takes into account the higher order statistics and, as such, has more of a possibility to capture the behavior of the system. It turned out that this long way towards linearization leads to results that are closer to those obtained via exact solution, when the latter is available, or Monte Carlo simulation, when the exact solution is not available. Commenting on this method as exemplified on a Duffing oscillator, Anh (2006a) stresses that "in the [usual] linearization we go from X^3 [term] to X. That will yield some error, and we should do something to balance. For regulated Gaussian equivalent linearization (RGEL) we should go back [to balance the error]. Since [the difference of the powers of the original cubic and replacing linear terms is] 3-1=2, so we go back also 2 degrees, i.e. from X^3 to X^5 and come back to the first place X^3 but will regulated coefficient (7/9 in this case)".

For the details of implementation of RGEL for the Duffing oscillator the readers may consult with the study by Anh and Di Paola (1996). Anh (2006b) provides an additional justification of this method: "The natural explanation [of RGEL] is that when we want to go through a thing ahead we should move the hand back as how far ahead so far back. That is why we go first from X^3 to X^5." Since this method produces more accurate results than the classic linearization, one way metaphorically refer to as a "long shorter way," versus classical technique, that can be dubbed as a "short longer way." This metaphor is stemming from an folk story about a young boy who was asked by a stranger how to find the road to the big city. The boy asked: "Do you want a long shorter way, or a short longer way?" The stranger chose the latter, since the first adjective was a word "short." Yet, after several hours of wondering the man returned to the boy and told him: "The way is short, but there are unsurpassable rocks. Tell me the whereabouts of the long shorter way." This time the stranger succeeded to get to his destination.

Anh & Di Paola's (1996) derivation can be viewed as a " long shorter way " for it yields much more satisfactory results than the direct linearization technique ; the latter being a "short longer way." Recently, Elishakoff (2000) demonstrated that the expressions for the equivalent stiffness and damping coefficients adopted in the literature can be obtained by alternative means, namely via modified orthogonality criterion.

These two ideas, those by Anh & Di Paola (1996) and by Elishakoff (2000) are combined in this study. We first apply the Anh & Di Paola procedure to the Atalik & Utku oscillator with attendant dramatic decrease in error in comparison with the classical stochastic linearization. Then we extend the Anh & Di Paola methodology to two-step regulation. The latter extension shows considerable improvement of results in comparison with both classical scheme as well as the single step regulation, in the Lutes & Sarkani oscillator.

In this section we will derive the results of Anh and Di Paola by the orthogonality requirement. Anh and Di Paola (1996) studied the following nonlinear random vibration problcm

$$\ddot{X}+2h\dot{X}+\omega_0^2X+\varepsilon g\left(X,\dot{X}\right)=f(t) \tag{4.1}$$

where $X(t)$ is the displacement, $\dot{X}(t)$ is the velocity, $\ddot{X}(t)$= acceleration of a single degree of freedom system, h = damping coefficient, ω_0= natural frequency of the system obtained when $h\equiv 0, \varepsilon\equiv 0$, $f(t)\equiv 0$; $g\left(X,\dot{X}\right)$ is a nonlinear function, ε= amplitude of nonlinearity, $f(t)$= random excitation.

a polynomial expression of X and $\dot{X}$. The nonlinear function $g\left(X,\dot{X}\right)$ was taken as

$$g\left(X,\dot{X}\right)=\sum_{k=0}^{N}\sum_{j=0}^{N}\left(\alpha_{kj}\dot{X}^{2k}X^{2j+1}+\beta_{kj}X^{2k}\dot{X}^{2j+1}\right)$$

Classical linearization would perform following replacements of the nonlinear terms by the linear ones:

$$\alpha_{kj} \dot{X}^{2k} x^{2j+1} \rightarrow \lambda_{kj} X \tag{4.3}$$

$$\beta_{kj} X^{2k} \dot{X}^{2j+1} \rightarrow \mu_{kj} \dot{X} \tag{4.4}$$

Instead, most unusually, at least at the first glance, Anh and Di Paola (1996) suggested to replace non-linear terms by *higher-order nonlinear* ones

$$\begin{aligned} \alpha_{kj} \dot{X}^{2k} X^{2j+1} &\rightarrow c_{kj} \left(\dot{X}^{2k} X^{2j+1} \right) \left(\dot{X}^{2k} X^{2j} \right) \\ &= c_{kj} \dot{X}^{4k} X^{4j+1} \end{aligned} \tag{4.5}$$

$$\begin{aligned} \beta_{kj} X^{2k} \dot{X}^{2j+1} &\rightarrow d_{kj} \left(X^{2k} \dot{X}^{2j+1} \right) \left(X^{2k} \dot{X}^{2j} \right) \\ &= d_{kj} X^{4k} \dot{X}^{4j+1} \end{aligned} \tag{4.6}$$

where authors used the mean-square criterion for obtaining the coefficients d_{kj} and c_{kj}:

$$c_{kj} = \alpha_{kj} E\left[X^{6k} \dot{X}^{6j+2} \right] \Big/ E\left[X^{8k} \dot{X}^{8j+2} \right] \tag{4.7}$$

$$d_{kj} = \beta_{kj} E\left[X^{6k} \dot{X}^{6j+2} \right] \Big/ E\left[X^{8k} \dot{X}^{8j+2} \right] \tag{4.8}$$

Anh and Di Paola (1996) then replaced higher-order non-linear terms into the original non-linear terms

$$c_{kj}\dot{X}^{4k}X^{4j+1} \rightarrow q_{kj}\dot{X}^{4k}X^{4j+1} \tag{4.9}$$

$$d_{kj}X^{4k}\dot{X}^{4j+1} \rightarrow b_{kj}X^{2k}\dot{X}^{2j+1} \tag{4.10}$$

Where

$$b_{kj} = d_{kj}E\left[X^{6k}\dot{X}^{6j+2}\right] \Big/ E\left[X^{4k}\dot{X}^{4j+2}\right] \tag{4.11}$$

$$q_{kj} = c_{kj}E\left[X^{6k}\dot{X}^{6j+2}\right] \Big/ E\left[X^{4k}\dot{X}^{4j+2}\right] \tag{4.12}$$

This step is followed by the conventional replacement

$$b_{kj}X^{2k}\dot{X}^{2j+1} \rightarrow h_{kj}\dot{X} \tag{4.13}$$

$$q_{kj}\dot{X}^{2k}X^{2j+1} \rightarrow l_{kj}X \tag{4.14}$$

Where

$$h_{kj} = b_{kj}E\left[X^{6k}\dot{X}^{6j+2}\right] \Big/ E\left(\dot{X}^{2}\right) \tag{4.15}$$

$$l_{kj} = q_{kj}E\left[\dot{X}^{6k}X^{6j+2}\right] \Big/ E[X^{2}] \tag{4.16}$$

Let us show that the procedure by Anh & Di Paola (1996) can be directly obtained via modified stochastic linearization technique.

Indeed, we demand statistical orthogonality of the difference of the left and right hand sides in Eq.(4.6):

$$e_1 = \beta_{kj} X^{2k} \dot{X}^{2j+1} - d_{kj} X^{4k} \dot{X}^{4j+1} \tag{4.17}$$

with $\dot{X}^{4k} X^{4j+1}$, i.e. we require

$$\left(e_1, \dot{X}^{4k} X^{4j+1} \right) = 0 \tag{4.18}$$

where (. , .) is the inner product defined as

$$(\varphi, \psi) = E[\varphi, \psi] \tag{4.19}$$

Thus, Eq. (4.18) becomes:

$$E\left[\left(\beta_{kj} X^{2k} \dot{X}^{2j+1} - d_{kj} X^{4k} \dot{X}^{4j+1} \right) \dot{X}^{4k} X^{4j+1} \right] = 0 \tag{4.20}$$

yielding the expression (4.8) for d_{kj}. Analogously, the orthogonality requirement

$$\left(e_2, \dot{X}^{4k} X^{4j+1} \right) = 0 \tag{4.21}$$

where e_2 is the difference between the left and right hand sides in Eq. (4.5)

$$e_2 = \alpha_{kj} \dot{X}^{2k} X^{2j+1} - c_{kj} \dot{X}^{4k} X^{4j+1} \tag{4.22}$$

yields Eq. (4.8).

The results of the second step is likewise deducible from the requirements

$$\left(e_3, X^{2k}\dot{X}^{2j+1}\right) = 0 \tag{4.23}$$

$$\left(e_4, \dot{X}^{2k} X^{2j+1}\right) = 0 \tag{4.24}$$

where e_3 is the difference between the left and the right sides of Eq. (4.10):

$$e_3 = d_{kj} X^{4k}\dot{X}^{4j+1} - b_{kj} X^{2k}\dot{X}^{2j+1} \tag{4.25}$$

and e_4 is the difference between the left and the right hand sides of Eq.(4.9):

$$e_4 = c_{kj}\dot{X}^{4k} X^{4j+1} - q_{kj}\dot{X}^{2k} X^{2j+1} \tag{4.26}$$

Eqs. (4.23) and (4.24) lead to Eqs. (4.11) and (4.12), respectively. In perfect analogy, Eqs. (4.15) and (4.10) are obtained by postulating following conditions

$$\left(e_5, \dot{X}\right) = 0 \tag{4.27}$$

$$\left(e_6, X\right) = 0 \tag{4.28}$$

Where

$$e_5 = b_{kj} x^{2k} \dot{X}^{2j+1} - h_{kj} \dot{X} \quad (4.29)$$

$$e_6 = q_{kj} \dot{X}^{2k} X^{2j+1} - l_{kj} X \quad (4.30)$$

As is seen, Equations (4.11), (4.12), (4.15) and (4.16) are obtained by stochastic Galerkin-type orthogonality conditions. As a result, the final, linear replacement takes place:

$$g\left(X, \dot{X}\right) = \sum_{k=0}^{N} \sum_{j=0}^{N} \left(h_{kj} \dot{X} + l_{kj} X\right) \quad (4.31)$$

Where

$$h_{kj} = \frac{E\left[X^{2k} \dot{X}^{2j+2}\right]}{E[X^2]} \frac{E\left[X^{6k} \dot{X}^{6j+2}\right]}{E\left[X^{4k} \dot{X}^{4j+2}\right]} \frac{E\left[X^{6k} \dot{X}^{6j+2}\right]}{E\left[X^{8k} \dot{X}^{8j+2}\right]} \beta_{kj} \quad (4.32)$$

$$l_{kj} = \frac{E\left[\dot{X}^{2k} X^{2j+2}\right]}{E[X^2]} \frac{E\left[\dot{X}^{6k} X^{6j+2}\right]}{E\left[\dot{X}^{4k} X^{4j+2}\right]} \frac{E\left[\dot{X}^{6k} X^{6j+2}\right]}{E\left[\dot{X}^{8k} X^{8j+2}\right]} \alpha_{kj} \quad (4.33)$$

Anh and Di Paola (1996) evaluated by their method several oscillators. For the Duffing oscillator in the investigated numerical range, the numerical results led to roughly half the pourcentagewise error, than that resulted by the conventional stochastic linearization technique, i.e. without recourse to amending the original system by the higher non-linearity degree. As noted before, Anh & Di Paola (1996) call their method as "a regulated Gaussian equivalent linearization (RGEL)." As is seen, RGEL can be interpreted as a multiple orthogonalization technique.

Consider the following nonlinear system, usually called as Atalik and Utku (1976) oscillator:

$$\ddot{X}(t)+\beta\dot{X}(t)+\alpha X^3(t)=F(t) \tag{4.34}$$

where β is the damping constant, α is the nonlinear stiffness constant and $F(t)$ is a Gaussian white noise process with

$$E[F(t)]=0\ ,\ E[F(t)F(t+\tau)]=2d\beta\delta(\tau) \tag{4.35}$$

The exact stationary probability density function of the above system, obtained by the Fokker-Planck approach, is

$$p(x)=p_0\exp\left(-\frac{\alpha}{4d}x^4\right) \tag{4.36}$$

where p_0 is the normalization constant. To obtain the exact mean square displacement,

$$\sigma_x^2=E[X^2]=\int_{-\infty}^{+\infty}x^2p(x)dx \tag{4.37}$$

we use the integration formula

$$\int_0^{+\infty}x^{s-1}\exp\left(-ax^h\right)dx=\left(h^{-1}\right)\left(a^{-s/h}\right)\Gamma(s/h) \tag{4.38}$$

where $\Gamma(\bullet)$ is the Gamma function. The mean square displacement becomes

$$\sigma_x^2 = \frac{(1/4)(\alpha/4d)^{-3/4}\Gamma(3/4)}{(1/4)(\alpha/4d)^{-1/4}\Gamma(1/4)} \approx 0.6760(\alpha/d)^{1/2} \tag{4.39}$$

The equivalent linear system to Eq. (4.34) can be written as

$$\ddot{X}(t) + \beta\dot{X}(t) + k_{eq}X(t) = F(t) \tag{4.40}$$

where the equivalent linear spring constant k_{eq} is found by processing the conventional linearization, as equal

$$k_{eq} = E\left[\frac{d}{dx}(\alpha X^3)\right] = 3\alpha E\left[X^2\right] \tag{4.41}$$

The mean-square value of the displacement of the linearized system is

$$E[X^2] = d/k \tag{4.42}$$

Thus, we obtain the approximate solution as

$$\sigma_{x_e}^2 = (d/3\alpha)^{1/2} \approx 0.5776\,(d/\alpha)^{1/2} \tag{4.43}$$

The percentagewise error committed by using the classical equivalent linearization technique in evaluating the mean-square displacement is

$$(0.6760 - 0.5776)/0.6760 = 14.6\,\% \tag{4.44}$$

Let us apply the RGEL method, proposed by Anh and Di Paola. The scheme of the process can be read as follows

$$\alpha\, X^3(t) \rightarrow k_1\, X^5(t) \rightarrow k_2\, X^3(t) \rightarrow k_{eqI}\, X(t) \tag{4.45}$$

We can readily utilize the results obtained by Anh and Di Paola for the Duffing oscillator of which the Atalik and Utku oscillator is a particular case:

$$\alpha\, X^3(t) \rightarrow \frac{\alpha}{9E[X^2(t)]} X^5(t) \rightarrow \frac{7\alpha}{9} X^3(t) \rightarrow \frac{7\alpha}{3} X(t) \tag{4.46}$$

One gets the equivalent linearized equation

$$\ddot{X}(t) + \beta\,\dot{X}(t) + \frac{7}{3}\alpha\, E[X^2(t)]\, X(t) = F(t) \tag{4.47}$$

The mean-square value of $X(t)$ is evaluated by the following expression

$$E[X^2(t)] = \sqrt{\frac{3d}{7\alpha}} \approx 0.6546\left(\frac{d}{\alpha}\right)^{1/2} \tag{4.48}$$

Now, the percentagewise error found by using the RGEL linearization technique to calculate the mean-square displacement is

$$(0.6760 - 0.6546)/0.6760 = 3.17\,\% \tag{4.49}$$

We note a significant, over four-fold improvement, which demonstrates the extreme efficiency of the RGEL method. Naturally, the question of continuing the process to greater order arises. However calculation of such process beyond the first step in Eq. (4.45), namely,

$$\alpha X^3(t) \to \frac{\alpha}{9E[X^2(t)]} X^5(t) \to \frac{\alpha}{117\sigma_x^2 E[X^2(t)]} X^7(t)$$
$$\to \frac{11\alpha}{117E[X^2(t)]} X^5(t) \to \frac{77\alpha}{117} X^3(t) \to \frac{77\alpha}{39} E[X^2(t)] X(t) \tag{4.50}$$

leads to numerically worse result than the one previously found in Eq. (4.48). Hence, for the Atalik and Utku oscillator, the optimum number of regulation step is unity. It is important to note that the evaluation of the two steps in Duffing oscillator leads to the same conclusion. Question arises if there is an oscillator where the optimum number of regulation steps is greater than one. The reply to this question is shown in the next section to be affirmative.

Consider now the nonlinear oscillator by Lutes and Sarkani (2004)

$$\dot{X}(t) + k X^a(t) \operatorname{sgn}[X(t)] = F(t) \tag{4.51}$$

where a is a real number, $F(t)$ is a zero – mean, stationary Gaussian white noise with spectral density S_0. Lutes and Sarkani (2004) derive the probability density of the response

$$p_{X(t)} = A \exp\left[-\frac{k u^{a+1}}{(a+1)\pi S_0} \right] \tag{4.52}$$

Where

$$A = \left(\frac{k}{(a+1)\pi S_0} \right)^{\frac{1}{a+1}} \left(\frac{a+1}{2} \right) \left[\Gamma\left(\frac{1}{a+1} \right) \right]^{-1} \tag{4.53}$$

The variance of the response

$$\sigma_X^2 = 2A\int_0^\infty u^2 \exp\left[-\frac{k\,u^{a+1}}{(a+1)\pi\,S_0}\right] du \tag{4.54}$$

is obtained exactly

$$\sigma^2_{X,exact} = \left(\frac{\pi\,S_0}{k}\right)^{\frac{2}{a+1}} (a+1)^{\frac{2}{a+1}}\,\Gamma\left(\frac{3}{a+1}\right)\left[\Gamma\left(\frac{1}{a+1}\right)\right]^{-1} \tag{4.55}$$

Lutes and Sarkani (2004) also derived the approximate response via the classical stochastic linearization technique as follows:

$$\sigma^2_{X,approx} = \left(\frac{\pi\,S_0}{k}\right)^{\frac{2}{a+1}} \left[\frac{(2\pi)^{1/2}}{2^{a/2}\,a\Gamma(a/2)}\right]^{\frac{2}{a+1}} \tag{4.56}$$

The error η between exact and approximate solutions defined as

$$\eta = \frac{|\,\sigma^2_{X,exactx} - \sigma^2_{X,approx}\,|}{\sigma^2_{X,exactx}} \times 100\% \tag{4.57}$$

is shown in the Fig. 1. Lutes and Sarkani (2004) concluded that "the statistical linearization gives a good approximation of the response variance only when a is relatively near unity." Indeed, for a = 1 the error equals zero. For a = 2, the error constitutes $\eta = 5.6\%$; for a = 3, the error equals $\eta = 14.6\%$; for a = 4, the error is $\eta = 22.8\%$; for a = 5, the error reaches $\eta = 29.9\%$. It appears to be of interest to investigate this oscillator via the modified Anh and Di Paola (1996) approach.

For simplicity we will limit ourselves by considering the case when a is a positive integer; then instead of $X^a(t)\mathrm{sgn}[X(t)]$ we have $X^a(t)]$. We intend to replace the power oscillator with nonlinear term X^a by a linear oscillator with the term $k_{eq}\,X$, the difference of powers being a-1. During the "regulation" procedure by Anh and Di Paola (1996) we are recommended to increase the nonlinearity, i.e. power a by original power plus the increment a-1, i.e. to use the new regulation power of a+(a-1)=2a-1. Hence the procedure can be represented schematically as follows:

$$k\,X^a(t) \to k_1\,X^{2a-1}(t) \to k_2\,X^a(t) \to k_{eq}\,X(t) \tag{4.58}$$

We form a difference $k\,X^a(t) - k_1\,X^{2a-1}(t)$ and demand the statistical orthogonality of this expression to $X^{2a-1}(t)$,

$$E\{[k\,X^a(t) - k_1\,X^{2a-1}(t)]X^{2a-1}(t)\} = 0 \tag{4.59}$$

which leads to

$$k_1 = k\frac{E[X^{3a-1}(t)]}{E[X^{4a-2}(t)]} \tag{4.60}$$

General expression for $E[X^a(t)]$ is:

$$E[X^a(t)] = \sigma_X^a \int_{-\infty}^{\infty} \frac{\xi^a}{\sqrt{2\pi}} \exp(-\xi^2/2)\,d\xi \tag{4.61}$$

We obtain the first coefficient k_1 as follows

$$k_1 = k\,\sigma_X^{a-1} \frac{\int_{-\infty}^{\infty} \xi^{3a-1} \exp\left(-\xi^2/2\right) d\xi}{\int_{-\infty}^{\infty} \xi^{4a-1} \exp\left(-\xi^2/2\right) d\xi} \tag{4.62}$$

Proceeding in perfect analogy we obtain the following equivalent coefficients,

$$k_2 = k \frac{\left(\int_{-\infty}^{\infty} \xi^{3a-1} \exp\left(-\xi^2/2\right) d\xi\right)^2}{\int_{-\infty}^{\infty} \xi^{4a-1} \exp\left(-\xi^2/2\right) d\xi \int_{-\infty}^{\infty} \xi^{2a} \exp\left(-\xi^2/2\right) d\xi} \tag{4.63}$$

$$k_{eq} = k \frac{\sigma_X^{a+1}}{E[X^2(t)]} \frac{1}{\sqrt{2\pi}} \frac{\left(\int_{-\infty}^{\infty} \xi^{3a-1} \exp\left(-\xi^2/2\right) d\xi\right)^2 \int_{-\infty}^{\infty} \xi^{a+1} \exp\left(-\xi^2/2\right) d\xi}{\int_{-\infty}^{\infty} \xi^{4a-1} \exp\left(-\xi^2/2\right) d\xi \int_{-\infty}^{\infty} \xi^{2a} \exp\left(-\xi^2/2\right) d\xi} \tag{4.64}$$

$$= \frac{k\, E\left[X^2(t)\right]^{\frac{a-1}{2}}}{R} \tag{4.65}$$

Where

$$R = \sqrt{2\pi} \frac{\int_{-\infty}^{\infty} \xi^{4a-1} \exp\left(-\xi^2/2\right) d\xi \int_{-\infty}^{\infty} \xi^{2a} \exp\left(-\xi^2/2\right) d\xi}{\left(\int_{-\infty}^{\infty} \xi^{3a-1} \exp\left(-\xi^2/2\right) d\xi\right)^2 \int_{-\infty}^{\infty} \xi^{a+1} \exp\left(-\xi^2/2\right) d\xi} \tag{4.66}$$

Equation of motion becomes

$$\dot{X}(t)+\frac{k\,E\left[X^2(t)\right]^{\frac{a-1}{2}}}{R}X(t)=F(t) \tag{4.67}$$

We deduce the mean-square displacement

$$E\left[X^2_{regulated}(t)\right]_I=\left(\frac{\pi S_0}{k}\right)^{\frac{2}{a+1}}R^{\frac{2}{a+1}} \tag{4.68}$$

In order to express R as a function of Gamma–functions, note that

$$\int_{-\infty}^{\infty}\xi^{2a}\exp\left(-\xi^2/2\right)d\xi=2\int_{0}^{\infty}\xi^{2a}\exp\left(-\xi^2/2\right)d\xi \tag{4.69}$$

Then we make a change in the variable $\eta=\xi^2/2$, to get

$$\int_{-\infty}^{\infty}\xi^{2a}\exp\left(-\xi^2/2\right)d\xi=2^{a+\frac{1}{2}}\int_{0}^{\infty}\eta^{a-\frac{1}{2}}\exp\left(-\eta\right)d\eta \tag{4.70}$$

According to the definition of the Gamma-function,

$$\Gamma(z)=\int_{0}^{\infty}t^{z-1}\exp\left(-t\right)dt \tag{4.71}$$

we finally obtain,

$$\int_{-\infty}^{\infty}\xi^{2a}\exp\left(-\xi^2/2\right)d\xi=2^{a+\frac{1}{2}}\Gamma\left(a+\frac{1}{2}\right) \tag{4.72}$$

By applying this process to the other integrals, we get:

$$E\left[X_{regulated}^{2}(t)\right]_{I}=\left(\frac{\pi S_{0}}{k}\right)^{\frac{2}{a+1}}\left[\sqrt{\pi}\,2^{-\frac{a+1}{2}}\frac{\Gamma\left(\frac{4a-1}{2}\right)\Gamma\left(\frac{2a+1}{2}\right)}{\Gamma\left(\frac{3a}{2}\right)^{2}\Gamma\left(\frac{a+2}{2}\right)}\right]^{\frac{2}{a+1}} \quad (4.73)$$

Table 1 below presents the percentagewise error due to approximate nature of the solutions (with both conventional linearization and RGEL method) in comparison to the exact solution provided by Eq. (38), for different integer values of *a*.

a	$\sigma_{X,exact}^{2}$	$\sigma_{classical\ X,approx}^{2}$	Error, %	$E[X_{regulated}^{2}(t)]_{I}$	Error, %
1	1	1.0000	0	1.0000	0
2	0.7765	0.7323	5.6877	0.7824	0.7713
3	0.6760	0.5774	14.5904	0.6547	3.1546
4	0.6175	0.4764	22.8490	0.5620	8.9861
5	0.5786	0.4055	29.9225	0.4917	15.0206
6	0.5505	0.3529	35.8981	0.4367	20.6846
7	0.5291	0.3124	40.9630	0.3925	25.8224

Table 1: Error incurred by using a single-step in the Anh & Di Paola regulation

We can observe that there is an important improvement in the performance of the stochastic linearization when we utilize the RGEL method. Namely, whereas for a=2, the classical linearization is in error of about 5.69%, the regulated linearization has an error which is over 7 times less, namely 0.77%. For large value of *a* namely, a=5, the regulated linearization has about half the error of that classical linearization namely 15% vs. 29.9%. For even larger values of *a*, error is much less than that in the classical scheme but still quite large: for a=7, Anh and Di Paola approach leads to 25.8% error, whereas the classical approach is associated with error of about 41%. Still, the regulation reduces the error in this case by about 15%.

Natural question arises: What is the effect of additional steps in regulation? Anh and Di Paola (1996) considered only a single-step regulation. Here, the two step regulation is performed, as illustrated schematically below:

$$\begin{aligned} &k\, X^{a}(t) \to k_1\, X^{2a-1}(t) \to k_2\, X^{3a-1}(t) \\ &\to k_3\, X^{2a-1}(t) \to k_4\, X^{a}(t) \to k_{eq,II}\, X(t) \end{aligned} \tag{4.74}$$

Proceeding in perfect analogy with a single step procedure, we get

$$k_1 = k\,\sigma_X^{a-1} \frac{\int_{-\infty}^{\infty} \xi^{3a-1} \exp\left(-\xi^2/2\right) d\xi}{\int_{-\infty}^{\infty} \xi^{4a-1} \exp\left(-\xi^2/2\right) d\xi} \tag{4.75}$$

$$k_2 = k_1\,\sigma_X^{1-a} \frac{\int_{-\infty}^{\infty} \xi^{5a-3} \exp\left(-\xi^2/2\right) d\xi}{\int_{-\infty}^{\infty} \xi^{6a-4} \exp\left(-\xi^2/2\right) d\xi} \tag{4.76}$$

$$k_3 = k_2\,\sigma_X^{a-1} \frac{\int_{-\infty}^{\infty} \xi^{5a-3} \exp\left(-\xi^2/2\right) d\xi}{\int_{-\infty}^{\infty} \xi^{4a-2} \exp\left(-\xi^2/2\right) d\xi} \tag{4.77}$$

$$k_4 = k_3\ \sigma_X^{a-1} \frac{\int_{-\infty}^{\infty} \xi^{3a-1} \exp\left(-\xi^2/2\right) d\xi}{\int_{-\infty}^{\infty} \xi^{2a} \exp\left(-\xi^2/2\right) d\xi} \tag{4.78}$$

$$k_{eq,II} = k_4\ \sigma_X^{a+1} \frac{\int_{-\infty}^{\infty} \xi^{a+1} \exp\left(-\xi^2/2\right) d\xi}{\sqrt{2\pi}\ E[X^2(t)]} \tag{4.79}$$

After expressing $k_{eq,II}$ via k, the initial equation of motion is replaced by

$$\dot{X}(t) + \frac{k\ E\left[X^2(t)\right]^{\frac{a-1}{2}}}{Q} X(t) - F(t) \tag{4.80}$$

where

$$Q = \sqrt{2\pi} \frac{\left(\int_{-\infty}^{\infty} \xi^{4a-2} \exp\left(-\xi^2/2\right) d\xi\right)^2 \int_{-\infty}^{\infty} \xi^{6a-4} \exp\left(-\xi^2/2\right) d\xi \int_{-\infty}^{\infty} \xi^{2a} \exp\left(-\xi^2/2\right) d\xi}{\left(\int_{-\infty}^{\infty} \xi^{5a-3} \exp\left(-\xi^2/2\right) d\xi\right)^2 \left(\int_{-\infty}^{\infty} \xi^{3a-1} \exp\left(-\xi^2/2\right) d\xi\right)^2 \int_{-\infty}^{\infty} \xi^{a+1} \exp\left(-\xi^2/2\right) d\xi} \tag{4.81}$$

We arrive at the mean-square displacement,

$$E\left[X^2_{regulated}(t)\right]_{II} = \left(\frac{\pi S_0}{k}\right)^{\frac{2}{a+1}} Q^{\frac{2}{a+1}} \tag{4.82}$$

or, via the Gamma–functions

$$E[X^2_{regulated}(t)]_{II} = \left(\frac{\pi S_0}{k}\right)^{\frac{2}{a+1}} \left[\sqrt{\pi}\, 2^{-\frac{a+1}{2}} \frac{\Gamma\left(\frac{6a-3}{2}\right) \Gamma\left(\frac{2a+1}{2}\right) \Gamma\left(\frac{4a-1}{2}\right)}{\Gamma\left(\frac{5a-2}{2}\right)^2 \Gamma\left(\frac{3a}{2}\right)^2 \Gamma\left(\frac{a+2}{2}\right)}\right]^{\frac{2}{a+1}} \tag{4.83}$$

The Roman subscript II indicates that the result is obtained in the second step of the regulation process. The Table 2 presents a comparison of the two-step procedure with the exact solution on one hand, a single-step procedure, and classical stochastic linearization.

a	$\sigma^2_{X,exact}$	$E[X^2_{regulated}(t)]_I$	Error,%	$E[X^2_{regulated}(t)]_{II}$	Error, %
1	1	1.0000	0	1	0
2	0.7765	0.7824	0.7713	0.8205	5.6693
3	0.6760	0.6547	3.1546	0.7117	5.2820
4	0.6175	0.5620	8.9861	0.6251	1.2229
5	0.5786	0.4917	15.0206	0.5554	4.0131
6	0.5505	0.4367	20.6846	0.4988	9.4038
7	0.5291	0.3925	25.8224	0.4521	14.5548

Table 2: Error incurred by using an extended two-step regulation

We note that two-step regulation provides an additional improvement in comparison to the single-step regulation; for the moderate value of a=4, the classical stochastic lineariztion is associated with the error of about 23%; the single-step regulation results in error of about 9% whereas the two-step regulation leads to the error of 1.23%. Thus, the error in two-step regulation is about 18 times less than in the classical scheme, and about 7 times less than in a single-step regulation. For larger values of a, though still much better than the classical of single-step regulation linearization, the two-step regulation reduces the error by about additional 10% in comparison with a single-step regulation: The classical linearization yields about 41% for a=7; the single-step regulation leads to 26% whereas the two-step regulation results in

14.6% of error. We should note that the two-step regulation turns out to be the optimal one for the Lutes & Sarkani oscillator, since it turns that the three-step regulation yields greater errors, than the two-step version. The optimum number of steps needed should be established for each oscillator at hand.

Figure Captions

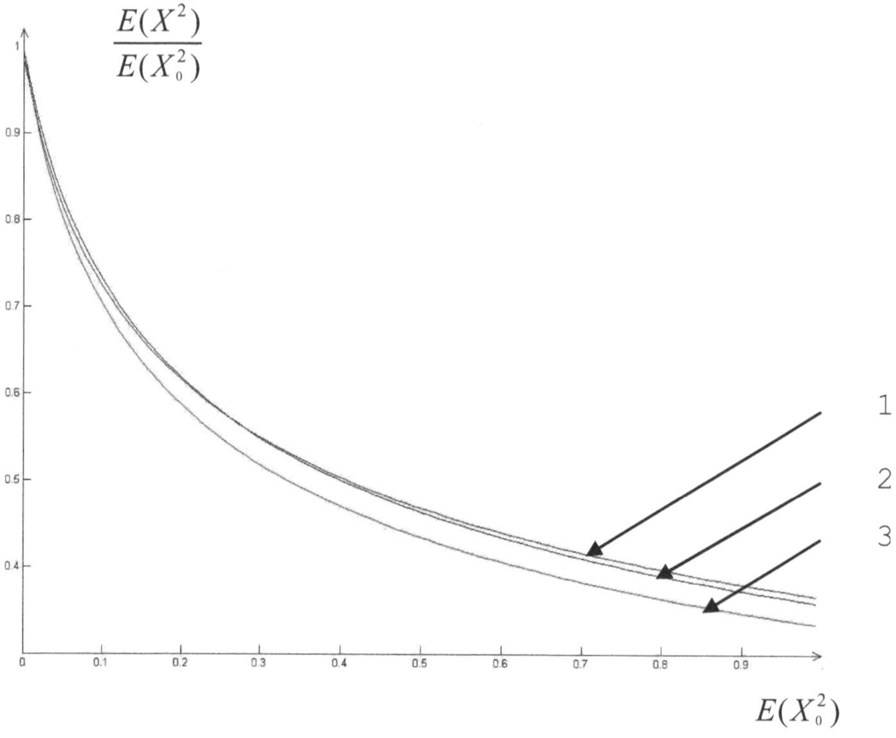

Figure 4.1. Comparison of approximate solutions with exact results: (1) exact solution; (2) potential energy criterion in Eq. (4.10); (3) classical stochastic linearization results

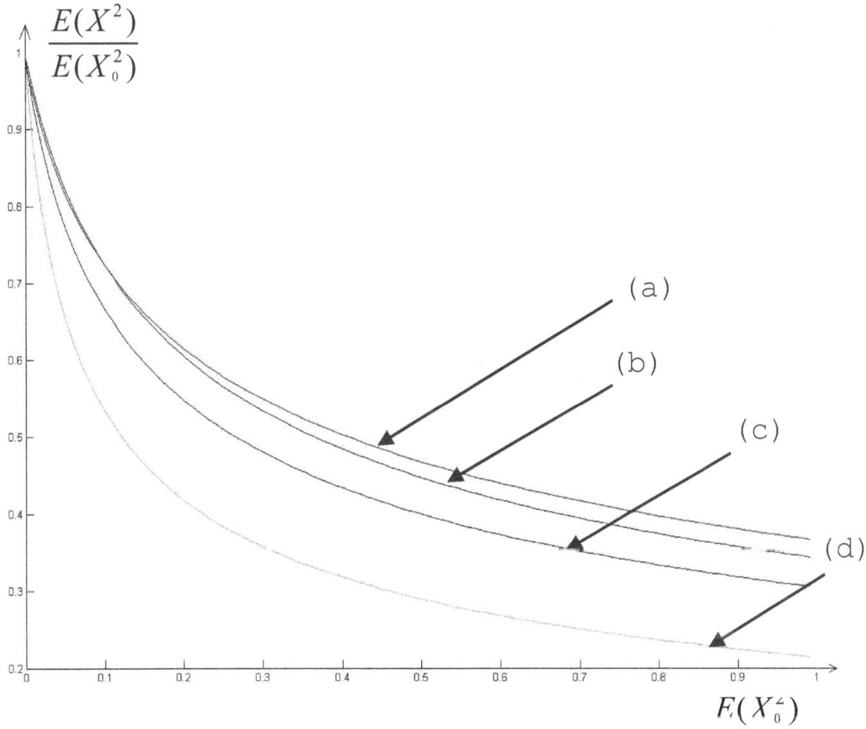

Figure 4.2. Contrast of potential and complementary energy criteria with exact solution: (a) exact solution, (b) criterion based on equality of mean-squares of potential energies, in Eq. (27), (c) criterion based on complementary energy in Eq. (13), (d) criterion based on equality of mean-squares of complementary energies.

Conclusion

This review cannot be concluded without mentioning that Crandall (2001) in his tellingly entitled paper “Is stochastic Equivalent Linearization a Subtly Flawed Procedure ?” commented: “It must be admitted that the literature … has been confusing.” Elsewhere, in the same study, he wrote: “… although there has been some confusions concerning the standard procedure, it does *not* possess a subtle flaw. “In another paper, Crandall (2006) noted: “… the inconsistency of applying recipes based on non-linear response statistics independent of k to linear system statistics which were function of k was recognized and corrected by Socha and Pawleta (1994, 1999) and by Elishakoff and Colajanni (1997, 1998), abd Colojanni and Elishakoff (1998). On the same topic Proppe, Schueller and Pradwarter (2000), p. 1210

“… the minimization procedure is carried out with respect to the probability density function of the non-linear system and that at the last step, this probability density function is approximated by the probability density function of the linearized system. the approximation error us thus not minimized with respect to the probability density function of the linearized system!”

Thus Proppe, Schueller and Pradlwarter (2000) do not recognize that the stochastic linearization is associated with either confusion or inconsistency, i.e. the terms mentioned by Crandall (2001, 2006). This topic appears to be in need to be addressed additionally . It is hoped that future, unbiased and impartial investigators will find the topic to be fascinating enough to deal with this issue in more details.

Acknowledgement

Authors express their sincere gratitude to Mr. Yohann Miglis and Mr. Baptiste Ducreux of the Florida Atlantic University for editorial help.

References

Abaoui-Ismaili M. and Bernard P, 1997, "Asymptotic Analysis and Linearization of the Randomly Perturbed Two-Wells Duffing Oscillator," *Probabilistic Engineering Mechanics*, Vol. 12, pp. 171-178.

Anh, N.D., Krause, R., and Schiehlen, W., 1992, "Statistical Linearization and Large Excitation of Nonlinear Stochastic Mechanical Systems," *Nonlinear Stochastic Mechanics*, (N. Bellomo and F. Casciati, Springer Verlag, eds.) Berlin, pp. 1-12.

Anh, N.D. and Schiehlen W., 1998, "A Technique for Obtaining Approximate Solutions in Gaussian Equivalent Linearization," *Computer Methods in Applied Mechanics and Engineering*, Vol. 168, pp. 113-119.

Anh N.D (2006a) Private Communication to I.E., May 25.

Anh N.D (2006b) Private Communication to I.E., May 24.

Anh N.D and Di Paola M., 1995, "Some Extensions of Gaussian Equivalent Linearization", *International Confernce on Nonlinear Stochastic Dynamics*, Hanoi, Vietnam, pp. 5-16, Dec. 7-10.

Atalik, T.S., and Utku S., 1976, "Stochastic Linearization of Multi-Degree-of-Freedom Non-Linear Systems," *Earthquake Engineering and Structural Dynamics*, Vol. 4, pp. 411-420.

Beaman, J.J., 1980, "Accuracy of Statistical Linearization," *New Approaches to Nonlinear Problems in Dynamics*, (P. J. Holmes, eds.), SIAM, Philadelphia, pp. 195-207.

Beaman, J.J., and Hedrick J.K., 1981, "Improved Statistical Linearization for Analysis and Control of Nonlinear Stochastic Systems," *Journal of Dynamical Systems, Measurement, and Control*, Vol. 103, pp. 22-27.

Bellizi S. and Bouc R., 1999, "Analysis of Multi-Degree of Freedom Strongly Non-Linear Mechanical Systems with Random Input. Part II: Equivalent Linear Systems with Random Matrices and Power Spectral Density Matrix," *Probablistic Engineering Mechanics*, Vol. 14 pp. 245-256.

Bernard, P., 1992, "About Stochastic Linearization," *Nonlinear*

Stochastic Mechanics, (N. Bellomo and F. Casciati, eds.) Springer Verlag, Berlin, pp. 61-70.

Bernard, P., 1998, "Stochastic Linearization: What Is Available and What Is Not," *Computers and Structures*, Vol. 67, pp. 9-18.

Bernard P., and Wu L.,1998, "Stochastic Linearization: The Theory," *Journal of Applied Probability*, Vol. 35, pp. 718-730.

Bolotin, V.V., 1984, *Random Vibrations of Elastic Systems*, Martinus Nijhoff Publishers, The Hague, pp. 240-292.

Booton, R.C., 1953, "The Analysis of Nonlinear Control Systems with Random Inputs," *Proceedings Symposium on Nonlinear Circuit Analysis*, Vol. 2, pp. 341-344.

Bouc R. and Defelippi M., 1997, "Multi-Modal Nonlinear Spectral Response of a Beam with Impact under Random Load," *Probabilistic Engineering Analysis*, Vol. 12,pp. 163-170.

Busby, H.R., Jr., and Weingarten, V.I., 1973, "Response of Nonlinear Beam to Random Excitation," *Journal of Engineering Mechanics Division*, Vol. 99, pp. 55-68.

Casciati, F., 1986, "Equivalent Linearization Technique in the Analysis of Seismic Excited Structures," Proceedings, *Euro-China Joint Seminar of Earthquake Engineering,* Beijing, pp. 398-410.

Casciati, F., Faravelli L. and Hasofer A. M., 1993, "A New Philosophy for Stochastic Equivalent Linearization," *Probabilistic Engineering Mechanics*, Vol. 8, pp. 179-185.

Caughey, T.K., 1963, "Equivalent Linearization Techniques," *Journal of Acoustical society of America*, Vol. 35 (11), pp. 1706-1711.

Caughey, T.K., 1998, July 27, and an undated one. Personal Communications.

Caughey, T.K., and Dienes, J.K., 1961, "Analysis Nonlinear First-Order System with a White Noise Input," *Journal of Applied Physics*, Vol. 23, pp. 2476-2479.

Caughey, T.K., 1953, "Response of Nonlinear Systems to Random Excitation", Lecture Note, California Institute of Technology (unpublished, quoted in Caughey, T.K., 1963)

Chang, R.J., 1990, "Non-Gaussian Linearization Method for Stochastic Parametrically and Externally Excited Nonlinear Systems," ASME Paper 90-WA/DSC-16, Presented at the ASME Winter Annual Meeting, Dallas, TX, Nov. 25-30, 1990.

Chang, R.J., and Young, G.E., 1989, "Methods and Gaussian Criterion for Statistical Linearization of Stochastic Parametrically and Externally Excited Nonlinear Systems," *Journal of Applied Mechanics*, Vol. 56, pp. 179-185.

Chen G., 1999, "Cascade Linearization of Nonlinear Systems Subjected to Gaussian Excitations," *Stochastic Structural Dynamics* (B.F. Spencer, Jr., and E. Johnson, eds.), Balkema, Rotterdam, pp. 69-76.

Crandall, S.H., 1974, "Nonlinearities in Structural Dynamics," *The Shock and Vibration Digest*, Vol. 6(8), pp. 1-13.

Crandall, S.H., 1977, " On Statistical Linearization for Nonlinear Oscillators," *Problems of the Asymptotic Theory of Nonlinear Oscillators*, Academy of Sciences of the Ukrainian SSR, Naukova Dumka, pp. 115-122. Reprinted, 1980, *Nonlinear System Analysis and Synthesis*, Vol. 2, *Techniques and Applications*, (R.V. Ramnath, J.K. Hedrick and H.M. Paynter, eds.) ASME, New York, pp. 199-209.

Crandall S.H., 2001, "Is Stochastic Equivalent Linearization a Subtly Flawed Procedure?", Probabilistic Engineering Mechanics, Vol. 16. 169-176.

Crandall S.H., 2004, "Fifty Years of Stochastic Equivalent

Linearization." *Seminar*, Center for Applied Stochastics Research, College of Engineering, Florida Atlantic University, 12 February.

Crandall S.H., 2006, "A Half Century of Stochastic Equivalent Linearization", Structural Control and Health Monitoring, Vol. 13, 27-40.

Colajanni, I. and Elishakoff, I., 1998a, "A Subtle Error in the Stochastic Linearization Technique," *Chaos, Solitons and Fractals*, Vol. 9, pp. 479-491.

Colajanni, P. and Elishakoff, I., 1998b, "A New Look at the Stochastic Linearization Technique for Hyperbolic Tangent Oscillator," *Chaos, Solitons and Fractals*, Vol. 9, pp. 1611-1623.

Constantinou, M.C., and I.G. Tadjbakhsh, 1984, "Response of a Sliding Structure to Filtered Random Excitation," *Journal of Structural Mechanics*, Vol. 12, pp. 401-418.

Constantinou, M.C., 1985, "Vibration Statistics of the Duffing Oscillator," *Soil Dynamics and Earthquake Engineering*, Vol. 4, pp. 221-223.

Cunha A. A. M. F., 1994, "The Role of the Stochastic Equivalent Linearization Method in the Analysis of the Nonlinear Seismic Response of Building Structures," *Earthquake Engineering and Structural Dynamics*, Vol. 23, 837-857.

Dobson, S., Noori M., Hou Z. and Dimentberg M., 1998, "Direct Implementation of Stochastic Linearization for SDOF Systems with General Hysteresis," *Structural Engineering and Mechanics,* Vol. 6, 473-484.

Dogan V. and Vaicaitis R., 1999, "Nonlinear Response of Cylindrical Shells to Random Excitation," *Nonlinear Dynamics,* Vol. 20, pp. 33-53.

Donley, M.G., and Spanos, P.D., 1990, *Dynamic Analysis of Non-Linear Structures by the Method of Statistical Quadratization*, Springer Verlag, New York.

Donley, M.G., and Spanos, P.D., 1992, "Equivalent Statistical Quadratization for Multi-Degree-of-Freedom Nonlinear Systems," *Nonlinear Stochastic Mechanics*, (N. Bellomo and F. Casciati, eds.) Springer Verlag, Berlin, pp. 185-200.

Elishakoff, I., 1991, "Method of Stochastic Linearization Revisited and Improved," *Computational Stochastic Mechanics*, (P.D. Spanos and C.A.Brebbia, eds.), Computational Mechanics Publications, Southampton, pp. 101-111.

Elishakoff I., 1995, "Some Results in Stochastic Linearization of Nonlinear Systems," *Nonlinear Dynamics and Stochastic Mechanics*, (W. Kliemann and N. Sri Namachchivaya, eds.), pp. 259-281 CRC Press, Boca Raton.

Elishakoff I., 2000, Stochastic Linearization Technique: A New Interpretation and a Selective Review, *The Shock and Vibration Digest*, Vol. 32(3), 179-188.

Elishakoff I. and Andriamasy L., 2010, "Nonclassical Linearization Criteria in Nonlinear Stochastic Dynamics," *Journal of applied Mechanics,* Vol. 77(4), paper 04501.

Elishakoff I., Andriamasy L. and Dolley M., 2009, "Application and Extension of the Stochastic Linearization by Anh and Di Paola," *Acta Mechanica,* Vol. 204, 89-98.

Elishakoff, I. and Bert, C.W., 2000, "Complementary Energy Criterion in Nonlinear Stochastic Dynamics", *Applications of Statistics and Probability*, Balkema, Rotterdam, pp. 821-825.

Elishakoff, I., and Colombi, P., 1993, "Successful Combination of the Stochastic Linearization and Monte-Carlo Methods," *Journal of Sound and Vibration*, 160 (3), pp. 554-558.

Elishakoff, I. and Colajanni P., 1997, "Stochastic Linearization

Critically Re-examined," *Chaos, Solitons and Fractals*, Vol. 8, pp. 1957-1972.

Elishakof, I. and Colajanni P., 1998, "Booton's Problem Critically Re-Examined," *Journal of Sound and Vibration*, Vol. 210, pp. 683-691.

Elishakoff I. and Falsone G., 1993, "Some Recent Developments in Stochastic Linearization Technique," *Computational Stochastic Mechanics* (A. H.-D. Cheng and C.Y. Yang, eds.), Computation Mechanics Publications, pp. 175-194, Southampton.

Elishakoff, I., and Zhang, R., 1992.b, "Comparison of the New Energy-Based Versions of the Stochastic Linearization Technique," *Nonlinear Stochastic Mechanics* , (N. Bellomo and F. Casciati, eds.), Springer Verlag, Berlin, pp. 201-212.

Elishakoff, I., and Zhang, X.T., 1992.a, "An Appraisal of Different Stochastic Linearization Techniques," *Journal of Sound and Vibration*, Vol. 153, pp.370-375.

Emam H.H., Pradlwarter H.J., and Schüeller G.I., 2000, "A Computational Procedure for the Implementation of Equivalent Linearzation on Finite Element Analysis," *Earthquake Engineering and Structural Dynamics*, Vol. 29, pp. 1-17.

Falsone, G., 1992.a, "Stochastic Linearization for the Response of MDOF Systems Subjected to External and Parametric Gaussian Excitations," *Computational Stochastic Mechanics*" (P.D. Spanos and C.A. Brebbia, eds.), Computational Mechanics Publications, Southampton.

Falsone, G., 1992.b, "Stochastic Linearization of MDOF Systems under Parametric Excitations," *International Journal of Non-Linear Mechanics*, Vol. 27, pp. 1025-1037.

Falsone, G., and Elishakoff, I., 1994, "Modified Stochastic Linearization Technique for Colored Noise Excitation of Duffing Oscillator," *International Journal of Non-Linear Mechanics*, submitted for publication.

Falsone, G. and Ricciardi, G., 2003, "Stochastic Linearization: Classical Approach and New Developments", *Recent Research Developments in Structural Dynamics* (Luongo, A., ed.), pp. 81-106.

Fan, F.-G., and Ahmadi, G., 1990, "On Loss of Accuracy and Non-Uniqueness of Solutions Generated by Equivalent Linearization and Cumulant-Neglect Methods," *Journal of Sound and Vibration*, Vol. 137, pp. 385-401.

Fang, J., and Fang, T.S., 1991, " A Weighted Equivalent Linearization Method in Random Vibration," *Chinese Journal of Applied Mechanics*, Vol. 8(3), pp. 114-120, (in Chinese).

Fang, J., Elishakoff, I. and Caimi, R., 1995, "Nonlinear Response of a Beam under Stationary Random Excitation by Improved Stochastic Linearization Method", *Applied Mathematical Modeling*, **19**, pp. 106-111.

Fang, T., Zhang, T. S. and Fang, J., 1991, "Nonstationary Response of Nonlinear Systems Under Random Excitation," *Proceedings, Third National Conference on Random Vibrations,* Taiwan, pp. 147-155, (in Chinese).

Fang, J.J. and Elishakoff, I., 1995, "Nonlinear Response of a Beam under Stationary Random Excitation by Improved Stochastic Linearization Method," *Applied Mathematical Modelling*, Vol. 19, 106-111.

Gao S., Niemann H.J. and Ou K., 1997, "Analysis of Stochasitc Response of Rotational Shell Considering Its Constructive Nonlearity," *Journal of Bejing Institute of Technology*, Vol. 6,pp. 180-186.

Giannini R.and Pinto P.E., 1997, "Stochastic Analysis Methods," *Computer Analysis and Design of Earthquake Resistant Structures*, Vol.3, pp. 103-152.

Grigoriu, M., 1999, "Sample Properties of Some Equivalent Non-

Gaussian Processes," *13th ASCE Engineering Mechanics Conference*, The Johns Hopkins University, Baltimore, June 13-16.

Grunddmann, H., Hartmann Ch. and Waubke H., 1998, "Structures Subjected to Stationary Stochastic Loadings. Preliminary Assessment by Statistical Linearization Combined with an Evolutionary Algorithm," *Computers and Structures*, Vol. 67, pp. 53-64.

Hurtado, J.E., 1996, "Improved Stochastic Linearzation Method using Mixed Distributions," *Structural Safety*, Vol. 18, pp. 49-62.

Iwan, W.D., and Mason, A.B., Jr., 1980, "Equivalent Linearization for Systems Subjected to Non-Stationary Random Excitation," *International Journal of Non-Linear Mechanics*, Vol. 15, pp. 71-82.

Iyengar R.N. and Roy D., 1998, "Extensions of the Phase Space Linearization Method," *Journal of Sound and Vibration*, Vol. 211, pp. 877-906.

Izumi, M., Zaiming, L., and Kimura, M., 1989, "A Stochastic Linearization Technique and Its Application to Response Analysis of Nonlinear Systems Based on Weighted Least-Square Minimization," *Journal of Structural and Construction Engineering, Transaction of AIJ*, Vol. 395, pp. 72-81.

Jiminez J.C., Shoi I. and Ozaki T. 1999, Simulation of Stochastic Differential Equations Through the Local Linearization Method, A Comparative Study," *Journal of Statistical Physics*, Vol. 94, pp. 587-602.

Kang J.W. and Havichandran R.S., 1999, "Random Vibration of Laminated FRP Plates with Material Nonlinearity using High-Order Shell Theory," *Journal of Engineering Mechanics*, Vol. 125, pp. 1081-1088.

Kazakov, I.E., 1954, "An Approximate Method for the Statistical Investigation for Nonlinear Systems," *Trudi Voenno-Vozdushnoi Inzhenernoi Akademii imeni Prof. N.E. Zhukovskogo*, Vol. 399, pp. 1-52, (in Russian).

Kazakov I.E., 1998, "An Extension of the Method of Statistical Linearzation," *Automation and Remote Control*, Vol. 59, pp. 1682-1685.

Khabbaz, G.R., 1964 "Power Spectral Density of the Response of a Nonlinear System to Random Excitation," *Journal of Acoustical Society of America*, Vol. 38, pp. 847-850.

Kimura K., Komada M. and Sakata M., 1995, "NonGaussian Equivalent Linearzation for Estimation for Response Distribution of Nonlinear Systems," *Transitions JSME, Part C,* Vol. 61, No: 583, pp. 831-835.

Kolovskii, M.Z., 1966, "Estimating the Accuracy of Solutions Obtained by the Method of Statistical Linearization," *Automation and Remote Control*, Vol. 27, pp. 1692-1701.

Kozin F., "Parameter Estimation and Statistical Linearization", 1987, *Stochastic Structural Mechanics*, (Y.K. Lin and G.I. Schuëller, eds.), Springer Verlag, Berlin, pp. 259-267.

Kozin, F., 1965, "Comments Upon the Technique of Statistical Linearization," *Proceedings Japan Association of Automatic Control Engineers*, Annual Convention, pp. 15.1-15.10.

Kozin, F., 1987, "The Method of Statistical Linearization for Non-Linear Stochastic Vibrations," *Nonlinear Stochastic Dynamic Engineering Systems*, (F. Ziegler and G.I. Schuëller, eds.), Springer, Berlin, pp. 45-56.

Langley, R.S., 1988, "An Investigation of Multiple Solutions Yielded by the Equivalent Linearization Method, *Journal of Sound and Vibration*, Vol. 127, pp. 271-281.

Li, W. Chang, C.W. and Tseng, S., 2006, "The Linearization Method Based on the Equivalence of Dissipated Structural Systems", *Journal of Sound and Vibration*, **295**, pp. 797-809.

Liang, J.W. and Feeny, B., 2006, "Balancing Energy to Estimate

Damping Parameters in Forced Oscillators", *Journal of Sound and Vibrations*, **295**, pp. 988-998.

Lee J., 1995, "Improving the Equivalent Linearzation Technique for Stochastic Duffing Oscillators," *Journal of Sound and Vibration*, Vol. 186, pp. 845-855.

Lee, X.X., and Chen, J.Q., 1992, "A Linearization Technique for Random Vibrations of Nonlinear Systems," *Mechanics Research Communications*, Vol. 19, pp. 1-6.

Lutes L. D. and Sarkani S., 2004, *Random Vibration: Analysis of Structural and Mechanica Systems*, pp. 423-424 and 437-438, Elseview, Amsterdam.

Mei, C., and Wentz, K. R. 1982, "Large Amplitude Random Response of Angle Ply Laminated Composite Plates," *AIAA Journal*, Vol. 20, pp. 1450-1458.

Mei, C., and Wolfe, H.F., 1986, "On Large Deflection Analysis in Acoustic Fatigue Design," *Random Vibration-Status and Recent Developments*, (I. Elishakoff and R.H. Lyon, eds.), Elsevier, Amsterdam, pp. 279-302.

Mochio, T., Samaras, E. and Shinozuka, M., 1985, "Stochastic Linearization in a Finite Element Based Reliability Analysis," *Proceedings, Fourth International Conference on Structural Safety and Reliability*, (I. Konishi, A.H. S. Ang and M. Shinozuka, eds.), pp. 1375-1384.

Muravyov, A., Turner T., Robinson J. and Rizzi S, 1999, A new Stochastic Equivalent Linearization Implementation for Prediction of Geometrically Nonlinear Vibrations, *AIAA Paper 59-1376* AIAA/ASME/ASCE/AHS/ASC Structures, Structural Dynamics Conference, St. Louis,Vol. 2, pp.1489-1497.

Naess A., 1995, "Prediction of Extreme Response of Nonlinear

Structures by Extended Stochastic Linearzation, *Probablistic Engineering Mechanics*, Vol. 10, pp. 153-160.

Noori, M., and Davoodi, H., 1990, "Comparison between Equivalent Linearization and Gaussian Closure for Random Vibration Analysis of Several Nonlinear Systems," *International Journal of Engineering Sciences*, Vol. 28, pp. 897-905.

Piszczek, K., and Niziol, J., 1984, *Random Vibration of Mechanical Systems*, Ellis Horwood, Chichester, pp. 173-175.

Proppe C., Schueller G.I. and Pradlwarter H.J., 2000, Equivalent Linearization Revisisted," *Advanced in Structural Dynamics* (J.M Ko and Y.L. Xu, eds.), Vol. II, pp. 1207-1214.

Pradlwarter, H.J., 1991, "Non-Gaussian Linearization an Efficient Tool to Analyze Nonlinear MDOF-Systems," *Nuclear Engineering and Design*, Vol. 128, pp. 175-192.

Pradlwarter, H.J., and Schüeller, G.I., 1988, "Accuracy and Limitations of the Method of Equivalent Linearization for Hysterectic Multi-Store Structures," *Nonlinear Stochastic Dynamic Engineering Systems*, (F. Ziegler and G. I. Schüeller, eds.), Springer Verlag, Berlin, pp. 3-21.

Pradlwarter, H. J., and Schüeller, G.I., 1992, "A Practical Approach to Predict the Stochastic Response of Many-DOF Systems Modeled by Finite Elements," *Nonlinear Stochastic Mechanics*, (N. Bellomo and F. Casciati, eds.), Springer Verlag, Berlin, pp. 427-437.

Popov, E.P., and Paltov, I. N., 1960, *Approximate Methods of Investigation of Nonlinear Automatic Systems*, "Fizmatgiz" Publishers, Moscow (in Russian).

Reardon D. and Leithead W.E.,1990, "Statistical Linearzation A Comparative Study," *International Journal of Control,* Vol.52, pp.1083-1105.

Reddy, G.R., Suziki K., Watanabe T. and Mahajan S.C., "Linearization Techniques for Seismic Analysis of Piping System on Friction Support," *Journal of Pressure Vessel Technology*, Vol. 121, 103-108, 1999.

Roberts, J.B., 1989, "Response of Nonlinear Mechanical Systems to Random Excitation, Part 2: Equivalent Linearization and Other Methods," *The Shock and Vibration Digest*, Vol. 13, No. 5, pp. 15-29.

Roberts, J.B., 1981, Response of Nonlinear Mechanical Systems to Random Excitation, Part 2: Equivalent Linearization and Other Methods, *The Shock and Vibration Digest*, Vol. 13(5), pp. 15-29.

Roberts, J.B., 1990, "Statistical Linearization: Multiple Solutions and Their Physical Significance," *Structural Dynamics*, (W.B. Krätzig et al, eds.), Balkema Publishers, Rottterdam, pp. 671-681.

Roberts, J.B., and Spanos, P.D., 1991, *Random Vibration and Statistical Linearization*, John Wiley and Sons, Chichester.

Schuëller G.I., 1999, "On the Stochastic Response of Nonlinear FE Models," *Archive of Applied Mechanics*, Vol. 69,pp. 765-784.

Schuëller, G.I., and Pradlwarter H.J., 1999, "Advances in Stochastic Structural Dynamics under the Perspective of Reliability Estimation," *Structural Dynamics-EURODYN'99* (L. Fryba and J. Naprstek, eds), pp. 267-272, Balkema, Rotterdam.

Schuëller, G.I., Pradlwarter H.J., Vasta M. and Harnpornchai N., 1998, Benchamark Study on Non-Linear Stochastic Structural Dynamics, *Structural Safety and Reliability*, (Shiraishi N., Shinozuka M. and Wen Y.K., eds), pp. 355-362, Balkema, Rotterdam.

Seide, P., 1975, "Nonlinear Stresses and Deflections of Beams Subjected to Random Time Dependent Uniform Pressure," *Israel Journal of Technology*, Vol. 13, pp. 143-151.

Seide, P., 1976, "Nonlinear Stresses and Deflections of Beams Subjected to Random Time Dependent Uniform Pressure," *Journal of*

Engineering for Industry, Vol. 98, pp. 1014-1020.

Shinozuka, M., 1972, "Monte Carlo Solution in Structural Dynamics," *Computers and Structures*, Vol. 2, pp. 855-874.

Shinozuka, M., and Deodatis, G., 1991, "Simulation of Stochastic Process by Spectral Decomposition," *Applied Mechanics Reviews*, Vol. 44, pp. 191-203.

Sinitsin, I.N., 1974, "Methods of Statistical Linearization (Survey)", *Automation and Remote Control*, Vol. 35, pp. 765-8776.

Sireteanu, T., Bellizzi S. and Ursu I., 1999, "An Extension of Gaussian Equivalent Linearization for a Class of Nonlinear Oscillator," *13th ASCE Engineering Mechanics Conference*, The Johns Hopkins University, Baltimore, June 13-16, p. 336.

Socha, L., 2005, "Linearization in Analysis of Nonlinear Stochastic Systems: Recent Results – Part I: Theory", *Applied Mechanics Reviews*, **58**, pp. 178-204, 2005; Part II: Applications, **58**, pp. 303-315.

Socha L. and Pawleta M., 1999, "Some Remarks on Equivalent Linearization of Stochastic Dynamic Systems," *Stochastic Structural Dynamics* (B.F. Spencer, Jr. and E. A. Johnson, eds.), Balkema, Rotterdam, pp. 106-112.

Socha L. and Pawleta M., 1994, "Corrected Equivalent Linearization of Stochastic Dynamic Systems," *Machine Dynamic Problem*, Vol. 7, pp. 149-161.

Socha, L., and Soong, T. T., 1991, "Linearization in Analysis of Nonlinear Stochastic Systems," *Applied Mechanics Reviews*, Vol. 44, No. 1, 399-422.

Socha L., 2008, *Linearization Method for Stochastic Dynamic Problems,* Springer, Berlin.

Spanos, P.D., 1980, "Formulation of Stochastic Linearization for

Symmetric or Asymmetric M.D.O.F. Nonlinear Systems," *Journal of Applied Mechanics*, Vol. 47, pp. 209-211.

Spanos, P.D., 1981, Stochastic Linearization in Structural Dynamics, *Applied Mechanics Reviews*, Vol. 34(1), 1-8.

Spanos, P.D., 1998, discussion, April 8.

Stratonovich, R.L., 1961, *Selected Problems of Theory of Fluctuation in Radiotechnics*, "Sovetskoe Radio" Publishers, Moscow, pp. 368 and 545 (in Russian).

To, C.W.S., 1984, "The Response of Nonlinear Structures to Random Excitation," *The Shock and Vibration Digest*, Vol. 16, pp. 13-33.

Vanmarkcke, E.H., 1979, "Some Recent Developments in Random Vibration," *Applied Mechanics Reviews*, Vol. 32(10), 1197-1202.

Zhang, X.T., 1989, "Equivalent Potential Technique for Deterministic and Random Response Analysis of Nonlinear Systems," *Applied Mechanics*, pp. 236-241, (in Chinese).

Zhang, X., 1992, "Study of Weighted Energy Technique in Analysis of Nonlinear Random Vibration," *Nonlinear Vibration and Chaos*, Tianjin University Press, pp. 139-144.

Zhang, R. Ch., Elishakoff, I., and Shinozuka, M., 1992, "Analysis of Nonlinear Sliding Structures by Modified Stochastic Linearization Methods," *Probabilistic Mechanics and Structural and Geotechnical Reliability*, (Y.K.Lin, ed.), ASCE Press, New York, pp. 196-199. Extended version, *International Journal of Nonlinear Dynamics*,Vol. 5, pp.299-312.

Zhang, X.T., Elishakoff, I., and Zhang, R.Ch., 1991, "A Stochastic Linearization Technique Based on Minimum Mean Square Deviation of Potential Energies", *Stochastic Structural Dynamics - New Theoretical Developments*, (Y. K. Lin and I. Elishakoff, eds.), Springer Verlag, Berlin, pp. 327-338.

Zhang, X.T. and Zhang R. Ch., 1999, "A Stochastic Response-Consistent Equivalence Technique," *13th ASCE Engineering Mechanics Conference*, The Johns Hopkins University, Baltimore, June 13-16.

Wang, C., and Zhang, X.T., 1985, "An Improved Equivalent Linearization Technique in Nonlinear Random Vibration," *Proceedings, International Conference on Nonlinear Mechanics*, pp. 959-964.

Wen, Y.K., 1989, Methods of Random Vibration for Inelastic Structures, *Applied Mechanics Reviews*, 42(2), 39-52.

Wentz, K.R., Paul D.B., and Mei, C., 1982, "Large Deflection Random Response of Symmetric Laminated Composite Plates," *Shock and Vibration Bulletin*, Vol. 52, pp. 99-111.

Yu, Z.D., and Gao, G.A., 1988, *Theory of Random Vibration with Applications,* Tongji University Press (in Chinese).

Zhang, R.C., Elishakoff, I. and Shinozuka, M., 1992, “Analysis of Nonlinear Sliding Structures by Modified Stochastic Linearization Methods”, *International Journal of Nonlinear Dynamics*, **5**, pp. 299-312.

Zhang, X., and Zhang R., 1999, "Energy-Based Stochastic Equivalent Linearization with Optimized Power," *Stochastic Structural Dynamics*, (B.F. Spencer, Jr. and E. A. Johnson, eds.), Balkema, Rotterdam, pp. 113-117.

Part 2
Non-Stochastic Models

Comprehensive Modeling of Uncertain Systems Using Fuzzy Set Theory

Thomas Haag[*] and Michael Hanss[†]

[*] Institute of Applied and Experimental Mechanics, University of Stuttgart, Germany

[†] Institute of Engineering and Computational Mechanics, University of Stuttgart, Germany

Abstract Non-determinism in structural mechanics is ubiquitous. Whenever a mathematical model of a real world engineering system is developed, simplification techniques are employed which lead to systematic errors in the modeling procedure. The fuzzy arithmetical approach which is presented in this chapter is designed to handle epistemic uncertainties, as those systematic errors are also termed, in numerical simulations by the use of fuzzy numbers. The approach consists of the transformation method which is designed to compute fuzzy-valued output quantities and an inverse fuzzy arithmetical method which additionally determines the fuzzy-valued model parameters on the basis of measurement data in a second step. Additionally, a kind of a sensitivity analysis is provided along with a criterion to assess the quality of different competing mathematical models.

1 Introduction

A well-known problem in the numerical simulation of real-world systems is the question of how detailed the structure of the model has to be chosen in order to appropriately represent reality. If the structure of the model does not exhibit any simplifications of the reality, classical, crisp-valued model parameters are adequate to describe the real system, and a conventional model-updating procedure can be used for their identification. Due to computational limitations, however, it is often necessary to neglect certain phenomena in the modeling phase, such as high frequency dynamics or nonlinearities. This leads to simplified and, from a computational perspective, less expensive models, but obviously also to specific modeling errors when crisp parameters are used as the optimal ones. In order to cover these modeling errors in simplified models, uncertain parameters can be used,

representing a so-called comprehensive model and providing a conservative prediction of the real system behavior. Based on the fact that the inherent uncertainty is a consequence of model simplifications, it can be classified as epistemic, Oberkampf et al. (2002). As probability theory may not be appropriate to effectively represent epistemic uncertainties, the alternative strategy of quantifying epistemic uncertainties by fuzzy numbers (Zadeh (1975a)) is pursued in this work. The propagation of the uncertainty through the system, i.e. the evaluation of the model with fuzzy-valued parameters, is performed by the use of fuzzy arithmetic on the basis of the transformation method Hanss (2005).
As an extension of the transformation method of fuzzy arithmetic, an inverse fuzzy arithmetical method to estimate the uncertain parameters of a simplified model on the basis of the output of an advanced model, or based on measurement data is proposed in Hanss (2003a); Haag et al. (2009b, 2010, 2011) and described in the second part of this chapter. The proposed method uses inverse computations that are based on the data of the transformation method. This implies that an inversion of the model equations is not needed, which enables the method to be used with existing software, e.g. commercial finite element codes.

2 General fuzzy arithmetic

2.1 Modeling epistemic uncertainties using fuzzy numbers

This work concentrates on uncertainties of epistemic type and their numerical representation by fuzzy numbers. Fuzzy numbers are defined as a normal, convex fuzzy set with exactly one nominal value, which is usually the one value one would use for a single crisp-valued simulation, Hanss (2005). Fuzzy numbers can be interpreted as a set of nested intervals, realizing a smooth transition between a worst case interval and a crisp, nominal value. In order to distinguish between the nested intervals, a membership level μ is introduced which quantifies the possibility for the occurrence of a specific interval.

Triangular fuzzy numbers In the subsequent computations and definitions, the notion of a triangular fuzzy number is used frequently. For this reason, the following definition is given: A *triangular* or *linear fuzzy number* is denoted as

$$\widetilde{p} = \mathrm{tfn}\left(\overline{x}, \alpha_{\mathrm{l}}, \alpha_{\mathrm{r}}\right)$$

with the membership function given by

$$\mu_{\widetilde{p}}(x) = \begin{cases} 0 & \text{for} \quad x \le \overline{x} - \alpha_\mathrm{l}\,, \\ 1 + (x - \overline{x})/\alpha_\mathrm{l} & \text{for} \quad \overline{x} - \alpha_\mathrm{l} < x \le \overline{x}\,, \\ 1 - (x - \overline{x})/\alpha_\mathrm{r} & \text{for} \quad \overline{x} < x \le \overline{x} + \alpha_\mathrm{r}\,, \\ 0 & \text{for} \quad x > \overline{x} + \alpha_\mathrm{r}\,. \end{cases}$$

The value $\overline{x}$ is called the *nominal value* of the fuzzy number $\widetilde{p}$. The values α_l and α_r are called the *left-hand* and *right-hand worst-case deviation* from the nominal value $\overline{x}$, respectively. An example of a triangular fuzzy number is illustrated in Fig. 1.

As an alternative to triangular fuzzy numbers, any other membership

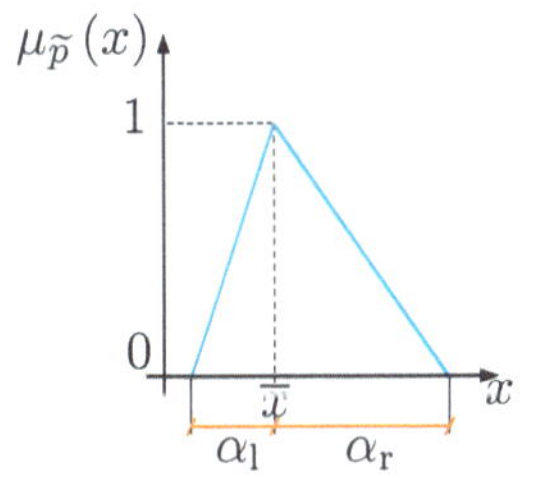

Figure 1. Triangular fuzzy number, $\widetilde{p} = \mathrm{tfn}(\overline{x}, \alpha_\mathrm{l}, \alpha_\mathrm{r})$.

function can be used which ensures the convexity of the corresponding fuzzy set. In literature, Gaussian, quadratic and exponential fuzzy numbers are found among many others. These alternative types of fuzzy numbers are not detailed here due the fact that, very often, their utilization is very vague for real engineering problems.

2.2 Zadeh's extension principle

In order to be able to perform computations with fuzzy numbers, classical arithmetic has to be extended from the ordinary computation with crisp numbers to this widened class of fuzzy-valued numbers. This extension to fuzzy arithmetic is commonly based on the extension principle which has been developed by Zadeh (1975a,b,c). The extension principle states that for a functional mapping $f : \mathbb{R}^n \to \mathbb{R}$, the membership function $\mu_{\widetilde{q}}(z)$ of the output $\widetilde{q} = f(\widetilde{p}_1, \widetilde{p}_2, \ldots, \widetilde{p}_n)$ is given by

$$\mu_{\widetilde{q}}(z) = \begin{cases} \sup_{z=f(x_1,x_2,\ldots,x_n)} & \min\{\mu_{\widetilde{p}_1}(x_1), \mu_{\widetilde{p}_2}(x_2), \ldots, \mu_{\widetilde{p}_n}(x_n)\} \\ & \text{if} \quad \exists z = f(x_1, x_2, \ldots, x_n) \\ 0 & \text{otherwise.} \end{cases} \tag{1}$$

Thus, the solution to any fuzzy arithmetical computation can be found through a multi-dimensional optimization problem, whose complexity and computational effort depends on the number of fuzzy variables and the function of interest. All modern approaches for the calculation with fuzzy numbers are based on the extension principle and try to solve the optimization problems in different ways. One way to solve it is given by the transformation method of fuzzy arithmetic (see Section 2.3) which is used in this work.

Alternative realizations, which are based on the elementary arithmetical operations addition, subtraction, multiplication and division only are not detailed in this work as they are irrelevant for practical engineering computations. The direct application of the extension principle by using standard optimization techniques is in principle possible but the related computational effort and convergence behavior is strongly dependent on the system under consideration and, very often, it is thus not practicable for real engineering systems.

2.3 The transformation method (TM) of fuzzy arithmetic

Theoretical background The transformation method is a practical implementation of Zadeh's extension principle without the typical drawbacks that characterize elementary fuzzy arithmetic like the overestimation of result uncertainties and the restriction to elementary arithmetical operations. In particular, the transformation method of fuzzy arithmetic can be used in conjunction with any mathematical function and thus also with black-box models of engineering systems. The method is available in a general, a reduced, and an extended form, Hanss (2005, 2002, 2003b). In the following, the notion *simulation of fuzzy-parameterized systems* denotes the procedure of computing the fuzzy-valued outputs $\widetilde{q}_r$, $r = 1, \ldots, k$, of the output functions $f_r : \mathbb{R}^n \to \mathbb{R}$, i.e.

$$\widetilde{q}_r = f_r(\widetilde{p}_1, \widetilde{p}_2, \ldots, \widetilde{p}_n)\,, \tag{2}$$

where $\widetilde{p}_i$, $i = 1, 2, \ldots, n$, are n independent fuzzy variables or parameters.

The basic idea of the transformation method is to choose crisp parameter values from the interval range of the decomposed fuzzy input parameters and to combine these values in well-defined parameter combinations, for which the problem of interest is evaluated. Assuming the uncertain system to be characterized by n fuzzy-valued model parameters $\widetilde{p}_i$, $i = 1, 2, \ldots, n$, the major five steps of the simulation tool based on the transformation method can be briefly described as follows.

1. Decomposition of the input fuzzy numbers At first, the μ-axis is subdivided into m equally-spaced intervals of length $\Delta\mu = 1/m$. The discrete values μ_j of the $(m+1)$ levels of membership are then given by

$$\mu_j = \frac{j}{m}, \quad j = 0, 1, \ldots, m\,.$$

Each fuzzy parameter $\widetilde{p}_i$, $i = 1, 2, \ldots, n$, is then decomposed into $(m+1)$ α-cuts leading to a set P_i of intervals $X_i^{(j)}$ of the form

$$P_i = \left\{X_i^{(0)}, X_i^{(1)}, \ldots, X_i^{(m)}\right\}$$

with

$$X_i^{(j)} = \left[a_i^{(j)}, b_i^{(j)}\right], \quad a_i^{(j)} \leq b_i^{(j)}, \quad i = 1, 2, \ldots, n\,, \quad j = 0, 1, \ldots, m\,. \quad (3)$$

The decomposition step is illustrated in Fig. 2.3. The parameter m is denoted as *decomposition number*.

2. Transformation of the input intervals The input intervals $X_i^{(j)}$ are transformed to arrays $\widehat{X}_i^{(j)}$. The structure of the arrays $\widehat{X}_i^{(j)}$ is different for the reduced and for the general transformation method.

- *Reduced transformation method.*
  In the case of the transformation method in its reduced form, the intervals $X_i^{(j)}$ are transformed into arrays $\widehat{X}_i^{(j)}$ of the form

$$\widehat{X}_i^{(j)} = \Big(\overbrace{(\alpha_i^{(j)}, \beta_i^{(j)}), (\alpha_i^{(j)}, \beta_i^{(j)}), \ldots, (\alpha_i^{(j)}, \beta_i^{(j)})}^{2^{i-1}\text{ pairs}}\Big)$$

  with

$$\alpha_i{}^{(j)} = \underbrace{(a_i^{(j)}, \ldots, a_i^{(j)})}_{2^{n-i}\text{ elements}} \quad \text{and} \quad \underbrace{\beta_i{}^{(j)} = (b_i^{(j)}, \ldots, b_i^{(j)})}_{2^{n-i}\text{ elements}}\,.$$

  Only the boundary values $a_i^{(j)}$ and $b_i^{(j)}$ of the intervals $X_i^{(j)}$ are considered in the transformation scheme for the reduced transformation method, as plotted in Fig. 2.3(left). This is sufficient if the objective function $f(\widetilde{p}_1, \widetilde{p}_2, \ldots, \widetilde{p}_n)$ is monotonic with respect to the variables $\widetilde{p}_1, \widetilde{p}_2, \ldots, \widetilde{p}_n$ or if the problem is characterized by only one uncertain variable, that is $n = 1$. For all other cases, the reduced transformation method cannot guarantee a correct propagation of the uncertain parameters, i.e. a proper estimation of the uncertain result, and the general transformation method should be used instead.

- *General transformation method.*
In case of the general transformation method, the intervals $X_i^{(j)}$ are transformed into arrays $\widehat{X}_i^{(j)}$ of the form

$$\widehat{X}_i^{(j)} = \Big(\overbrace{(\gamma_{1,i}^{(j)}, \gamma_{2,i}^{(j)}, \ldots, \gamma_{(m+1-j),i}^{(j)}), \ldots, (\gamma_{1,i}^{(j)}, \gamma_{2,i}^{(j)}, \ldots, \gamma_{(m+1-j),i}^{(j)})}^{(m-j+1)^{(i-1)}\,(m-j+1)-\text{tuples}}\Big)$$

with

$$\gamma_{l,i}^{(j)} = \underbrace{(c_{l,i}^{(j)}, \ldots, c_{l,i}^{(j)})}_{(m-j+1)^{(n-i)}\ \text{elements}}$$

and

$$c_{l,i}^{(j)} = \begin{cases} a_i^{(j)} & \text{for } l = 1 \\ & \text{and } j = 0, 1, \ldots, m\,, \\ \frac{1}{2}\left(c_{l-1,i}^{(j+1)} + c_{l,i}^{(j+1)}\right) & \text{for } l = 2, 3, \ldots, m-j \\ & \text{and } j = 0, 1, \ldots, m-2\,, \\ b_i^{(j)} & \text{for } l = m-j+1 \\ & \text{and } j = 0, 1, \ldots, m\,. \end{cases}$$

The transformation scheme of the general transformation method is illustrated on the right-hand side of Fig. 2.3.

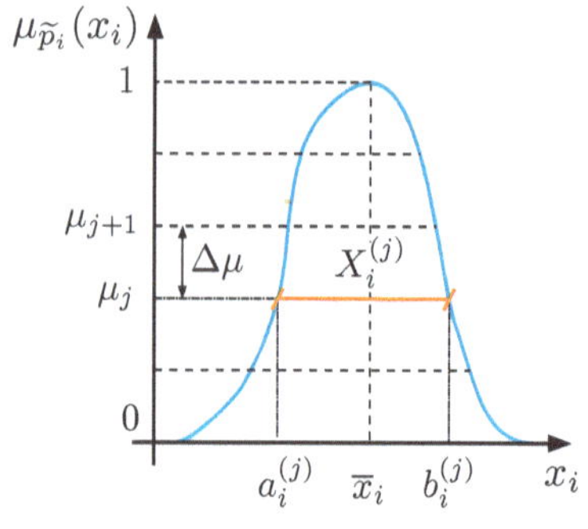

Figure 2. Decomposition of the ith fuzzy parameter $\widetilde{p}_i$.

3. Evaluation of the model The estimation of the fuzzy-valued outputs $\widetilde{q}_r$, $r = 1, \ldots, k$, is carried out by evaluating the expression of f_r separately for each column of the arrays $\widehat{X}_i^{(j)}$, using the classical arithmetic of crisp numbers. Practically, only the well-known concepts of crisp arithmetic are

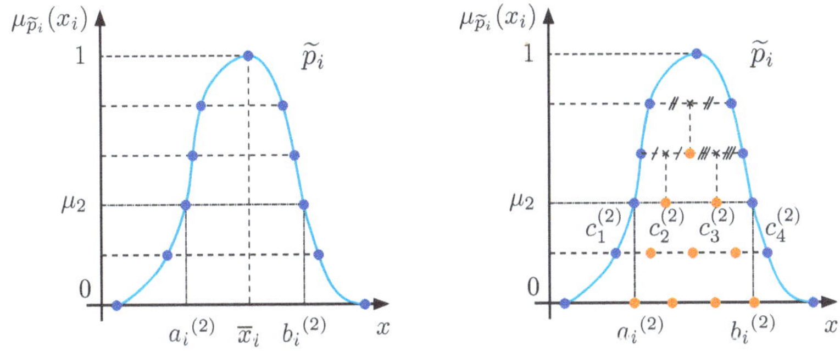

Figure 3. Transformation scheme according to the *reduced* (left) and the *general transformation method* (right).

needed for these evaluations. The outputs $\widetilde{q}_r$ can now be expressed in their decomposed and transformed form by the arrays $\widehat{Z}_r^{(j)}$, $j = 0, 1, \ldots, m$. The kth element of the arrays $\widehat{Z}_r^{(j)}$ is then given by

$$^k\hat{z}_r^{(j)} = f_r\left({}^k\hat{x}_1^{(j)}, {}^k\hat{x}_2^{(j)}, \ldots, {}^k\hat{x}_n^{(j)}\right)$$

where ${}^k\hat{x}_i^{(j)}$ denotes the kth element of the array $\widehat{X}_i^{(j)}$. In detail, the output arrays $\widehat{Z}_r^{(j)}$ for the reduced transformation method can be written as

$$\widehat{Z}_r^{(j)} = \left({}^1\hat{z}_r^{(j)}, {}^2\hat{z}_r^{(j)}, \ldots, {}^{2^n}\hat{z}_r^{(j)}\right)$$

and for the general transformation method as

$$\widehat{Z}_r^{(j)} = \left({}^1\hat{z}_r^{(j)}, {}^2\hat{z}_r^{(j)}, \ldots, {}^{(m-j+1)^n}\hat{z}_r^{(j)}\right).$$

4. Retransformation of the output arrays In the fourth step of the simulation tool, the result values of the arrays $\widehat{Z}_r^{(j)}$ are retransformed to the output intervals $Z_r^{(j)} = [c_r^{(j)}, d_r^{(j)}]$ for each membership level μ_j, where

$$Q_r = \left\{Z_r^{(0)}, Z_r^{(1)}, \ldots, Z_r^{(m)}\right\}$$

represents the fuzzy output $\widetilde{q}_r$ of the model in its decomposed form. The output intervals $Z_r^{(j)}$ are obtained by retransforming the arrays $\widehat{Z}_r^{(j)}$ according to the recursive formulas

$$c_r^{(j)} = \min_k \left(c_r^{(j+1)}, {}^k\hat{z}_r^{(j)}\right), \quad j = 0, 1, \ldots, m,$$
$$d_r^{(j)} = \max_k \left(d_r^{(j+1)}, {}^k\hat{z}_r^{(j)}\right), \quad j = 0, 1, \ldots, m,$$

and

$$c_r^{(m)} = \min_k({}^k\hat{z}_r^{(m)}) = \max_k({}^k\hat{z}^{(m)}) = d_r^{(m)} = \overline{z}_r \,.$$

5. Recomposition of the output intervals Finally, the intervals $Z_r^{(j)}$ are recomposed to the fuzzy-valued outputs $\widetilde{q}_r$ of the system.

If the problem $z_r = f_r(p_1, p_2, \ldots, p_n)$ is characterized by a function $f_r(\cdot)$ which is monotonic with respect to each variable $x_i \in \text{supp}(\widetilde{p}_i)$, $i = 1, 2, \ldots, n$, the intervals with bounds given by the minimum and the maximum value of the elements of $\widehat{Z}_r^{(j)}$, $j = 0, 1, \ldots, m$, do already represent the proper solution of the problem at every level of membership. Thus, for monotonic problems it is sufficient to use the TM in its reduced form. In case of local extrema of the functions $f_r(\cdot)$ within $\text{supp}(\widetilde{p}_i)$ this will not be the case any more. The TM, however, can also be applied for complex, non-monotonic problems. If the problem $z_r = f_r(p_1, p_2, \ldots, p_n)$ is non-monotonic with respect to only one of the n variables $x_i \in \text{supp}(\widetilde{p}_i)$, $i = 1, 2, \ldots, n$, it is still sufficient to use the TM in its reduced form. In the most general case, where the problem is non-monotonic with respect to an arbitrary number of variables, the TM has to be applied in its general or extended form where the problem is evaluated at more than just the boundary points of the decomposed input intervals.

With the *extended transformation method* only those n_g parameters that actually cause non-monotonic behavior – i.e. the parameters $\widetilde{p}_1, \widetilde{p}_2, \ldots, \widetilde{p}_{n_\text{g}}$ – are transformed using the general form of the transformation method; the others – i.e. the parameters $\widetilde{p}_{n_\text{g}+1}, \widetilde{p}_{n_\text{g}+2}, \ldots, \widetilde{p}_n$ – are transformed using the reduced form. For further details on the extended formulation of the transformation method and for some illustrative examples on the application of the transformation method, the reader is referred to Hanss (2005, 2003b).

The TM represents a practical way to implement the fuzzy arithmetical approach for the simulation of complex physical systems with uncertain parameters. The fuzzy-valued inputs of the system represent the uncertain parameters, and the fuzzy-valued outputs of the system result from the fuzzy arithmetical evaluation of the system. In this context, the simulation tool of the TM allows a quantification of the propagation of the uncertainty through the system. On the other hand, the analysis tool provides a kind of sensitivity analysis by quantifying the influence of each fuzzy input parameter on the fuzzy outputs. This finally leads to a better understanding of the investigated physical system. For example, it may be a practical way to reduce computational complexity by neglecting those uncertain parameters whose influence is small in comparison to the others.

Due to the fact that the TM is based on multiple crisp-valued computations, the method can be coupled to any software code, such as finite element software, without time and cost expensive rewriting of the program code for the use of fuzzy-valued parameters. This has been shown for various engineering applications in the publications Turrin et al. (2009); Haag et al. (2009b); Fleissner et al. (2009a,b); Haag and Hanss (2010a); Haag et al. (2010); Haag and Hanss (2010c); Hanss et al. (2011); Turrin et al. (2011); Pratellesi et al. (2011).

Practical implementation The steps of decomposition and transformation as well as of retransformation, recomposition and analysis can be coupled to any existing commercial software by a separate pre- and postprocessing tool without changes in the source code of the software. For this purpose, the in-house software package FAMOUS (**F**uzzy **A**rithmetical **M**odeling **O**f **U**ncertainty **S**ystem) has been developed, representing a very efficient implementation of the pre- and postprocessing steps, that works for different commercial FE solvers.

Time- or frequency-dependent fuzzy quantities In Eq. (2), the definition of a system with fuzzy-valued model parameters $\widetilde{p}_i$, $i = 1, 2, \ldots, n$, and fuzzy-valued outputs $\widetilde{q}_r$ is given. Even though Eq. (2) covers the following case implicitly, time or frequency dependence is introduced explicitly in this section in order to give a clearer and more widespread description of a system with fuzzy-valued model parameters and varying fuzzy-valued outputs. The most comprehensive generalization of Eq. (2) is

$$\widetilde{q}_r(s) = f_r(\widetilde{p}_1(s), \widetilde{p}_2(s), \ldots, \widetilde{p}_n(s), s), \tag{4}$$

with $f_r : \mathbb{R}^n + 1 \rightarrow \mathbb{R}$, $r = 1, \ldots, k$. Thereby, frequency or time are discretized at specific instances, and the variable s, $s = 1, 2, ..., S$, is used as an index for these instances. For a specific value of s, Eq. (4) reduces to Eq. (2). The s-dependent quantities $\widetilde{p}_i(s)$ and $\widetilde{q}_r(s)$ imply that also their interval bounds and nominal values are dependent on the independent variable s. Thus, the interval bounds and nominal values for inputs and outputs, respectively, become $a_i^{(j)}(s), b_i^{(j)}(s), \overline{x}_i(s), i = 1, 2, \ldots, n$, and $c_r^{(j)}(s), d_r^{(j)}(s), \overline{z}_r(s), r = 1, \ldots, k$.

When applying the transformation method or the inverse fuzzy arithmetical method, which is described in the Section 2.3, all fuzzy arithmetical computations are applied separately for each specific value of the variable s. In Eq. (4), the model parameter $\widetilde{p}_i(s)$, $i = 1, 2, \ldots, n$, may also vary over s, which can be found in many practical applications, for example in the frequency-dependent damping behavior of a mechanical system.

In order to illustrate the previously described issue, the function $f : x, A \rightarrow y$ with

$$y(x) = A \sin(x) \tag{5}$$

is used in the following, where the amplitude A is assumed to be the uncertain model parameter $\widetilde{p}_1$ and $x(s) = s\frac{2\pi}{S}$, $s = 1, 2, ..., S$, represents time which is discretized into S steps.
In a first step, the constant fuzzy-valued amplitude

$$\widetilde{p}_1 = \widetilde{A} = \mathrm{tfn}(1, 0.2, 0.2)$$

is used, being invariant with respect to s. The fuzzy-valued output

$$\widetilde{q}(s) = \widetilde{A} \sin(s)$$

of the function f, which can be computed by applying the transformation method to the problem, is illustrated in Fig. 4 for $x \in [0, 2\pi]$. For each specific value of s, a fuzzy number is obtained for the output $\widetilde{q}(s)$. The computation of these fuzzy numbers for a set of values of s leads to a three-dimensional output where time (or frequency) is the first dimension, the physical value of the parameter A is the second and the membership level of the fuzzy number is the third. The latter is visualized by the saturation of the color in all subsequent plots of a dependent fuzzy-valued output quantity. In other words, the contour plot of a s-dependent fuzzy number results from stacking S independent fuzzy numbers (like the one shown in Fig. 1) in the s-direction.

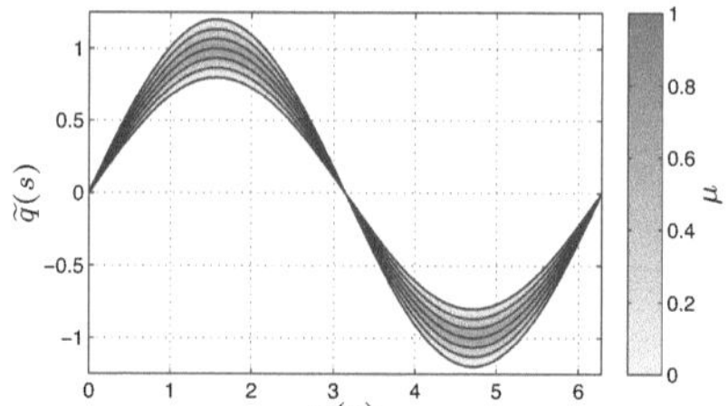

Figure 4. Fuzzy-valued output $\widetilde{q}(s)$.

In a second step, the fuzzy-valued amplitude

$$\widetilde{p}_1(s) = \widetilde{A}(s) = \mathrm{tfn}(1, 0.2\,(1 - \frac{x(s)}{2\pi}), 0.2\,(1 - \frac{x(s)}{2\pi}))$$

is implemented, which is now varying and depending explicitly on the variable s. The uncertain parameter $\widetilde{A}(s)$ is shown on the left-hand side of

Fig. 2.3 while the corresponding output $\widetilde{q}(s)$ is visualized on the right-hand side.

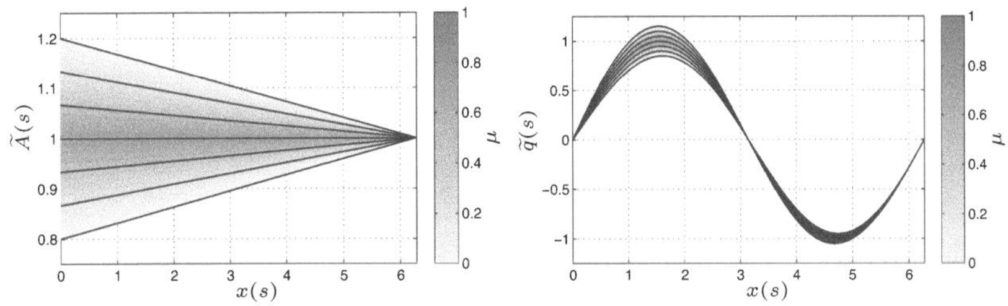

Figure 5. s-dependent model parameter $\widetilde{A}(s)$ (left) and corresponding fuzzy-valued output $\widetilde{q}(s)$ (right).

2.4 Application of fuzzy arithmetical uncertainty analysis

As an application area of high practical use and significant industrial importance, the proposed fuzzy-based approach to comprehensive modeling, simulation and analysis of systems with (predominantly) epistemic uncertainties is employed in automotive crash in the following. Explicitly, the crash of a vehicle against a rigid wall is studied, where, for the sake of simplicity, the vehicle is idealized by an elastic bumper with its finite element mesh shown in Fig. 6, and by a virtual mass located on the crossmember and representing the seat of the passenger. Besides other output quantities that are of general interest, such as the wall force and the absorbed energy, the acceleration (of the crossmember) shall exclusively be considered in the following. The following section is based on the publication Hanss and Turrin (2010).

Conventional modeling of automotive crash In conventional modeling of automotive crash, where no uncertainties are taken into account, the model parameters are quantified by appropriate nominal values only, and uncertainties due to subjectivity in implementation or due to simplification and idealization are not considered. In this case, the numerical simulation of the crash process provides exactly one crisp curve of the interesting output quantity over time, such as the acceleration transient plotted in Fig. 7(top left). Additionally, resulting from experimental crash tests, measured data for the acceleration of the passenger cabin may be available, and a possible realization of this experimentally obtained acceleration is given in Fig. 7(top right). The direct comparison of the results, provided by the numerical sim-

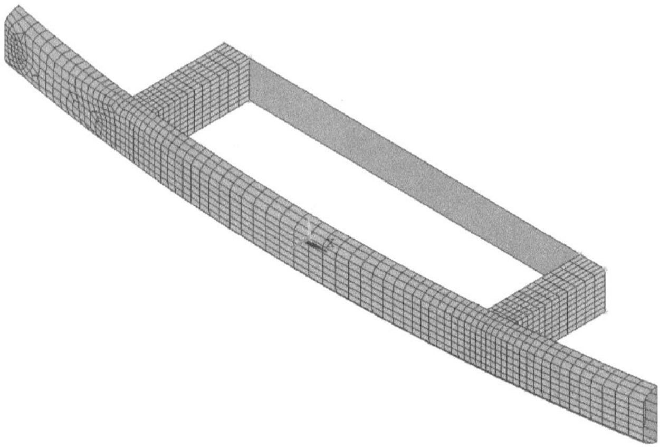

Figure 6. Finite element mesh of the bumber crashing against a rigid wall.

ulation on the one side and the experimental testing on the other, shows a significant discrepancy (Fig. 7(bottom)), which may be a decisive factor if the activation of an airbag is controlled by the exceedance of a certain threshold of the acceleration.

This discrepancy may be rated as unsatisfactory, and usually some practical remedy is supposed to be found in a further (finite element) model refinement of the vehicle components if the technological progress provides some significant increase of the computer power. However, the observed discrepancy may be absolutely satisfactory if the uncertainties inherent to the models are appropriately taken into account in the simulation procedure by the proposed concept of comprehensive modeling, as outlined in the following.

Uncertainties in automotive crash Besides the well-studied aleatory uncertainties, such as variability of material properties and geometrical parameters due to irregularities in manufacturing or assembly, the epistemic uncertainties play an important role in automotive crash simulation. Those

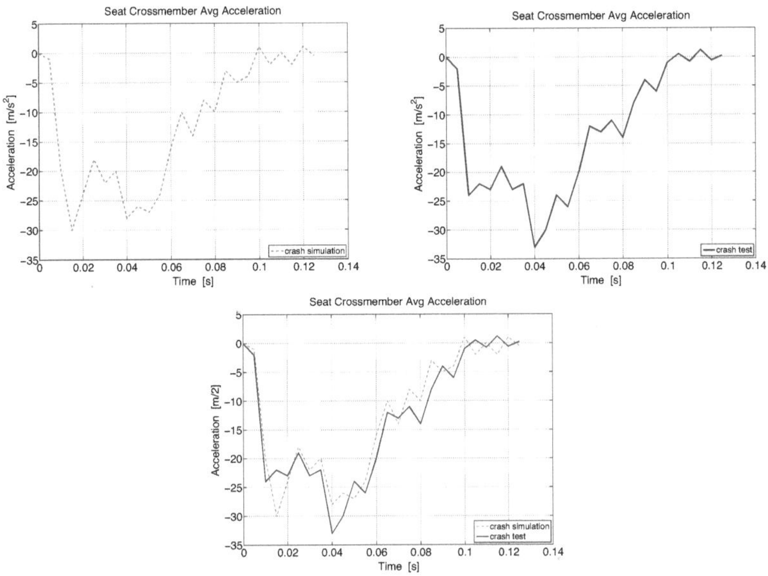

Figure 7. (top left): Acceleration of the cabin resulting from a numerical crash simulation, (top right): acceleration of the cabin resulting from an experimental crash test, (bottom): combined illustration of the accelerations of both crash simulation and crash test.

epistemic uncertainties can be classified on the basis of their origin and their nature in the following categories:

- Lack of knowledge or vagueness,
  e.g. a broad spectrum of possible operating conditions which are not known or only vaguely defined, such as the impact angle of a frontal crash, which can be assumed to be 'quasi frontal' in reality, rather than 'perfectly frontal' as ideally assumed, or the impact velocity, etc.
- Subjectivity in implementation,
  e.g. the selection of different methods of numerical evaluation by using different finite element solvers, or the mesh refinement, the element types, the calculation parameters, such as time step and contact parameters, etc.
- Errors or defects in modeling,
  e.g. idealization or simplification due to a linearization of the model equations or the assumption of linear model behavior to basically ensure the numerical evaluation, or idealized or simplified constitutive laws for material models or friction behavior, etc.

Usually, epistemic uncertainties of either of the three categories may occur in the modeling procedure of automotive crash. However, in the framework of comprehensive modeling as introduced above, these uncertainties can successfully be represented and quantified by fuzzy numbers, and the propagation of the uncertainties through the finite element model can be evaluated by the use of the TM of fuzzy arithmetic.

Comprehensive modeling of automotive crash In the example presented in this section, the following three parameters are assumed to be uncertain, reflecting uncertainties of epistemic type:

- The yield strength,
  a characteristic parameter of the material model assumed (material of the bumper) which quantifies the transition from elastic to plastic behavior;
- The frictional exponential decay coefficient,
  a characteristic parameter of the friction model assumed, which quantifies the dynamic behavior of the friction coefficient between bumper and wall;
- The shell thickness scale factor,
  a characteristic parameter of the finite element solver which controls the penetration of one element into another.

These uncertain model parameters are represented by triangular fuzzy numbers $\widetilde{p}_1$, $\widetilde{p}_2$ and $\widetilde{p}_3$, as shown exemplarily for the frictional exponential decay coefficient $\widetilde{p}_2$ in Fig. 8.

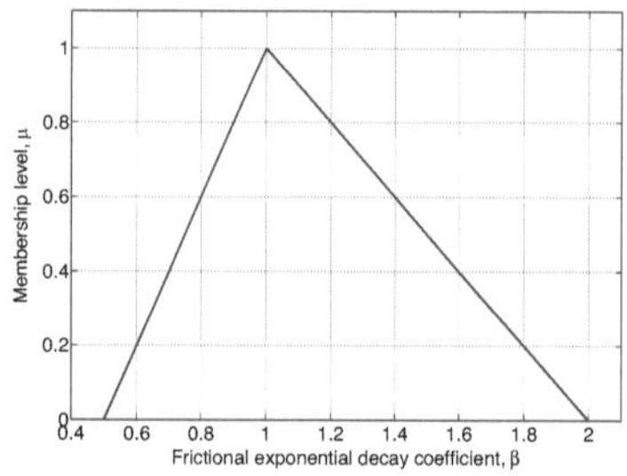

Figure 8. Uncertain frictional exponential decay coefficient represented by the (linear) triangular fuzzy number $\widetilde{p}_2$.

After applying the TM of fuzzy arithmetic, the fuzzy-valued acceleration of the cabin, representing the output fuzzy number $\widetilde{q}(t)$ of interest, can be determined over time. It is shown in Fig. 9(left) as a contour plot with the membership level expressed by a gray scale. By slicing through the contour plot at a specific time, the uncertain acceleration at this time can be obtained and appropriately visualized by a fuzzy number (Fig. 9(right)). At the membership level of unity, the nominal value can be read off, which is identical to the crisp value achieved if no uncertainty is assumed. At the membership level of zero, instead, the fuzzy number shows the worst-case interval of possible values and thus the maximum uncertainty that can be anticipated. Finally, the obvious loss of the initially assumed linear, triangular shape allows the drawing of conclusions about the degree of non-linearity of the transfer behavior of the system with respect to the uncertain parameters.

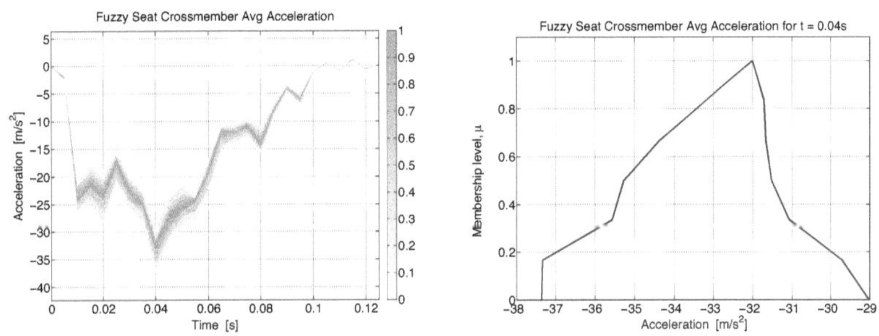

Figure 9. (left): Contour plot of the fuzzy-valued acceleration $\widetilde{q}(t)$ of the cabin, (right): fuzzy-valued acceleration $\widetilde{q}(t = 0.04 \text{ sec})$ at the specific time $t = 0.04$ sec.

In contrast to conventional, crisp modeling, the uncertain, fuzzy-valued acceleration in Fig. 9 can be interpreted as a comprehensive result of automotive crash simulation that does not pretend to be exact as the crisp curves do, but additionally contains the information about how the unavoidable uncertainties effect the simulation and about how exact a sincere and honest simulation can only be. If further information about the individual influences of the three uncertain model parameters $\widetilde{p}_i$, $i = 1, 2, 3$, on the overall fuzziness of the model output $\widetilde{q}(t)$ is required, the influence measures provided in the framework of the TM can be calculated, which shall, however, not be presented at this point.

Finally, if the experimental results of the crash tests from Fig. 7(top right) are now assembled into the results of comprehensive modeling, as shown in the direct comparison in Fig. 10, it is obvious that the results of the crash tests do not at all need to be rated as unsatisfactory; they are,

Figure 10. Contour plot of the fuzzy-valued acceleration $\widetilde{q}(t)$ of the cabin (crash simulation based on comprehensive modeling) as well as experimentally determined cabin acceleration (crash test).

in fact, absolutely acceptable, for they mostly range within the anticipated uncertainty bounds.

3 Inverse fuzzy arithmetic

The goal of the inverse fuzzy arithmetical approach is to identify a set of uncertain model parameters which lead to an uncertain output which, in turn, conservatively includes one or several measured outputs. In other words, the uncertainty of the output that results from a re-simulation with the identified uncertain parameters, has to be greater or equal than the measured output $\check{\widetilde{q}}_r$, see Section 3.1.
In this chapter, the two further theoretical aspects of applied fuzzy arithmetic are derived and explained in detail. The first one consists of the identification of uncertain model parameters on the basis of measurement values, the second one consists of the use of these identified uncertain parameters for the assessment of the quality of the model. Mathematically, these two tasks can be reduced to, firstly, the computation of a set of fuzzy input quantities $\widetilde{p}_1 ... \widetilde{p}_n$ of a given model, and secondly, the determination of a model validity criterion Λ.
The steps that are necessary in order to fullfil these tasks are presented in the following in the order of their occurrence.

This work is based on several publications on inverse fuzzy arithmetic and the TM. The principal idea of inverse fuzzy arithmetic has already been presented in Hanss (2003a, 2005), whereas in Haag et al. (2009b,a); Haag and Hanss (2010a) a generalization to time- or frequency dependent fuzzy-valued systems has been elaborated. The issue of model validation is addressed in Haag and Hanss (2010b); Haag et al. (2010); Haag and Hanss (2010c). In this work, all previously mentioned publications are restructured, unified and partly clarified.

3.1 Provision of fuzzy outputs from measured data

In order to identify the uncertain parameters of a model which is capable of representing a real system, a set of fuzzy-valued outputs $\tilde{\breve{q}}_r$ needs to be defined from the measured, crisp outputs of the real system. The nominal values $\bar{\breve{z}}_r = \bar{z}_r$ of the fuzzy values $\tilde{\breve{q}}_r$ must correspond to the nominal outputs $\bar{z}_r$ of the initial simulation due to the fact that the nominal values of the model parameters are also equal. The nominal values of the model parameters are equal because it is assumed that they are determined in some optimal way in the beginning of the analysis using an optimization procedure. Thus, they are not subjected to be identified within the inverse fuzzy arithmetical approach. The worst-case deviation from the nominal values is then defined by the measured output of the real system. Consequently, one value of the bounds of the support of $\tilde{q}_r$ is set to the measurement value z_r^* while the other one is set to the nominal value. For the other μ-levels, a linear interpolation between $\mu = 0$ and $\mu = 1$ is chosen. In the case of a single measurement value, this procedure leads to a single-sided, triangular fuzzy number, as visualized in Fig. 3.1(left). If multiple measurement values exist, the procedure described above is applied repeatedly for each measurement value. The final output fuzzy number $\tilde{\breve{q}}_r$ is then computed by applying the union operator to the acquired single-sided, triangular fuzzy numbers. In total, this way of proceeding leads to a triangular fuzzy number whose worst-case bounds are given by the minimum $z_{r,\min}^*$ over all measurement values and by the maximum $z_{r,\max}^*$, respectively, see Fig. 3.1(right). If the

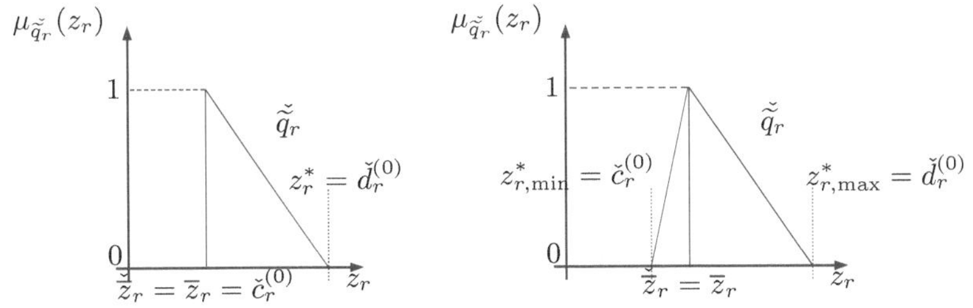

Figure 11. Definition of the output fuzzy number $\tilde{\breve{q}}_r$ based on a single (left) or multiple (right) measurement values.

output is time- or frequency-dependent, the procedure described above is applied successfully for each sampling instance of time or frequency. This implies that simulation and measurement values have to be available for the same instances of time or frequency which can be achieved through an interpolation in the one direction or the other.

3.2 System representation based on the transformation method

The inverse fuzzy arithmetical approach is based on the TM of fuzzy arithmetic. Before the inverse computation can be performed, an initial forward simulation has to be executed using the TM in order to generate a system representation and a set of preliminary output quantities.
For the initial simulation, a practical set of uncertain model parameters has to be chosen. The nominal values of these parameters are determined in the same way as the parameters of a classical crisp-valued simulation. Usually, this is carried out through a best-fit procedure which minimizes the quadratic deviation between simulated and measured output.

Recalling the motivation for introducing a membership level as a smooth transition between worst-case interval and nominal value, the nominal values of the uncertain parameters may not be changed throughout the inverse fuzzy arithmetical approach, as they represent the optimal set of parameters that would be used for a conventional crisp-valued simulation. Due to the fact that the nominal values of the uncertain model parameters represent the optimal parameters, they are not changed throughout the inverse fuzzy arithmetical approach. In other words, the optimal set of parameters must be independent of any fuzzy-arithmetical considerations as fuzzy numbers are explicitly based on these optimal parameters and a corresponding perturbation.

For the amount of uncertainty in the uncertain model parameters, i.e. for the worst-case bounds, some initial educated guesses have to be assumed. These initial guesses have to be reasonable with respect to two properties. Firstly, they have to be within the range of physically possible values, and secondly, they have to be as close as possible to the actual uncertainty of the system. The second of these two requirements results from the fact that for real engineering problems, relatively small discretization numbers m are used, which usually are $m = 3$ or $m = 4$. Otherwise, computational costs would exceed the permissible range. For those small discretization numbers, though, nonlinearities of the model are not covered accurately, if the worst-case intervals of the uncertain parameters for the initial simulation are in a totally different range as those that will be identified.
Moreover, the set of input parameters used in the initial simulation is used later on for an interpolation of the set of parameters that is identified. For

this reason, numerical accuracy is improved if rather large uncertainties are used for the initial simulation as interpolating data is more accurate than extrapolating it. Through the initial simulation, where the uncertainties of the model parameters have been chosen according to the best possible guesses, a set of fuzzy-valued input quantities and a corresponding set of fuzzy-valued output quantities is available. Based on these fuzzy-valued input and output quantities of the system under consideration, a system representation $\mathbf{H}^{(j)}(s)$ is derived where the system is considered as a linear mapping between the uncertainties of the inputs and the uncertainties of the outputs. This linear mapping , however, can be different for each level of membership and is given by the equation

$$\mathbf{H}^{(j)}(s) \begin{bmatrix} a_1^{(j)}(s) - \bar{x}_1(s) \\ b_1^{(j)}(s) - \bar{x}_1(s) \\ a_2^{(j)}(s) - \bar{x}_2(s) \\ b_2^{(j)}(s) - \bar{x}_2(s) \\ \vdots \\ a_n^{(j)}(s) - \bar{x}_n(s) \\ b_n^{(j)}(s) - \bar{x}_n(s) \end{bmatrix} = \begin{bmatrix} c_1^{(j)}(s) - \bar{z}_1(s) \\ -(d_1^{(j)}(s) - \bar{z}_1(s)) \\ c_2^{(j)}(s) - \bar{z}_2(s) \\ -(d_2^{(j)}(s) - \bar{z}_2(s)) \\ \vdots \\ c_k^{(j)}(s) - \bar{z}_k(s) \\ -(d_k^{(j)}(s) - \bar{z}_k(s)) \end{bmatrix} \tag{6}$$

with

$$\mathbf{H}^{(j)}(s) = \begin{bmatrix} \mathbf{H}_{11}^{(j)}(s) & | & \mathbf{H}_{12}^{(j)}(s) & | & \cdots & | & \mathbf{H}_{1n}^{(j)}(s) \\ --- & & --- & & --- & & --- \\ \mathbf{H}_{21}^{(j)}(s) & | & \mathbf{H}_{22}^{(j)}(s) & | & & | & \mathbf{H}_{2n}^{(j)}(s) \\ --- & & --- & & --- & & --- \\ \vdots & & \vdots & & \vdots & & \vdots \\ --- & & --- & & --- & & --- \\ \mathbf{H}_{k1}^{(j)}(s) & | & \mathbf{H}_{k2}^{(j)}(s) & | & \cdots & | & \mathbf{H}_{kn}^{(j)}(s) \end{bmatrix} \tag{7}$$

and

$$\mathbf{H}_{ri}^{(j)}(s) = \frac{1}{2} \begin{bmatrix} \eta_{ri-}^{(j)}(s)(1 + \mathrm{sign}(\eta_{ri-}^{(j)}(s))) & \eta_{ri+}^{(j)}(s)(1 - \mathrm{sign}(\eta_{ri+}^{(j)}(s))) \\ -\eta_{ri-}^{(j)}(s)(1 - \mathrm{sign}(\eta_{ri-}^{(j)}(s))) & -\eta_{ri+}^{(j)}(s)(1 + \mathrm{sign}(\eta_{ri+}^{(j)}(s))) \end{bmatrix}$$

$$i = 1, 2, \ldots, n \,, \quad r = 1, 2, \ldots, k \,, \quad j = 0, 1, \ldots, m-1 \,.$$

Obviously, the system representation $\mathbf{H}^{(j)}(s)$ is exclusively based on the single-sided gain factors $\eta_{ri-}^{(j)}(s)$ and $\eta_{ri+}^{(j)}(s)$. According to Hanss (2003a), the single-sided gain factors can be determined on the basis of the coefficients $\eta_{ri}^{(j)}$ of the TM, which are computed through

$$\eta_{ri}^{(j)} = \frac{1}{2^{n-1}(b_i^{(j)} - a_i^{(j)})} \sum_{k=1}^{2^{n-i}} \sum_{l=1}^{2^{i-1}} \left({}^{s_2}\hat{z}_r^{(j)} - {}^{s_1}\hat{z}_r^{(j)} \right) \tag{8}$$

with

$$s_1(k,l) = k + (l-1)2^{n-i+1}, \quad s_2(k,l) = k + (2l-1)2^{n-i}$$

for the reduced TM. The single-sided gain factors $\eta_{ri-}^{(j)}(s)$ and $\eta_{ri+}^{(j)}(s)$ can then be computed by using only the left branch, and the right branch, respectively, of the membership function of the ith fuzzy parameter $\widetilde{p}_i$ while the other fuzzy parameters remain unmodified. In other words, in order to compute the left-sided gain factor $\eta_{ri-}^{(j)}(s)$, the ith parameter $\widetilde{p}_i$ is modified in such a way that the right-hand deviation is zero while the left-hand side deviation and all other model parameters remain unmodified. For this modified set of model parameters, the output array $\widehat{Z}_{i-,r}^{(j)}$ has to be computed, and then, the left-sided gain factor $\eta_{ri-}^{(j)}(s)$ can be determined according to Eq. (8). Then, the whole process is repeated for the right-sided gain factor, where the left-hand deviation of the current parameter is zero. Finally, the entire procedure has to be repeated for each input parameter $\widetilde{p}_i, i = 1, 2, \ldots, n$. Consequently, the whole analysis with its costly model evaluations would have to be repeated twice for each input parameter $\widetilde{p}_i$ in order to derive all new sets of input and output quantities, corresponding to the right and left branch of each input fuzzy number.
Obviously, this way of proceeding would involve a very large amount of computational effort which is undesirable for a postprocessing step like the inverse fuzzy arithmetical method. In order to overcome this drawback, the output arrays $\widehat{Z}_r^{(j)}$, i.e. the results of the initial simulation, are used for the interpolation of the new output arrays $\widehat{Z}_{i+,r}^{(j)}$ and $\widehat{Z}_{i-,r}^{(j)}$.
In detail, the interpolation-based computation of the single-sided gain factors consists of the following steps. Exemplary, the computation of the single-sided gain factors is explained for one specific input $\widetilde{p}_\lambda$, $\lambda \in \{1, 2, ..., n\}$.

1. Right-zeroing of one input The right-hand branch of the fuzzy number $\widetilde{p}_\lambda$ is set to zero:
Modified input arrays $\widehat{X}_{\lambda-,i}^{(j)}, i = 1, 2, \ldots, n$, are determined according to

$$\widehat{X}_{\lambda-,i}^{(j)} = \begin{cases} \widehat{X}_i^{(j)} & \forall i \neq \lambda \\ \underbrace{\left((\alpha_\lambda^{(j)}, \beta_\lambda^{(j)}), (\alpha_\lambda^{(j)}, \beta_\lambda^{(j)}), \ldots, (\alpha_\lambda^{(j)}, \beta_\lambda^{(j)})\right)}_{2^{\lambda-1} \text{ pairs}} & i = \lambda \end{cases} \tag{9}$$

$$\text{with} \quad {\alpha_\lambda}^{(j)} = \underbrace{(a_\lambda^{(j)}, \ldots, a_\lambda^{(j)})}_{2^{n-\lambda}\text{ elements}} \quad \text{and} \quad {\beta_\lambda}^{(j)} = \underbrace{(\overline{x}_\lambda, \ldots, \overline{x}_\lambda)}_{2^{n-\lambda}\text{ elements}}.$$

2. Interpolation of the output arrays $\widehat{Z}_{\lambda-,r}^{(j)}$ Modified output arrays $\widehat{Z}_{\lambda-,r}^{(j)}$ that correspond to the input array $\widehat{X}_{\lambda-,i}^{(j)}$ are computed based on an interpolation of the data corresponding to the initial simulation. Throughout the initial simulation, output values have been computed for each column of the input array $\widehat{X}_i^{(j)}$, which correspond to points in the n-dimensional parameter space. Now, these samples are used to interpolate the output data at different points in the parameter space, which are defined by $\widehat{X}_{\lambda-,i}^{(j)}$. The interpolation is performed with the MATLAB function `griddatan` in its default form which uses a linear interpolation on the basis of the convex hull algorithm published in Barber et al. (1996).

3. Computation of the left-sided gain factors $\eta_{r\lambda-}^{(j)}$ Based on Eq. (8), the left-sided gain factors $\eta_{r\lambda-}^{(j)}$ are determined by using the elements from the modified input array $\widehat{X}_{\lambda-,i}^{(j)}$ and the corresponding modified output array $\widehat{Z}_{\lambda-,r}^{(j)}$.

4. Left-zeroing of one input Now, the left-hand branch of the fuzzy number $\widetilde{p}_\lambda$ is set to zero:
Modified input arrays $\widehat{X}_{\lambda+,i}^{(j)}, i = 1, 2, \ldots, n$, are determined according to

$$\widehat{X}_{\lambda+,i}^{(j)} = \begin{cases} \widehat{X}_i^{(j)} & \forall i \neq \lambda \\ \underbrace{\left((\alpha_\lambda^{(j)}, \beta_\lambda^{(j)}), (\alpha_\lambda^{(j)}, \beta_\lambda^{(j)}), \ldots, (\alpha_\lambda^{(j)}, \beta_\lambda^{(j)})\right)}_{2^{\lambda-1}\text{ pairs}} & i = \lambda \end{cases} \tag{10}$$

$$\text{with} \quad {\alpha_\lambda}^{(j)} = \underbrace{(\overline{x}_\lambda, \ldots, \overline{x}_\lambda)}_{2^{n-\lambda}\text{ elements}} \quad \text{and} \quad {\beta_\lambda}^{(j)} = \underbrace{(b_\lambda^{(j)}, \ldots, b_\lambda^{(j)})}_{2^{n-\lambda}\text{ elements}}.$$

5. Interpolation of the output arrays $\widehat{Z}_{\lambda+,r}^{(j)}$ Based on the data $\widehat{X}_i^{(j)}, \widehat{Z}_r^{(j)}$ from the initial simulation, the modified output arrays $\widehat{Z}_{\lambda+,r}^{(j)}$ are computed as above (see Step 2.).

6. Computation of the right-sided gain factors $\eta_{r\lambda+}^{(j)}$ The right-sided gain factors $\eta_{r\lambda+}^{(j)}$ are computed through Eq. (8) and on the basis of the modified input and output arrays $\widehat{X}_{\lambda+,i}^{(j)}, \widehat{Z}_{\lambda+,r}^{(j)}$.

Finally, steps 1 to 6 have to be repeated for each input parameter $\widetilde{p}_\lambda$, $\lambda = 1, 2, \ldots, n$. However, due to the fact that no further model evaluations are needed for these steps, the computational effort related to those is negligible.

3.3 Parameter identification

Formulation of the optimization problem The goal of the inverse fuzzy arithmetical approach is to identify a set of uncertain model parameters which lead to an uncertain output which, in turn, conservatively includes one or several measured outputs. In other words, the uncertainty of the output that results from a re-simulation with the identified uncertain parameters, has to be greater or equal than the measured output $\breve{\tilde{q}}_r$, see Section 3.1. Mathematically, this leads to the inequality

$$\mathbf{H}^{(j)}(s)\mathbf{U}^{(j)}(s) \leq \mathbf{M}^{(j)}(s),\ j = 0, 1, \ldots, m. \tag{11}$$

Thereby,

$$\mathbf{U}^{(j)}(s) = \begin{bmatrix} \breve{a}_1^{(j)}(s) - \breve{\bar{x}}_1(s) \\ \breve{b}_1^{(j)}(s) - \breve{\bar{x}}_1(s) \\ \breve{a}_2^{(j)}(s) - \breve{\bar{x}}_2(s) \\ \breve{b}_2^{(j)}(s) - \breve{\bar{x}}_2(s) \\ \vdots \\ \breve{a}_n^{(j)}(s) - \breve{\bar{x}}_n(s) \\ \breve{b}_n^{(j)}(s) - \breve{\bar{x}}_n(s) \end{bmatrix} = \begin{bmatrix} \breve{a}_1^{(j)}(s) - \bar{x}_1(s) \\ \breve{b}_1^{(j)}(s) - \bar{x}_1(s) \\ \breve{a}_2^{(j)}(s) - \bar{x}_2(s) \\ \breve{b}_2^{(j)}(s) - \bar{x}_2(s) \\ \vdots \\ \breve{a}_n^{(j)}(s) - \bar{x}_n(s) \\ \breve{b}_n^{(j)}(s) - \bar{x}_n(s) \end{bmatrix} \tag{12}$$

contains the unknown input uncertainties which are to be identified and

$$\mathbf{M}^{(j)}(s) = \begin{bmatrix} \breve{c}_1^{(j)}(s) - \breve{\bar{z}}_1(s) \\ -(\breve{d}_1^{(j)}(s) - \breve{\bar{z}}_1(s)) \\ \breve{c}_2^{(j)}(s) - \breve{\bar{z}}_2(s) \\ -(\breve{d}_2^{(j)}(s) - \breve{\bar{z}}_2(s)) \\ \vdots \\ \breve{c}_n^{(j)}(s) - \breve{\bar{z}}_n(s) \\ -(\breve{d}_n^{(j)}(s) - \breve{\bar{z}}_n(s)) \end{bmatrix} = \begin{bmatrix} \breve{c}_1^{(j)}(s) - \bar{z}_1(s) \\ -(\breve{d}_1^{(j)}(s) - \bar{z}_1(s)) \\ \breve{c}_2^{(j)}(s) - \bar{z}_2(s) \\ -(\breve{d}_2^{(j)}(s) - \bar{z}_2(s)) \\ \vdots \\ \breve{c}_n^{(j)}(s) - \bar{z}_n(s) \\ -(\breve{d}_n^{(j)}(s) - \bar{z}_n(s)) \end{bmatrix} \tag{13}$$

represents the measured uncertain output, see Fig. 3.1. Please note that $\mathbf{M}^{(j)}(s)$ only contains negative values. Thus, Eq. (11) indeed demands a conservative coverage of the measured uncertainties by the simulated uncertainties.

Equation (11) is solved for the unknown quantities $\breve{a}_i^{(j)}(s), \breve{b}_i^{(j)}(s)$. If the number of outputs k is smaller than the number of inputs n, the system of equations in (11) still possesses some degrees of freedom which are used to minimize the resulting uncertainty in the input parameters that are to be identified. Thus, Eq. (11) is solved by additionally minimizing

$$u(s) = \mathbf{U}(s)^{\mathrm{T}}\mathbf{W}(s)\mathbf{U}(s). \tag{14}$$

The weighting matrix

$$\mathbf{W}(s) = \operatorname{diag}\left(\left[\frac{1}{(\overline{x}_1(s))^2}, \frac{1}{(\overline{x}_1(s))^2}, \ldots, \frac{1}{(\overline{x}_n(s))^2}, \frac{1}{(\overline{x}_n(s))^2}\right]\right)$$

performs a rescaling of the entries of $\mathbf{U}(s)$ with respect to their dimensions. The goal of this rescaling is to equally weight the relative uncertainty of all model parameters as the scale of the parameters is not supposed to influence the result and is thus eliminated by the rescaling.
An alternative way of proceeding is needed if any of the nominal values is zero. In this case, a normalization of the data with respect to the width of the worst case interval might be indicated.

Solution of the optimization problem By applying the TM, the task of computing the uncertainties $\mathbf{U}^{(j)}(s)$ of the model parameters $\widetilde{p}_i$ has been formulated as a quadratic optimization problem (Eq. (14)) under linear inequality constraints (Eq. (11)). For this type of optimization problems, standard solvers are available. The solver which is used in this work is based on the work of Gill et al. (1984, 1991) and is implemented in the MATLAB command `quadprog`. The algorithm computes the optimal solution $\mathbf{U}$ which minimizes the quality function

$$J = \frac{1}{2}\mathbf{U}^{\mathrm{T}}\mathbf{W}\mathbf{U} + c^{\mathrm{T}}\mathbf{U}. \tag{15}$$

Additionally, the solution $\mathbf{U}$ must satisfy the inequality constraints

$$\mathbf{H}\mathbf{U} \leq \mathbf{M} \tag{16}$$

and the equality constraints

$$\mathbf{H}_{\mathrm{eq}}\mathbf{U} = \mathbf{M}_{\mathrm{eq}}. \tag{17}$$

For the problem at hand, the quantities $c^{\mathrm{T}}, \mathbf{M}_{\mathrm{eq}}$ and $\mathbf{H}_{\mathrm{eq}}$ are neglected, which means they are set to zero.
Besides the system theoretic constraints Eq. (11, additional constraints may be defined in order to keep the solution, namely the identified uncertainties, within a permissible range. In reality, many model parameters are governed by certain restrictions which can be added to the problem by constraints of

the form

$$\begin{bmatrix} -\alpha_{\mathrm{l,min},1} \\ 0 \\ -\alpha_{\mathrm{l,min},2} \\ 0 \\ \vdots \\ -\alpha_{\mathrm{l,min},n} \\ 0 \end{bmatrix} \leq \mathbf{U}^{(j)}(s) \leq \begin{bmatrix} 0 \\ \alpha_{\mathrm{r,max},1} \\ 0 \\ \alpha_{\mathrm{r,max},2} \\ \vdots \\ 0 \\ \alpha_{\mathrm{r,max},n} \end{bmatrix} \tag{18}$$

In the solution process, these additional constraints are appended to the constraints that result from the fuzzy arithmetical problem (Eq. 11).

Condensation of time- or frequency-dependent solutions For the general case of a fuzzy system where time- or frequency-dependence is present, see Eq. (4), a set of model parameters which also depends on s is obtained from the inverse fuzzy arithmetical approach. Very often, this dependency is undesired as static models, that means models with constant parameters, are aspired. For this reason, an approach to reduce the obtained, s-dependent set of fuzzy-valued model parameters to a static set of fuzzy parameters is presented in this section.

First of all, a specific step s_t is excluded from the overall solution if the optimization problem (Eq. (14)) is infeasible for the given constraints (Eq. (11)) for this step. This test is performed for all $s = 1, 2, \ldots, S$.

The next issue that has to be addressed in the process of reducing the s-dependent fuzzy-valued model parameters to a static set is the exclusion of unreliable solutions. In order to extract those inverse problems where a reliable solution can be expected, the magnitudes of the elements of the matrix $H^{(j)}(s)$ are considered as a feasibility criterion, Haag et al. (2009b,a). For very small elements in the matrix $H^{(j)}(s)$, this matrix becomes ill-conditioned and the numerical solution of Eq. (11). cannot be considered reliable due to numerical errors, even though the numerical solver might still produce some results. For this reason, those steps are excluded from the identification procedure where the values of all absolute measures of influence $\kappa_i(s)$, $i = 1, 2, \ldots, n$, remain below a certain threshold. According to Hanss (2005), the measure $\kappa_i(s)$, quantifying the influence of the uncertainty of the ith model parameter on the uncertainty of the output of interest, is defined by

$$\kappa_i(s) = \frac{\sum\limits_{j=1}^{m-1} \mu_j \left| \eta_i^{(j)}(s) \left(\tilde{\breve{a}}_i^{(j)}(s) + \tilde{\breve{b}}_i^{(j)}(s) \right) \right|}{2 \sum\limits_{j=1}^{m-1} \mu_j} . \tag{19}$$

For the reasons explained previously, the solution for one specific s_t is rated as reliable if for all output quantities the following inequality holds:

$$\kappa_i(s_t) > \epsilon > 0 \quad \forall i = 1, 2, \ldots, n. \tag{20}$$

The actual condensation of the time- or frequency-dependent set of fuzzy parameters to a static set of fuzzy parameters is based on those instances of time or frequency where a solution exists and is considered reliable with respect to Eq. (20). For the definition of the one set of uncertain input parameters $\tilde{\breve{p}}_i$ that is considered as representative for all values of s, for each level of membership μ_j, two extreme cases can be specified: the maximum uncertainty, i.e. the union of all intermediate uncertain input parameters over all s, or the average uncertainty, i.e. the mean of all intermediate uncertain input parameters over all s. The upper and lower bounds of the uncertain input parameters resulting from the application of the union operator are defined by

$$\begin{aligned} \breve{a}_{i,\text{union}}^{(j)} &= \min_s \left(\breve{a}_{i,\text{union}}^{(j)}(s),\, \breve{a}_{i,\text{union}}^{(j+1)} \right) ,i = 1, 2, \ldots, n;\ j = 0, 1, \ldots, m-1 \\ \breve{b}_{i,\text{union}}^{(j)} &= \max_s \left(\breve{b}_{i,\text{union}}^{(j)}(s),\, \breve{b}_{i,\text{union}}^{(j+1)} \right) \end{aligned} \tag{21}$$

whereas those obtained by applying the mean operator are defined by

$$\begin{aligned} \breve{a}_{i,\text{mean}}^{(j)} &= \min_s \left(\frac{1}{S} \sum_{s=1}^{S} \breve{a}_{i,\text{mean}}^{(j)}(s),\, \breve{a}_{i,\text{mean}}^{(j+1)} \right) ,i = 1, 2, \ldots, n;\ j = 0, 1, \ldots, m-1 \\ \breve{b}_{i,\text{mean}}^{(j)} &= \max_s \left(\frac{1}{S} \sum_{s=1}^{S} \breve{b}_{i,\text{mean}}^{(j)}(s),\, \breve{b}_{i,\text{mean}}^{(j+1)} \right) . \end{aligned} \tag{22}$$

Both Eq. (21) and Eq. (22) ensure the convexity of the resulting fuzzy parameters, which is a crucial property for further computations, especially for the re-simulation of the identified system.

The application of the union operator leads to the most uncertain set of fuzzy input parameters as the largest intervals over all s are transfered

to the final set of input parameters for each level of membership. For the union operator, a conservative coverage of the measured outputs can be guaranteed for all steps with a feasible solution.
In contrast to the union operator, the mean operator performs an averaging of the intervals which can be considered as a compromise over different operating regimes. The conservatism of the union operator is thus abandoned for the benefit of a smaller uncertainty. In practice, a suitable trade-off between conservatism on the one side and a small model uncertainty on the other side is recommended. For this reason, a degree of conservatism γ may be introduced which describes the transition between the uncertainty that is averaged over all steps $s = 1, \ldots, S$ (see Eq. (22)) and the most conservative uncertainty (see Eq. (21)). Based on the degree of conservatism γ, the bounds for the input parameters $\widetilde{p}_i$ are then computed through

$$\begin{aligned} \breve{a}_i^{(j)} &= \breve{a}_{i,\text{mean}}^{(j)} + \gamma \left(\breve{a}_{i,\text{union}}^{(j)} - \breve{a}_{i,\text{mean}}^{(j)} \right) \\ \breve{b}_i^{(j)} &= \breve{b}_{i,\text{mean}}^{(j)} + \gamma \left(\breve{b}_{i,\text{union}}^{(j)} - \breve{b}_{i,\text{mean}}^{(j)} \right) . \end{aligned} \tag{23}$$

For practical applications, experience is still needed in order to use an adequate value for the degree of conservatism. For problems with a very large number of steps, it is adviseable to use a very small number for the degree of conservatism γ or even to use zero. The reason for this recommendation is that the union operator is dominated by a single step s and thus ignores the largest number of stepwise solutions. Moreover, a complete coverage can usually not be achieved in any way as the models, which are used for the coverage of the measurements, cannot be driven to any possible value. In such cases, the optimization fails, but it leads to unrealistically large results on the boarders of the feasible regions, which are not supposed to be included in the final solution.
If multiple different models are compared in order to select the most appropriate model, the same value for the degree of conservatism γ has to be used for all models. This value does not necessarily have to be one.

To verify the identified model parameters $\breve{\widetilde{p}}_1, \breve{\widetilde{p}}_2, \ldots, \breve{\widetilde{p}}_n$, the model equations (2) can be re-simulated by means of the TM, using $\breve{\widetilde{p}}_1, \breve{\widetilde{p}}_2, \ldots, \breve{\widetilde{p}}_n$ as the fuzzy input parameters. The degree of conformity of the so-obtained output fuzzy numbers $\breve{\widetilde{q}}_1, \breve{\widetilde{q}}_2, \ldots, \breve{\widetilde{q}}_n$ with the measurement output can serve as a measure of the quality of the identification.

3.4 Model validation

In order to provide a quantitative measure for the quality of the identified model, the model validity criterion Λ is defined in the following. Its definition is based on the idea that the ideal model minimizes both the uncertainty of the output covering the measured output, and the uncertainty of the identified parameters of the model.
In Hanss (2005), the relative imprecision of a general fuzzy number $\widetilde{v}$ with nominal value $\bar{v}$ is defined as

$$\mathrm{imp}_{\bar{v}}(\widetilde{v}) = \frac{1}{2m\bar{v}} \sum_{j=0}^{m-1} \left[b_i^{(j)} - a_i^{(j)} + b_i^{(j+1)} - a_i^{(j+1)} \right] . \tag{24}$$

The relative imprecision $\mathrm{imp}_{\bar{v}}(\widetilde{v})$ corresponds to the area between the membership function $\mu_{\widetilde{v}}(x)$ and the x-axis, normalized by the nominal value $\bar{v}$. It quantifies the amount of the uncertainty that is inherent to a fuzzy number $\widetilde{v}$. Based on the quantification of uncertainty given above, the model validity criterion Λ is defined as

$$\Lambda = \sum_{i=1}^{n} \left| \mathrm{imp}_{\bar{\tilde{p}}_i}(\tilde{\tilde{p}}_i) \right| \cdot \sum_{r=1}^{k} \left[\frac{1}{S} \sum_{s=1}^{S} \left(\left| \mathrm{imp}_{\bar{q}_r(s)}(\widetilde{q}_r(s)) \right| \right) \right] . \tag{25}$$

The sum in the first factor captures the amount of uncertainty inherent to all input parameters whereas the sum in the second factor accounts for the average uncertainty of the outputs. For practical problems, the model validity criterion Λ can be used to assess the quality of a model, as it does not only focus on the norm of the output deviation, but it additionally takes into account the amount of the input deviation that causes the output deviation. The most appropriate model is thus characterized by a possibly small value of the model validity criterion Λ.

3.5 Illustrative examples to inverse fuzzy arithmetic

In the following, the inverse fuzzy arithmetical approach, including the model selection feature, is illustrated for two mathematical examples. A more practical example of high industrial relevance can be found in Haag et al. (2011). The presented results are based on Haag and Hanss (2010b).

Polynomial fit In this section, the presented method is applied to the rather simple mathematical problem of approximating the fourth-order polynomial

$$y(x) = 10 + x + x^2 + x^3 + x^4 \tag{26}$$

by a polynomial of lower degree $d < 4$

$$\widetilde{y}(x) = \sum_{l=0}^{d} \widetilde{a}_l x^l, \tag{27}$$

but with fuzzy-valued coefficients $\widetilde{a}_l$, $l = 0, 1, \ldots, d$. The nominal values of the fuzzy coefficients $\widetilde{a}_l$ are computed through a best fit to the reference output $y(x)$, their uncertainties by the application of inverse fuzzy arithmetic. The output $\widetilde{y}(x)$ of the fuzzy-valued model in Eq. (27) with uncertain fuzzy-valued coefficients $\widetilde{a}_l$ is capable of entirely covering the reference output $y(x)$.

Figures 12 to 15 show the approximations of the fourth-order polynomial in Eq. (26) with three lower-order polynomials ($d = 1, 2, 3$) as well as with a polynomial of proper order ($d = 4$). On the left-hand side, the fuzzy coefficients $\widetilde{a}_l$ are shown. On the right-hand side, the uncertain outputs are plotted, where the degree of membership is visualized by a contour plot. Each of the subfigures is labeled with the approximating model type and the value of the model validity measure Λ. For the latter, the contribution of each element is given in detail. The first factor corresponds to the uncertainty of the fuzzy model parameters $\widetilde{a}_0, ..., \widetilde{a}_d$, whereas the second factor corresponds to the output uncertainty.

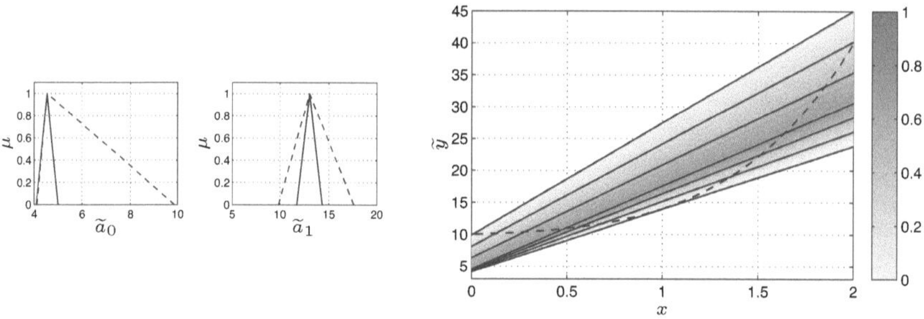

Figure 12. Linear polynomial: $\Lambda = (0.6335 + 0.2973) \cdot 0.4094 = 0.3811$

The fourth-order polynomial is shown in Fig. 15 and, obviously, no uncertainty is needed to cover the reference output. In Figs. 12 to 14, the degree of the approximation polynomial increases while the uncertainty in the coefficients decreases. Obviously, the quality of the model increases with increasing order of the polynomial as a smaller uncertainty in the model parameters is needed to cover the modeling error. The introduced measure

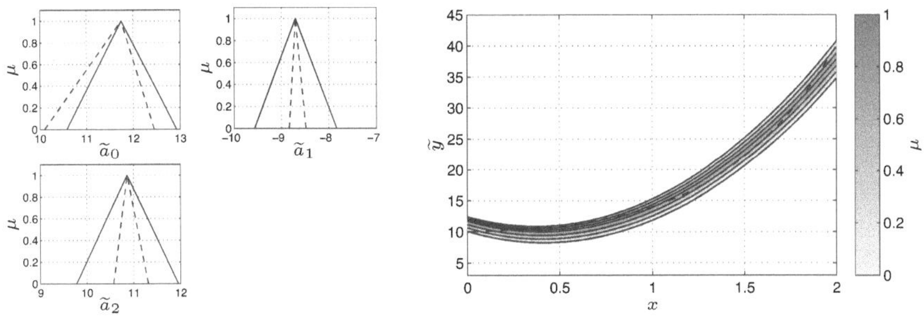

Figure 13. Quadratic polynomial: $\Lambda = (0.0999 + 0.0201 + 0.0338) \cdot 0.1123 = 0.0173$

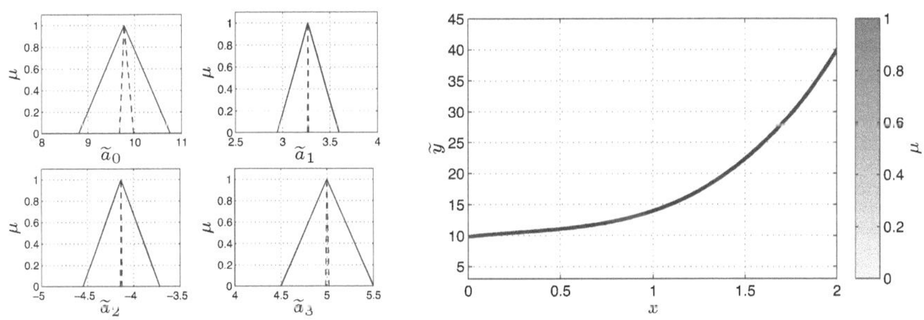

Figure 14. Cubic polynomial: $\Lambda = (0.0152 + 0.0016 + 0.0020 + 0.0037) \cdot 0.0124 = 0.0003$

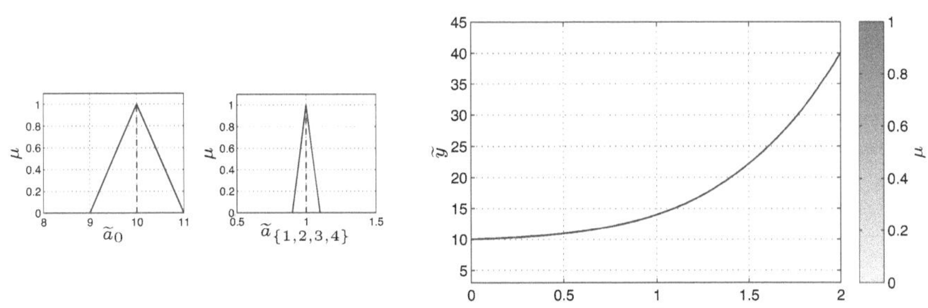

Figure 15. Fourth-order polynomial: $\Lambda = (5 \cdot 0) \cdot 0 = 0$

Λ of model validity of the model, which is given below the plots, is obviously capable of quantifying this fact in a systematic way as the values of Λ decrease continously.

Linearization of a quadratic function As a further example, five different fuzzy-valued linearizations of the fourth-order polynomial described in the previous section (see Eq. (26)) are compared. All these linearizations are of the form

$$\widetilde{y} = \widetilde{a}_0 + \widetilde{a}_1 x, \tag{28}$$

where $\widetilde{a}_0$, $\widetilde{a}_1$ are fuzzy parameters. The five different linearizations are labeled L_1 to L_5 and differ in the way the nominal values of the parameters $\widetilde{a}_0$, $\widetilde{a}_1$ are determined. The nominal values of the linearizations L_1 to L_3 are generated by computing the Jacobian linearizations of the reference output in Eq. (26) for the three points with $x = 0$, $x = 0.5$ and $x = 1$, respectively. For the linearization L_4, the best fit in terms of the mean-squared error of the output is computed and nominal values of the fuzzy parameters $\widetilde{a}_0$, $\widetilde{a}_1$ is derived therefrom. The nominal values of the linearization L_5 are computed by an optimization which minimizes the model validity criterion Λ of the model.

The uncertainty of the two fuzzy parameters $\widetilde{a}_0$, $\widetilde{a}_1$ are determined by inverse fuzzy arithmetic and the resulting input parameters lead to an uncertain output which in all cases covers the reference output in a conservative way. The identified uncertain input parameters are shown on the left-hand side of the Figs. 16 to 20, whereas the uncertain outputs with the reference output are shown on the right-hand side.

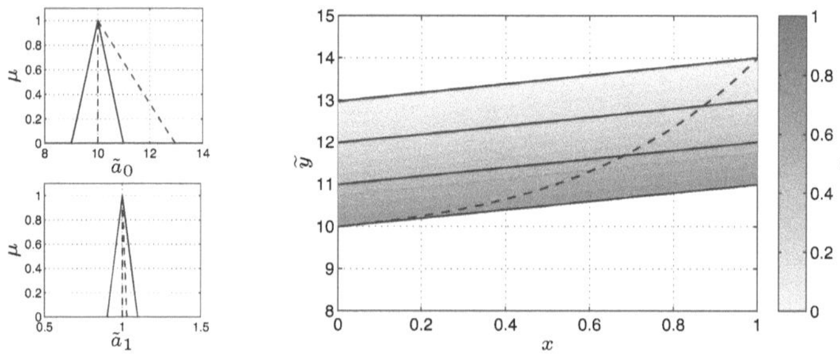

Figure 16. L_1: $\Lambda = (0.1485 + 0.0149) \cdot 0.1422 = 0.0232$

For each of the models L_1 to L_5, the corresponding measure Λ of the model validity criterion is given below the plots. Thus, based on the model

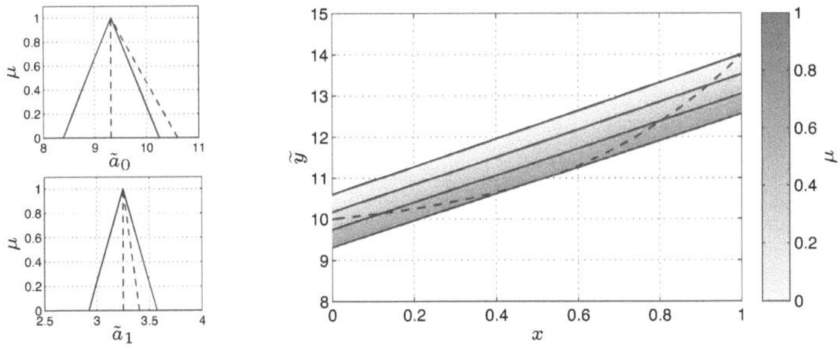

Figure 17. L_2: $\Lambda = (0.0688 + 0.0240) \cdot 0.0624 = 0.0058$

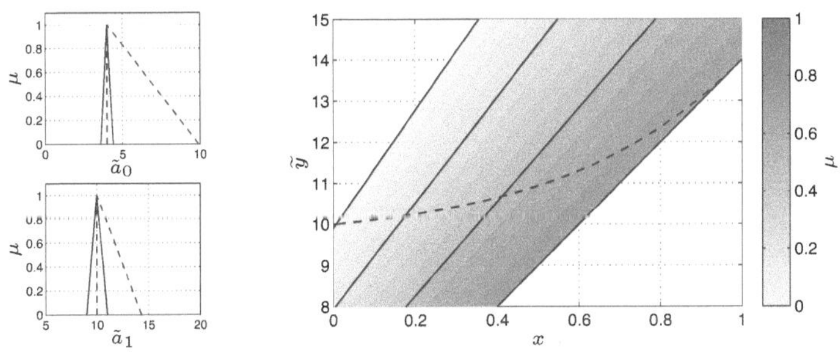

Figure 18. L_3: $\Lambda = (0.7383 + 0.2160) \cdot 0.4785 = 0.4566$

validity criterion Λ and its minimization, a new way of practically linearizing a nonlinear function can be formulated.

4 Conclusions

An interdisciplinary methodology to comprehensive modeling of uncertain systems using fuzzy set theory has been presented. It allows for the inclusion of uncertainties – in particular of those of epistemic type – from the very beginning of the modeling procedure and thus leads to a somehow more sincere and honest numerical simulation which reflects both the benefits and the limitations of the available information about the system model. Moreover, making use of an inverse fuzzy arithmetical approach in conjuction with measured output data, the fuzzy-valued parameters of the model can be identified in such a way that the re-simulated output of the fuzzy-parameterized model conservatively covers the range of possible output values. Finally, a novel criterion to assess the quality of models is

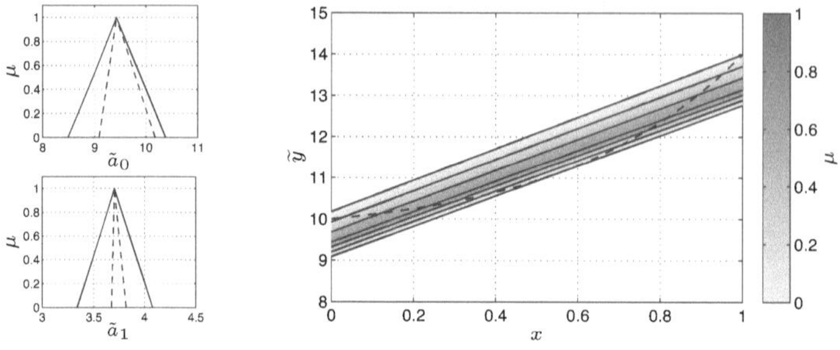

Figure 19. L_4: $\Lambda = (0.0577 + 0.0196) \cdot 0.0518 = 0.0040$

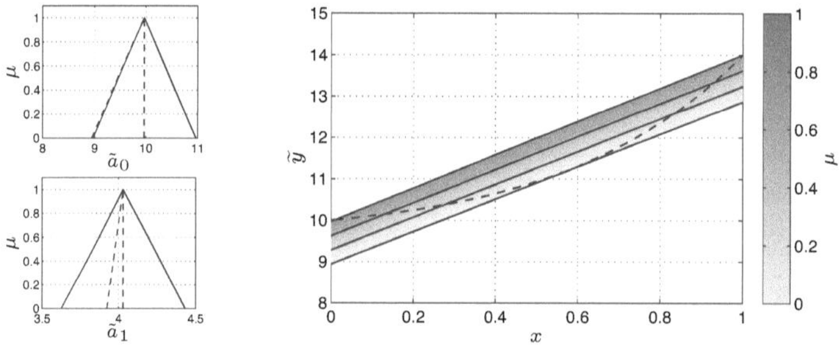

Figure 20. L_5: $\Lambda = (0.0515 + 0.0134) \cdot 0.0454 = 0.0029$

defined and verified on two mathematical examples. Unlike the conventional way of proceeding, which focuses on the L_2 norm of the output deviation only, the presented quality criterion also takes into account the uncertainty of the model parameters which are the source of the output deviation assuming a special model structure. By this method, the quality of models with different type or structure for the same real system can effectively be quantified and rated. Furthermore, the models can be optimized more comprehensively than this is possible with the rather narrow view of optimizing the output deviation only. For engineering applications, for example, the quantification of the model uncertainties, which cause the output deviation, enables the engineer to launch adequate measures in the actuator domain, rather than in the output domain, where this is impossible.

Bibliography

C. B. Barber, D. P. Dobkin, and H. T. Huhdanpaa. The Quickhull Algorithm for Convex Hulls. *ACM Transactions on Mathematical Software*, 22(4):469–483, 1996.

F. Fleissner, T. Haag, M. Hanss, and P. Eberhard. Uncertainty Analysis for a Particle Model of Granular Chute Flow. *Computer Modeling in Engineering & Sciences*, 52(2):181–196, 2009a.

F. Fleissner, T. Haag, M. Hanss, and P. Eberhard. Analysis of Granular Chute Flow Based on a Particle Model Including Uncertainties. In *Proceedings of the 1st International Conference on Computational Contact Mechanics (ECCOMAS)*, Lecce, Italy, 2009b.

P. E. Gill, W. Murray, M. A. Saunders, and M. H. Wright. Procedures for Optimization Problems with a Mixture of Bounds and General Linear Constraints. *ACM Transactions on Mathematical Software*, 10:282–298, 1984.

P. E. Gill, W. Murray, and M. H. Wright. *Numerical Linear Algebra and Optimization, Vol. 1.* Addison Wesley, 1991.

T. Haag and M. Hanss. Inverse Fuzzy Arithmetic for the Identification of Simplified Friction Models. In *IV European Conference on Computational Mechanics - ECCM 2010*, Paris, France, 2010a.

T. Haag and M. Hanss. Model assessment using inverse fuzzy arithmetic. In *Proc. of the IPMU 2010*, Dortmund, Germany, 2010b.

T. Haag and M. Hanss. An Inverse Fuzzy-Arithmetical Method for the Validation, Selection and Optimization of Models for Mechanical Systems. In *Proc. of the ISMA 2010*, Leuven, Belgium, 2010c.

T. Haag, P. Reuß, and M. Hanss. An Approach to the Identification of Uncertain Surrogate Models for Complex Systems. In R. Hoffmann and E. Hüllermeier, editors, *Proc. of the 19th Workshop Computational Intelligence*, pages 50–60, Dortmund, Germany, 2009a.

T. Haag, P. Reuß, S. Turrin, and M. Hanss. An inverse model updating procedure for systems with epistemic uncertainties. In *Proc. of the 2nd International Conference on Uncertainty in Structural Dynamics*, pages 116–125, Sheffield, UK, 2009b.

T. Haag, J. Herrmann, and M. Hanss. Identification procedure for epistemic uncertainties using inverse fuzzy arithmetic: Special issue: Isma 2010. *Mechanical Systems and Signal Processing*, 24(7):2021–2034, 2010.

T. Haag, S. Carvajal Gonzalez, and M. Hanss. Model validation and selection based on inverse fuzzy arithmetic. *Mechanical Systems and Signal Processing*, 2011. Accepted.

M. Hanss. The transformation method for the simulation and analysis of systems with uncertain parameters. *Fuzzy Sets and Systems*, 130(3): 277–289, 2002.

M. Hanss. An approach to inverse fuzzy arithmetic. In *Proc. of the 22nd International Conference of the North American Fuzzy Information Processing Society – NAFIPS 2003*, pages 474–479, Chicago, IL, USA, 2003a.

M. Hanss. The extended transformation method for the simulation and analysis of fuzzy-parameterized models. *International Journal of Uncertainty, Fuzziness and Knowledge-Based Systems*, 11(6):711–727, 2003b.

M. Hanss. *Applied Fuzzy Arithmetic – An Introduction with Engineering Applications.* Springer-Verlag, Berlin, New York, 2005. ISBN 3540242015.

M. Hanss and S. Turrin. A fuzzy-based approach to comprehensive modeling and analysis of systems with epistemic uncertainties. *Structural Safety*, 32(6):433 – 441, 2010.

M. Hanss, J. Herrmann, and T. Haag. Vibration analysis of fluid-filled piping systems with epistemic uncertainties. In A. K. Belyaev and R. S. Langley, editors, *IUTAM Symposium on the Vibration Analysis of Structures with Uncertainties*, volume 27 of *IUTAM Bookseries*, pages 43–56. Springer Netherlands, 2011. ISBN 978-94-007-0289-9.

W. L. Oberkampf, S. M. DeLand, B. M. Rutherford, K. V. Diegert, and K. F. Alvin. Error and uncertainty in modeling and simulation. *Reliability Engineering & System Safety*, 75(3):333 – 357, 2002. ISSN 0951-8320.

A. Pratellesi, S. Turrin, T. Haag, A. Scippa, and N. Baldanzini. On the effect of testing uncertainties in the homologation tests of motorcycle helmets according to ece 22.05. *International Journal of Crashworthiness*, 2011. Accepted.

S. Turrin, M. Hanss, and A. P. S. Selvadurai. An Approach to Uncertainty Analysis of Rockfall Simulation. *Computer Modeling in Engineering & Sciences*, 52(3):237–258, 2009.

S. Turrin, T. Haag, and M. Hanss. A Fuzzy Arithmetical Approach to the Quantification of Nonlinearity. In *Proc. of the 8th International Conference on Structural Dynamics*, Leuven, Belgium, 2011.

L. A. Zadeh. The concept of a linguistic variable and its application to approximate reasoning–i. *Information Sciences*, 8(3):199 – 249, 1975a.

L. A. Zadeh. The concept of a linguistic variable and its application to approximate reasoning–ii. *Information Sciences*, 8(4):301 – 357, 1975b.

L. A. Zadeh. The concept of a linguistic variable and its application to approximate reasoning-iii. *Information Sciences*, 9(1):43 – 80, 1975c.

Bounding Uncertainty in Civil Engineering: Theoretical Background and Applications

Alberto Bernardini[1], and Fulvio Tonon[2]

[1] Department of Structural and Transportation Engineering, University of Padova, Padova, Italy

[2] Department of Civil Engineering, University of Texas, Austin, TX, USA

Abstract. The design of civil engineering constructions frequently involves a great uncertainty about loading conditions, material properties and their degradation in time, human errors in modeling, construction and successive management. These uncertainties seldom can be described by mapping probabilistic input variables through a deterministic model, to obtain the precise expectation of output parameters. In the last years researchers referred to the more general idea of imprecise probabilities to derive lower/upper bounds of output expectations. In this field, the theory of random sets appears as the most appropriate and relatively simple approach for many typical engineering problems, containing probabilistic methods, interval analyses and fuzzy sets as particular cases. The theoretical background and its connection to the more general theory of imprecise probabilities are briefly summarized. Finally the results of real world field applications of the theory to large scale urban building constructions in order to evaluate their seismic vulnerability are presented[1].

1. Motivations

Modern computers allow to evaluate with very great precision the response to loading of complex structural systems. Non linear material and geometric behaviors can be easily simulated, once the applied loads, constraints and material constitutive laws are known.

However, particularly in the design, construction and management of civil engineering works, this ability is in many cases disappointed by the deficiency in

---

[1] Researches on seismic vulnerability summarized in this paper have been carried out with the support by the GNDT (Gruppo Nazionale per la Difesa dai Terremoti; "Damage scenarios in the Veneto-Friuli area" Project), and the Research Unit at Genova University of ReLUIS (Rete dei Laboratori Universitari per l'Ingegneria Sismica). The stimulating leadership, respectively of Marcello Riuscetti (University of Udine) and Sergio Lagomarsino (University of Genova), are here acknowledged

available information about expected loading conditions in the life time, material properties and their degradation over time; for example properties of soils are frequently greatly uncertain. Of course experimental testing procedures allow for both destructive and non-destructive evaluations and monitoring of such parameters. But their cost is not decreasing over time at same speed as the cost of computing numerical procedures. Hence output numbers with 16 or 32 significant digits are often computed by the engineer, starting with input data of only 1 or 2 maximum reliable digits.

In the ordinary design procedures this uncertainty is generally covered by partial safety factors suggested by Codes. Their calibration has been justified by considerations of the probability of exceeding serviceability or ultimate limit states in the expected lifetime of the structure. Sometimes, for special works, this probability is explicitly computed and bounded to a politically acceptable low value. In both cases a probabilistic paradigm seems to have been assumed: i.e. the hypothesis that uncertainties can be fully described by probability distributions for the design variables.

In recent decades researchers have risen some objections to this probabilistic paradigm. For example:

- the choice of the probability distributions and also of their parameters (mean value, variance,...) for the random variables is frequently arbitrary, not supported by data;
- the sensitivity of the final failure probabilities to these choices is frequently very great (many orders of magnitude when very low probabilities are computed) (Ben-Haim and Elishakoff, 1990);
- it seems impossible the probabilistic modeling of some critical uncertainties: e.g. uncertainties related to human errors in the design, construction or management.

Other criticism derives from more specific technical arguments. For example in the case of highly non linear response, the relation (explicitly or implicitly given by a numerical model) between input and output variables could be not a point-valued function, but a multi-valued function $X \rightarrow Y$. See case c) in Figure 1. The probabilities of the singletons $y_{c,1}$ and $y_{c,2}$ on the space Y cannot be determined, although their sum is equal to the probability of the singleton x_2 on X.

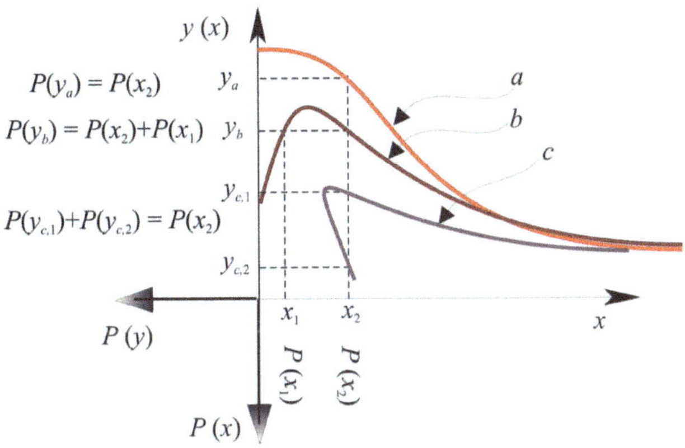

Figure 1. Extension of probabilities through a point-valued (a, b) and a multi-valued (c) mapping.

The same under-determination could be generated in the linear range when, for example, a stiffness parameter E is not exactly known but only estimated to be in a real interval, as shown in Figure 2.

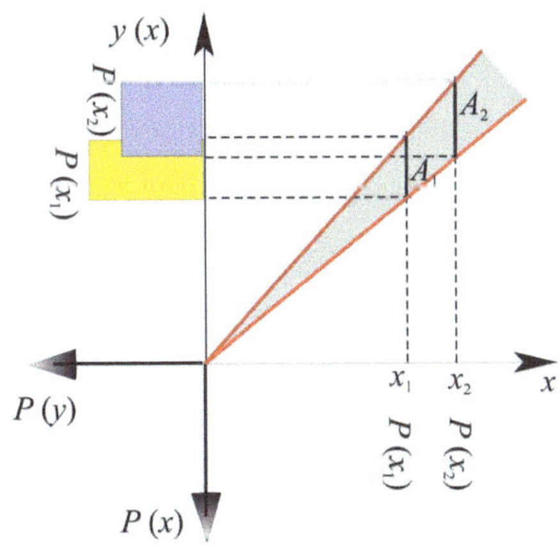

Figure 2. Extension of probabilities through a set-valued linear mapping

The probabilities of the singletons x_1 and x_2 are extended to the intervals A_1 and A_2 respectively, but the precise probabilities of the singletons y on Y cannot be determined. Notice that the intervals A_1 and A_2 could be jointed, as displayed in

Figure 2. To solve the problem preserving the probabilistic paradigm, the stiffness parameter E could be modeled by a further random variable: the Principle of Indifference could suggest a uniform density distribution on the real interval $[E_1, E_2]$ and 0 outside; moreover, stochastic independence of the variables x and E could be assumed. However the information that E belongs on $[E_1, E_2]$ does not imply a precise probability distribution, with a precise mean $(E_1+E_2)/2$ and variance $(E_1-E_2)^2/12$. Many other distributions are compatible with the given information: for example, Dirac distributions on E_1or E_2, with mean E_1 and E_2 respectively and variance equal to 0.

The recognition, besides the random uncertainty, of a set uncertainty not reducible to the probabilistic paradigm is a part of the engineer's daily experience; for example set uncertainty appears when information is obtained from experts, or is collected from published reports or papers, or when data are obtained by means of experimental instruments with finite precision (of course no instrument has infinite precision), or when statistical data are summarized by intervals of confidence.

Finally a considerable set uncertainty is specifically related to computational procedures. One of the first uses of digital computers was to simulate physical systems by numerically solving differential equations:

1. One models a physical phenomenon using a system of differential equations or a variational principle.
2. Then, one obtains the algebraic forms of the differential equations or variational principle by forcing them into the mold of discrete time and space; and
3. Finally, in order to commit those algebraic forms to algorithms, one projects real-valued variables onto finite computer words, thus introducing round-off during computation and truncation.

This approach led to numerical computations that could divert from the physical world. Errors included in Steps 1 through 3 should be addressed during verification and validation of numerical models (Oberkampf et al., 2003). A large body of literature has been devoted to estimating the discretization errors introduced in Step 2. Errors involved in Step 3 have been measured using interval analysis initiated by Moore (1966) and well developed in recent years.

The following Sections briefly summarize theoretical advancements combining random and set uncertainties in a more general framework. Researchers have proposed these advancements in the field of robust statistical methods, generalized information theory, computer science and artificial intelligence. The results described here derive from a more extended presentation proposed by the authors in a recently published monograph (Bernardini and Tonon, 2010). Applications of these results in the safety assessment of engineering systems are only at a beginning stage, e.g. in industrial risk reduction, design and retrofitting of civil engineering constructions, and particularly in soil-structures interactions (see, e. g., Aven, 2010 or Dubois, 2010). Some tentative real world applications in seismic vulnerability and damage evaluations of residential buildings in urban areas will be here presented.

2. Random Sets

A (finite) random set on the space S is a family $\mathcal{F}$ of n *focal elements* $A^i \subseteq S$ and attached weights of the *basic probabilistic assignment* $m(A^i)$ that satisfies the conditions: $m(\varnothing) = 0$; $\Sigma_i\, m(A^i) = 1$. The terms *focal elements* and *basic probabilistic assignment* were suggested by Shafer (1976) in his Evidence Theory.

A simple graphical representation of a random set on the space $S=\{s^1, s^2, s^3\}$ is displayed in Figure 3 as a pile of boxes. The width of each box covers a focal element, and the box height is equal to the relevant probabilistic assignment.

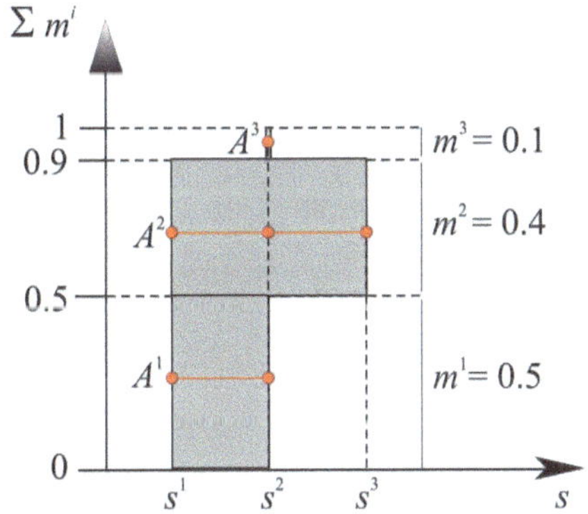

Figure 3. Random set $\mathcal{F}=\{(\{s^1, s^2\}, 0.5), (\{s^1, s^2, s^3\}, 0.4), (\{s^2\}, 0.1)\}$ (after Bernardini and Tonon, 2010; reprinted with permission).

Many different extensions to the classical probabilistic information lead to this equivalent mathematical structure.

The first extension was proposed by Dempster (1967) and is directly related to the examples above discussed in Section 1: a multi-valued mapping G: $X \rightarrow \mathcal{S}$ from a probability space to a measurable space $(S, \mathcal{S})$ is considered. To avoid mathematical complications, sets and spaces of finite cardinality, $|X|$ and $|S|$, are considered, $\mathcal{S}$ is the power set of the space S, with cardinality $2^{|S|}$, and the domain of G is equal to X. Clearly any singleton x on X with positive probability $P(x)$ gives a focal element $A = G(x)$ with probabilistic assignment $m(A) = P(x)$.

A second extension could be directly generated on the space S by statistics of set-valued data; for example the probabilistic assignments are given by the relative frequencies of interval measures experimentally obtained or of sets of arguments/values suggested by experts in a multi-choice questionnaire. Of course when the focal elements A^i define a disjoint partition of S, $m(A^i)$ is an ordinary probability distributions, precisely defining the probability of any event T in the σ-algebra generated by the singletons A^i. It is equivalent to information displayed by an histogram, when frequencies of point value data are summarized together in finite successive intervals, without recording the effective point in the interval. In this case also, as in the general case, it is a natural curiosity to ask what the available information can say about the probability of any other event T in the power set of S, and for example about the probability of the singletons of S. Taking into account that the probability $m(A^i)$ can possibly be attached to every point of the focal element A^i the additivity law gives the following bounds :

$$\sum_{A^i \subseteq T} m\left(A^i\right) \leq P(T) \leq \sum_{A^i \cap T \neq \varnothing} m\left(A^i\right) \tag{1}$$

Following Shafer the lower bound is the *Belief* of T ($Bel(T)$, while the upper bound is the *Plausibility* of T ($Pla(T)$). The set functions *Bel* and *Pla* clearly are monotone with respect to inclusion, as probabilities. For any pair of subsets of S, T^1 and T^2, *Bel* and *Pla* satisfy the constraints (equations in probability theory):

$$\begin{aligned} Bel\left(T^1\right)+Bel\left(T^2\right) &\leq Bel\left(T^1 \cup T^2\right)+Bel\left(T^1 \cap T^2\right) \\ Pla\left(T^1\right)+Pla\left(T^2\right) &\geq Pla\left(T^1 \cup T^2\right)+Pla\left(T^1 \cap T^2\right) \end{aligned} \tag{2}$$

Hence for any subset T of S , and its complementary set T^C:

$$\begin{aligned} Bel(T)+Bel\left(T^C\right) &\leq 1 \\ Pla(T)+Pla\left(T^C\right) &\geq 1 \end{aligned} \tag{3}$$

Moreover, considering that $A \subseteq T \Leftrightarrow A \cap T^C = \varnothing$:

$$Bel(T)+Pla\left(T^C\right)=1 \tag{4}$$

When $Bel(T)$ is assigned for all subsets T of the power set of S, $Pla(T)$ can be evaluated through the duality equation (4). However not every monotone (with respect to inclusion) set function can be assumed to be the Belief function of a random set. Further condition (for example Eq. 2) must be satisfied in order to obtain, through the inverse of the set function in the left bound of Eq. (1) zero or positive (for the focal elements) values of the probabilistic assignments summing to 1. The inverse function is the Möbius transform:

$$m(A) = \sum_{T \subseteq A} (-1)^{|A-T|} Bel(T) \quad (5)$$

Random sets are particular cases of a more general theory of monotone lower probabilities (Choquet Capacities (Choquet, 1954)): Belief functions are Choquet capacities of infinite order; lower probabilities simply satisfying Eq. 2 are 2-monotone Choquet capacities, while Belief functions satisfy similar relations for any set of subsets $T^1, T^2 \dots T^k$. Moreover Belief functions are particular cases of coherent lower probabilities and previsions (Walley, 1991). However they are general enough to describe uncertain information in a unified and simple framework; the framework contains both probability distributions and intervals or convex sets, and their combinations.

3. Consonant Random Sets and Fuzzy Sets

A random set is called *consonant* whenever the focal elements are nested, i.e. can be ordered in such a way that:

$$A^1 \subseteq A^2 \subseteq \dots \subseteq A^n \quad (6)$$

The definition was introduced by Shafer (1976). For example the random set displayed in Figure 3 is consonant, because the focal elements can be ordered as displayed in Figure 4.

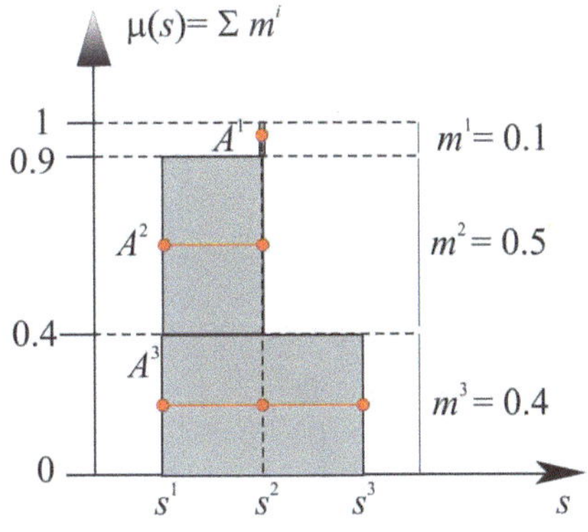

Figure 4. Consonant Random set $\mathcal{F}=\{(\{s^2\}, 0.1), (\{s^1, s^2\}, 0.5), (\{s^1, s^2, s^3\}, 0.4)\}$

The embodied information about the variable s is clearly centered around A^1, at least with probability $Bel(A^1) = m(A^1)$ and possibly with probability $Pla(A^1) = 1$, and surely contained in A^n ($Bel(A^n) = Pla(A^n) = 1$). Of course every random set with only one focal element (n=1) is consonant, and describes a fully non-probabilistic set uncertainty about a variable in the space S. It can be easily shown that, given two sets, T^1, T^2 both on S, consonant random sets satisfy the fundamental property

$$Pla\left(T^1 \cup T^2\right) = \max\left(Pla(T^1), Pla(T^2)\right) \tag{7}$$

This equation demonstrates that Plausibility of the union of any pair of disjointed sets is dependent solely on the individual sets' Plausibility ("*decomposability property*", as well as classical additive Probability measures). Therefore, in the case of a consonant random set, the point-valued *contour function* (Shafer, 1976) $\mu: S \rightarrow [0, 1]$:

$$\mu(s^j) = Pla\left(\left\{s^j\right\}\right) \tag{8}$$

allows to completely derive the Plausibility (and hence the Belief) of any subset T; similarly the point-valued probability distribution $P(s^j) = P(\{s^j\})$ defines the probability of every subset T in the algebra generated by the singletons. In actual fact:

$$\begin{aligned} Bel(T) &= 1 - \max_{s^j \in T^c} \mu\left(s^j\right) \leq \\ &\leq P(T) \leq Pla(T) = \max_{s^j \in T} \mu\left(s^j\right); \end{aligned} \tag{9}$$

There is a strong correspondence between consonant random sets and other decomposable measures of uncertainty: *fuzzy sets* and *possibility distributions*.

In a totally independent context, the idea of fuzzy set was proposed by Zadeh (Zadeh, 1965) as an extension of classical set theory. He suggested that the membership to a subset A could not always be a crisp property, absolutely verified or non-verified, but also, in some cases, partially verified. So the boundaries between A and A^C should be considered as separated by a fuzzy zone, where the singleton $\{s^j\}$ partly belongs both to A and A^C. Formally the extension was obtained through the generalization of the *indicator function* of a classical crisp set $A \subseteq S, \chi: S \rightarrow \{0, 1\}$, to the *membership function*: $\mu: S \rightarrow [0, 1]$. The *height* of a fuzzy set A is the largest membership grade : $h(A) = \max \mu(s) \mid s \in S$. A fuzzy set is called *normal* when its height is equal to 1; otherwise, it is called *subnormal*.

In *fuzzy logic* propositions exist with a degree of truth that goes from 0 (false) to 1 (true), every gradation being permitted in between. This not-Boolean logic seems to closely match human thinking when a true-false judgment has to be given about complex propositions relative to the real world.

But here we are only focused on the connection between the concept of fuzzy set and the concept of random set, as uncertain measures of a family of subsets. This connection can be clearly seen using the dual representation of a fuzzy set through their α-cuts ${}^{\alpha}A$. They are classical subsets of S defined, for any selected value α of membership, by the formula:

$$ {}^{\alpha}A = \{ s \in S \mid \mu(s) \geq \alpha \} \; ; \tag{10} $$

So it is clear that the α-cuts ${}^{\alpha}A$ of any given normal fuzzy set are a nested sequence of subsets of set S, and therefore the family of focal elements of an associated consonant random set: the membership function of normal fuzzy sets gives the contour function of the corresponding random sets, and the basic probabilistic assignment (for a finite sequence of α-cuts) is given by $m\left(A^i = {}^{\alpha^i}A\right) = \alpha^i - \alpha^{i+1}$. See for example Figure 5.

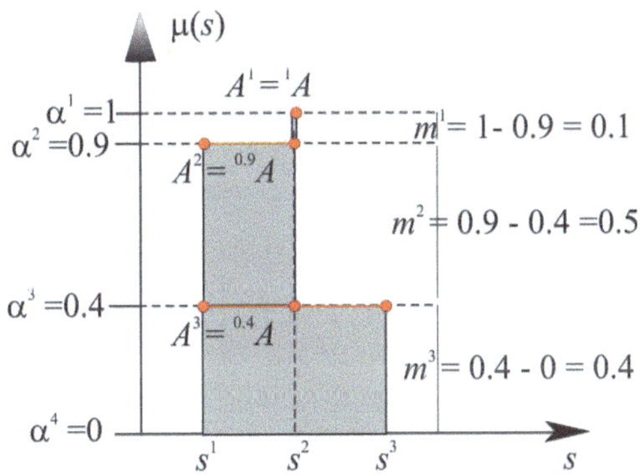

Figure 5. Consonant random set derived by the membership function of a fuzzy set.

From this point of view, the membership function of a fuzzy subset allows measures of Plausibility and Belief to be attached to every classical subset $T \subseteq S$; this very different interpretation of fuzzy sets was recognized by Zadeh (1978), as the basis of a theory of Possibilities, defined by a possibility distribution $\pi(s) = \mu_A(s)$, later extensively developed by other authors, in particular Dubois and Prade (1988). In a comparison between Evidence Theory and Possibility Theory, *Necessity* (*Nec*(*T*)) and *Possibility* (*Pos*(*T*)) measures of any classical subset *T* coincide respectively with Belief (*Bel*(*T*)) and Plausibility (*Pla*(*T*)) deriving from the associated consonant random set.

4. Probability Distributions Compatible with Random Sets

Alternatively it is possible to consider a random set as a class Ψ of probability measures $P(T)$ on the measurable space $(S, \mathcal{S})$, for the sets $T \in \mathcal{S}$. In a general way, let $\mathcal{S}$ be the algebra generated by a finite partition of S that contains only elementary events or *singletons*, s^j. For all $T \in \mathcal{S}$, $P(T) = \Sigma\,(P(s^j) = P(\{s^j\}) | s^j \in T)$, and $P(s^j)$ is the probability distribution attached to the singletons s^j.

Let us recall that the probability assignment m^i of any focal element A^i is attached to A^i without any other specification about its distribution on the singletons s^j contained in A^i. Then, the class Ψ implicit in a random set corresponds to the infinite ways in which m^i may be *selected* to be distributed on every focal element A^i (Fetz and Oberguggenberger, 2004). Formally, indicating with I^i the set of the indexes of the singletons in A^i and with Ψ^i the set of all probability distributions on the sub-space $A^i = \{s^j \mid j \in I^i\}$:

$$\Psi = \left\{ \sum_{i=1}^{n} \left(m^i \cdot P^{i,j} \right) | \, P^{i,j} = \begin{matrix} P^i\left(s^j\right) \in \Psi^i, j \in I^i \\ P^{i,j} = 0, j \notin I^i \end{matrix} \right\} \tag{11}$$

Clearly set Ψ is convex, because each member is the convex combination of distributions $P^{i,j}$ and sets Ψ^i are convex. Let us now underline the special meaning of some members of Ψ.

White distributions. According to the Principle of Indifference a uniform distribution could be selected along each focal element:

$$\forall j \in I^i : P^{i,j}\left(s^j\right) = \frac{1}{\left|I^i\right|} \;;\; i = 1 \text{ to } n \tag{12}$$

This particular distribution has been termed *white* (Bernardini, 1995) to underline some analogy with spectral uniformity. The same notion was previously considered by Dubois and Prade (1982).

Selectors., We assume in the following that S is a finite set of real numbers, so that the mass distribution of the simple real function $y_s = \Sigma_j\, s^j\, \chi(s^j)$ coincides with the probability distribution of the singletons $P(s^j)$. A random set can be considered as a random variable whose point-wise realizations on S cannot be precisely observed, so we can only say that each realization is a particular but unknown singleton of the focal element. The mass distributions corresponding to the different point-valued mapping, compatible with the multi-valued mapping that generates the focal elements, are called *selectors* of the random set: when focal element A^i is observed, the realization is an unknown but specific singleton $\{s^{k^i}\}$ of the focal element (see Figure 6). Therefore:

$$P^{i,j} = \begin{cases} 1 & \text{if} \quad j = k^i \in I^i \\ 0 & \text{if} \quad j \neq k^i \end{cases} \tag{13}$$

The set of selectors *SCT* is a subset of Ψ because it is generated by considering all *Dirac* probability distributions, i.e. probability distributions concentrated at one point in A^i. The number of selectors, is thus equal or lower than the product of the cardinalities $|I^i|$ (some selectors could coincide). In finite spaces, the number of focal elements and cardinalities $|I^i|$ are finite: therefore *SCT* is a finite (and hence non-convex) set included in the convex set Ψ.

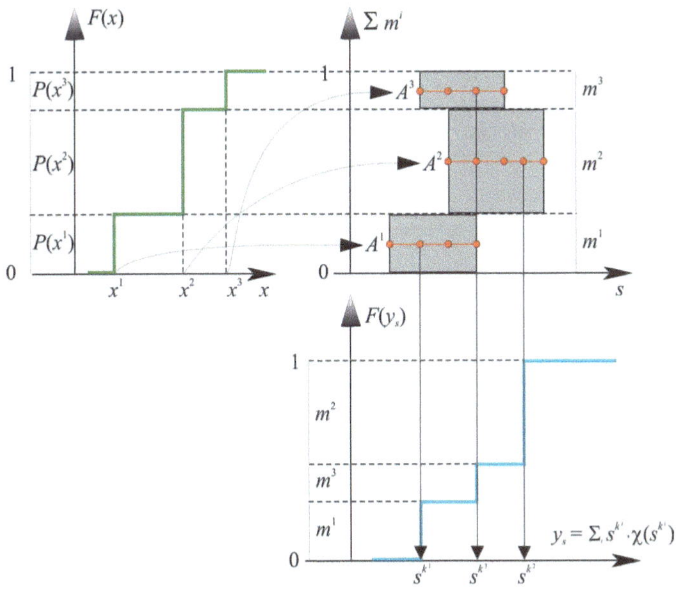

Figure 6. Selector from a random set(after Bernardini and Tonon, 2010; reprinted with permission).

Upper and lower distributions. Let us order each set I^i according to the increasing real values of the singletons. Two particular selectors can be generated by taking k^i equal to the upper bound of I^i ($k^i = \max(I^i)$), and to the lower bound of I^i ($k^i =$

$\min(I^i)$) respectively. The corresponding (discrete) cumulative distribution functions (CDF) of the simple function $y_s = \Sigma_j s^j \chi(s^j)$ are, respectively:

$$\begin{aligned} F_{UPP}(y_s) &= \sum_{s^{\min(I^i)} \le y_s} m(A^i) = Pla\left(\left\{s^j \mid s^j \le y_s\right\}\right) \\ F_{LOW}(y_s) &= \sum_{s^{\max(I^i)} \le y_s} m(A^i) = Bel\left(\left\{s^j \mid s^j \le y_s\right\}\right) \end{aligned} \tag{14}$$

They bound the CDF $F(y_s)$ given by any other probability distribution in Ψ.

Figure 7 shows upper and lower distributions for the random set displayed in Figure 3 (or Figure 4).

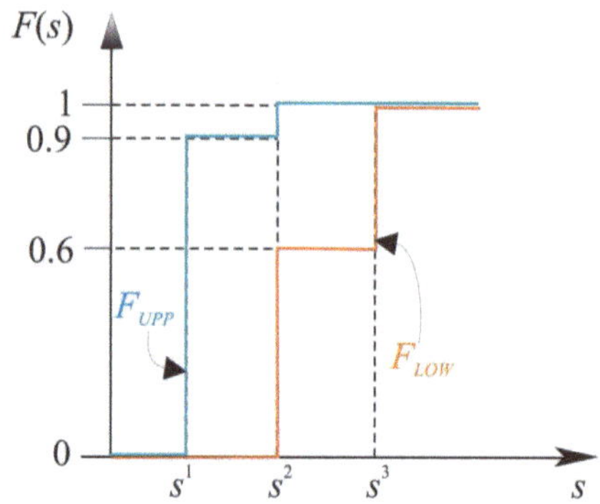

Figure 7. Upper and lower distributions for the random set in Figure 3.

If f is a point-valued monotonically increasing or decreasing mapping on S to $Y = \mathbb{R}$, the same upper and lower distribution functions extended to Y can be used to bound the CDF $F(y)$ (Figure 8) and the expectation $E[y]$.

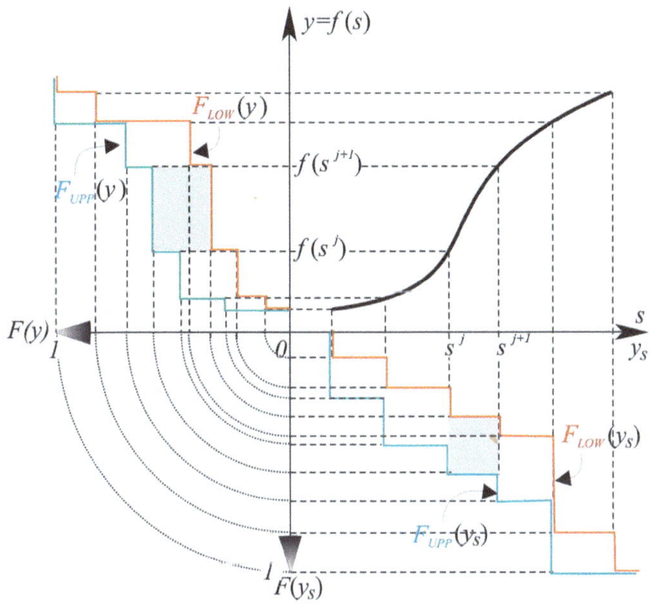

Figure 8. Extension of upper and lower distributions by a monotonic point-valued function(after Bernardini and Tonon, 2010; reprinted with permission).

Extreme distributions. Previous Section identified the two selectors that, for a given s^*, give upper and lower probabilities of the event (set) $\{s: s \leq s^*\}$. Given any set T, one may search for a distribution P^* in Ψ that maximizes (or minimizes) the probability of T, i.e. such that $P^*(T) = \max (\min)\{P(T)| P\in \Psi\}$. Such a distribution is called extreme distribution. One may then ask the same question for all sets $T \subseteq S$, i.e. determine all the distributions in Ψ that minimize or maximize the probability of some event T. Likewise, for a general non monotonic function $y = f(s)$, one may search for a distribution in Ψ that gives the upper (or lower) CDF of y.

Figure 9 shows that extreme distributions can be constructed by considering a permutation $\pi(j), j = 1$ to $|S|$ of the indexes j, in such a way as to obtain a monotonic (increasing or decreasing) function. By applying the maximum condition to the same permutation or, in an equivalent way, the minimum condition to the reversed permutation (i.e. the permutation $\pi_R(j) = |S| + 1 - \pi(j)$), a couple of dual extreme distributions are obtained.

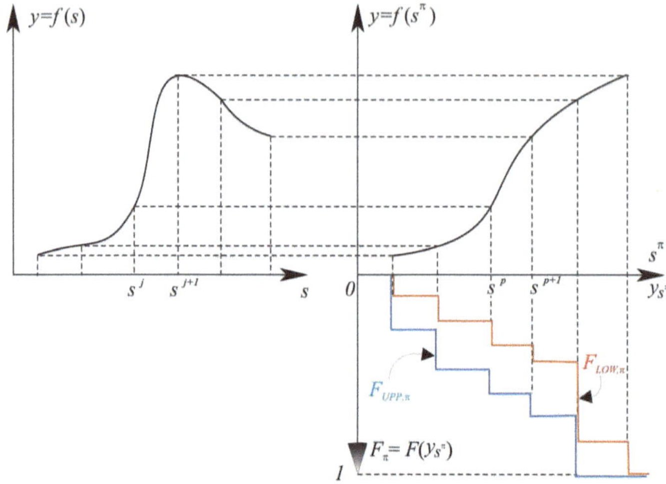

Figure 9. Extreme distributions and extension by a non monotonic point-valued function(after Bernardini and Tonon, 2010; reprinted with permission).

Formally, dual extreme cumulative distributions can be associated to any permutation $\pi(j), j = 1$ to $|S|$ of the indexes j, by extending Eq. (14) through the following formulas ($I^i_\pi = \{\pi(j) | j \in I^i\}$):

$$\begin{aligned} F_{UPP,\pi}(y_s) &= \sum_{s^{\min(I_\pi^i)} \le y_s} m(A^i) = Pla(\{s^{\pi(j)} \mid s^{\pi(j)} \le y_s\}) \\ F_{LOW,\pi}(y_s) &= \sum_{s^{\max(I_\pi^i)} \le y_s} m(A^i) = Bel(\{s^{\pi(j)} \mid s^{\pi(j)} \le y_s\}) \end{aligned} \tag{15}$$

Each extreme distribution $P_{EXT}(s^j)$ can finally be derived by differences of cumulative functions F_π and the inverse of the selected permutation:

$$\begin{aligned} P_\pi\left(s^{\pi(j)=k}\right) &= \left\langle \begin{array}{ll} F_\pi(y_s = s^{\pi(j)=1}) & \mid \pi(j)=1 \\ F_\pi(y_s = s^{\pi(j)=k}) - F_\pi(y_s = s^{\pi(j)=k-1}) & \mid \pi(j)>1 \end{array} \right. \\ P_{EXT}\left(s^j\right) &= P_\pi\left(s^{\pi^{-1}\pi(j)}\right) \end{aligned} \tag{16}$$

This algorithm shows that:

- extreme distributions are selectors (but not vice versa);

- upper and lower distributions are extreme dual distributions corresponding to the pair of reversed permutations $\pi(j) = j$ (identity) and $\pi(j) = |S|+1-j$;
- the number of extreme distributions is at most equal to the number of permutations, i.e. $|S|!$; however it can be much lower because different permutations could give the same extreme distribution (multiple extreme);
- let $y = f(s^j)$ be a point-valued mapping and $\pi_{LOW}(j)$, $\pi_{UPP}(j)$ a pair of reversed permutations so that $f(s^{\pi_{LOW}(j)})$ and $f(s^{\pi_{UPP}(j)})$ are an ordered list of monotonically increasing and decreasing values, respectively: the corresponding extensions to Y of the dual extreme distributions, obtained by formulae (16), give exact bounds to the CDF $F(y)$ and the expectation $E[y]$;
- since probabilities and expectations are linear functions of probabilities in Ψ, probabilities and expectations attain their maximum and minimum values at the vertices of Ψ (a $(|S|-1)$-dimensional polyhedron in the $|S|$-dimensional space of the probabilities of the singletons, taking into account that summation of the probabilities must be equal to 1). Therefore, the set of different extreme distributions is the set of vertices of Ψ and Ψ is the convex hull of the extreme distributions.

For the random set displayed in Figure 3 (or Figure 4), the set Ψ is the 2-dimensional shaded polyhedron shown in Figure 10 in the 3 dimensional space of the probabilities $P(s^j)$.

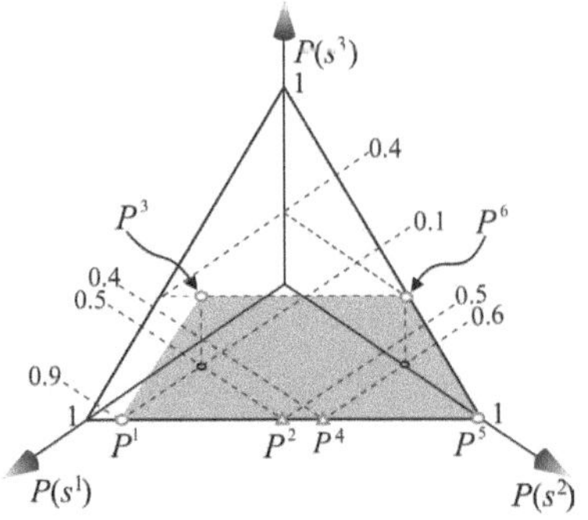

Figure 10. Set of probability distributions, selectors and extreme distributions of the random set shown in Figure 3 (after Bernardini and Tonon, 2010; reprinted with permission).

Figure 11 depicts the projection of Ψ in the 2-dimensional space of the probabilities $(P(s^1), P(s^2))$ (at every point, $P(s^3) = 1 - P(s^1) - P(s^2)$). Ψ is a quadrangle whose vertexes are the four extreme distributions. The 6 selectors and the white distributions are also displayed. Since the selectors P^2 and P^4 are not extreme distributions, they lie on the boundary of Ψ, precisely on the edge that connects extreme distributions P^1 and P^5. The white distribution is the centroid of both selectors and extreme distributions weighted by their multiplicities.

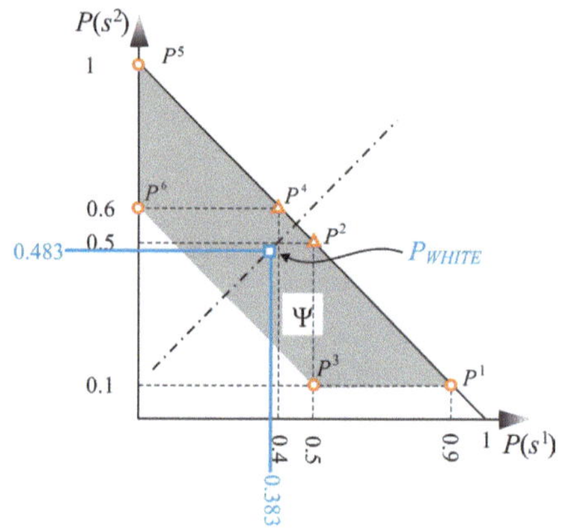

Figure 11. 2-dimensional projection of the set of probability distributions, selectors, extreme distributions and white distribution of the random set shown in Figure 3 (after Bernardini and Tonon, 2010; reprinted with permission).

5. Interval Probabilities and p-boxes

Interval Probabilities. In a finite space S, interval probabilities are a set of intervals attached to the singletons, defining a set Ψ of probability distributions:

$$IP_j = [l_j, u_j],\ j = 1 \text{ to } m \ ; \qquad \Psi = \left\{P : l_j \le P(s^j) \le u_j,\ j = 1 \text{ to } m\right\} \tag{17}$$

The set Ψ is non-empty if and only if the bounds satisfy the conditions: $\sum_{j=1}^{m} l_j \le 1 \ ; \ \sum_{j=1}^{m} u_j \ge 1$. However further stronger conditions are required for coherenc*e* (see e. g. Bernardini and Tonon, 2010 or Walley, 1991): each upper or lower bound should be *reachable* i.e. set Ψ should contain a particular distribution

P equally satisfying the bound. Numerically this condition requires that: $\forall j: \sum_{i \neq j} l_i + u_j \leq 1 \quad ; \quad \sum_{i \neq j} u_i + l_j \geq 1$.

When non-reachable interval probabilities are given, a set of equivalent coherent interval probabilities are given by:

$$\begin{aligned} \forall j: l_j' &= \max(l_j, u_j \ -(\sum_{i=1}^{m} u_i \text{-} 1)) \ ; \\ u_j' &= \min(u_j, l_j + (1 - \sum_{i=1}^{m} l_i)) \end{aligned} \tag{18}$$

These reachable interval probabilities can be used to evaluate bounds of the probability of any other set T in the algebra generated by the singletons:

$$\max\left(\sum_{s^i \in T} l_i', 1 - \sum_{s^i \in T^C} u_i'\right) \leq P(T) \leq \min\left(\sum_{s^i \in T} u_i', 1 - \sum_{s^i \in T^C} l_i'\right) \tag{19}$$

Generally the set of compatible probability distributions conveyed by reachable interval probabilities cannot be described by an equivalent random set. The bounds given by Eq. (19) are only 2-monotone Choquet capacities (Campos et al., 1994). However again extreme distributions and exact bounds of monotonic and non-monotonic functions can be derived by the permutation procedure above presented for random sets.

P-boxes. In a finite space S a set Ψ of probability distributions is implicitly defined by lower and upper bounds, $F_{LOW}(s^j)$ and $F_{UPP}(s^j)$, of the cumulative distribution functions $F(s^j)$:

$$\Psi = \left\{P : F_{LOW}\left(s^j\right) \leq F\left(s^j\right) \leq F_{UPP}\left(s^j\right), j = 1 \text{ to } |S|\right\} \tag{20}$$

The set Ψ is non-empty if, for any pair j, $k \leq j$ $F_{LOW}(s^k) \leq F_{UPP}(s^j)$. However coherence requires stronger conditions, e.g. monotonicity: see for example (Walley, 1991, § 4.6.6).

The lower probabilities derived from a p-box are Choquet capacities of infinite order; hence a p-box is equivalent to a (non consonant) random set. The focal elements and probabilistic assignment can be derived by the α-cuts of the p-box (see Figure 12) through a rule quite similar to the procedure for deriving an equivalent random set from a normal fuzzy set or a possibility distribution) (see Bernardini and Tonon, 2010, Section 3.4.2).

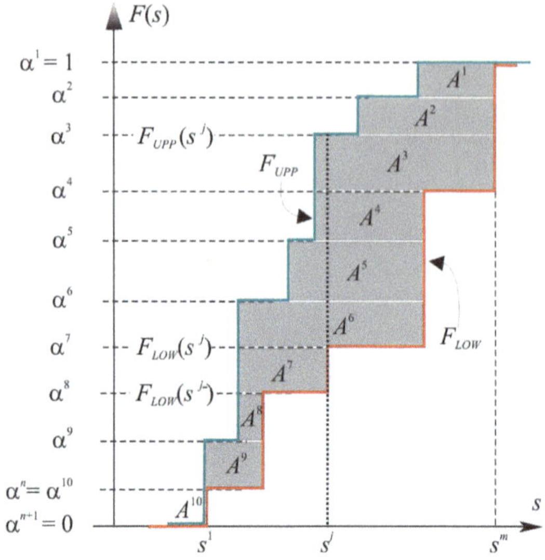

Figure 12. Random set equivalent to a p-box (after Bernardini and Tonon, 2010; reprinted with permission).

For example Table 1 gives the lower/upper bounds of the CDFs of a p-box and of the probability of singletons (columns 4 and 5, are computed by non negative cross differences of the CDFs in columns 2 and 3).

Table 1. Lower/upper CDF of a p-box and reachable bounds of the singletons (after Bernardini and Tonon, 2010; reprinted with permission)..

s^j	$F_{LOW}(s^j)$	$F_{UPP}(s^j)$	$l = Bel(\{s^j\})$	$u = Pla(\{s^j\}) = \mu(s^j)$
s^1	0	0.2	0	0.2
s^2	0	0.3	$\max(0, 0 - 0.2) = 0$	$0.3 - 0 = 0.3$
s^3	0.7	1.0	$\max(0, 0.7 - 0.3) = 0.4$	$1.0 - 0 = 1$
s^4	1.0	1.0	$\max(0, 1 - 1) = 0$	$1.0 - 0.7 = 0.3$

The upper bound in column 5 is the contour function of the equivalent random set displayed in the Figure 13a.

The set Ψ displayed in Figure 13b is the convex combination of 8 extremes. But it does not coincide with the set Ψ that could be computed starting from the interval probabilities of the singletons. P^1 and P^2 are the lower and upper distributions (coinciding of course with lower/upper bounds of the p-box).

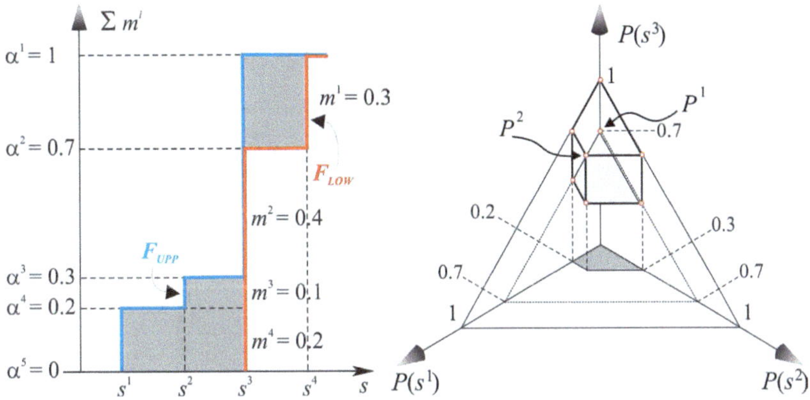

Figure 13. Random set equivalent to the p-box in Table 1 (a) and extreme points in the 3-dimensional space (b).

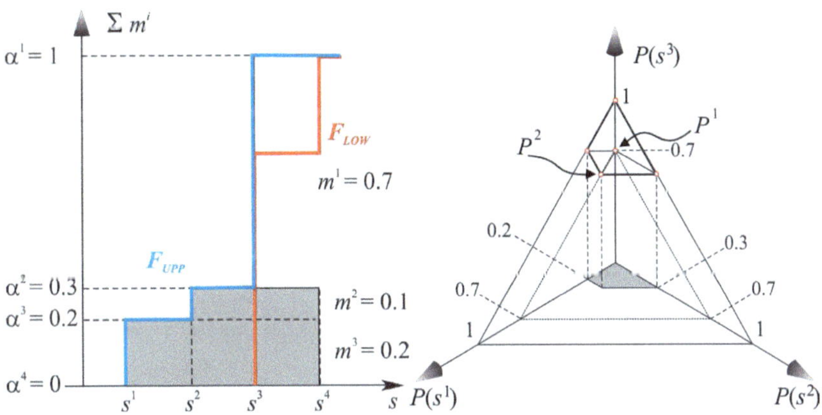

Figure 14. Focal elements, upper/lower distributions and probabilistic assignment (a) and extreme points in the 3-dimensional space (b).of a consonant random set equivalent to the contour function of the p-box in Table 1.

In this example the contour function of the p-box (last column in the table) is equivalent to the membership function of a normal fuzzy set

Hence for comparison an equivalent consonant random set has be derived by the α-cuts of the contour function and displayed in Figure 14. Observe that the upper/lower distributions coincide with upper/lower distributions of the p-box (and of the equivalent random set).

However the set Ψ is very different. It is a pyramid with vertex in P^1 and quadrangular base with the four extremes of upper face of the previous set Ψ of the p-box.

The comparison confirm again that random sets with the same upper/lower distributions can have a very different probabilistic content and give very different bounds to the expectation of non monotonic functions.

6. Random Relations, Inclusion and Mapping

The theory of random sets also consider multivariate problems and analyzes reciprocal interactions (dependence, independence, correlation) between different variables and their influence on the estimation of dependent functions. Joint random relations in multi-dimensional spaces determine, through projection of the focal elements and additivity law, focal elements and probabilistic assignments of marginal random sets. However, when marginal random sets are given, both focal elements and probabilistic assignment of the joint random relation are under-determined. Different hypotheses can be considered to derive solutions to better respect available information: unknown interaction, epistemic independence, strong independence and random set independence (see Bernardini and Tonon, 2010, Chapter 4).

The important rule of *inclusion* of random sets (i.e. inclusion of the corresponding sets of compatible probability distributions) appears to be of particular interest for reliability and risk evaluations of engineering systems. In fact, inclusion of random sets or relations, measuring input variables, implies inclusion of the intervals bounding the probability of any event T. E.g. non consonant random sets can be included in consonant random sets; in such a way that it is possible to profit from the computational advantages of fuzzy sets and relations in the mapping procedures (see Bernardini and Tonon, 2010, Chapter 5). Moreover the expectation of a point-valued or set-valued function of input variables can be similarly included.

7. Applications

Damage Probability Matrixes in EMS98 Scale. Statistical records of damage to building types after past earthquakes allow a probabilistic estimation of the degree of damage, conditional to a parameter measuring the ground motion intensity. Data regarding observed damage, summarized as Damage Probability Matrixes (DPM), have been used in Italy to analyze vulnerabilities and forecast expected damage, particularly learning by the damages observed after the1980 Irpinia earthquake

[18]. For different building types subjected to seismic input, measured by macro-seismic intensity, the DPMs give the probability of different degrees of damage occurrence. However the validation of these results is limited to a restricted area and to local building types. The extrapolation of these results to other contexts, such as different building types and/or different ground characteristics, is questionable.

From the beginning of the last century, the macro-seismic scales have defined and measured local intensities of earthquakes and included them in a report database. These more extensive reports describe the effects of observed earthquakes in different countries and/or the effects reported in old documents of past centuries. Definitions of the higher intensities take into account essentially the degree of damage to different building types: hence suggesting a rough qualitative DPM for some buildings types (masonry buildings, r.c. buildings, and so on).

The idea of deriving DPMs from macro-seismic scale definitions appears much more justified (Bernardini, 2007) when considering the European macro-seismic scale EMS98 (Grünthal, 1998). In this scale the DPMs are not attached to building types but to abstract vulnerability classes (A, B, C, D, E, F in order of decreasing vulnerability) and qualitative statistical correlation is given between classes and a list of building types containing both traditional non engineered buildings and modern construction built up according to rules of Earthquake Resistant Design (ERD). See Figure 15. So the vulnerability of a town, and scenarios of damage for prescribed local macro-seismic intensity can be evaluated from a complete inventory of the building types or at least from the survey of a representative sample. Of course it will be important a specific historical knowledge on the local buildings techniques and Code prescriptions in the design.

In the following we are going to see that the theoretical background of random sets and imprecise probabilities is particularly useful in the numerical treatment of the great uncertainties implicit in the definitions of the scale and allows to find robust bounds of the expected damages.

For each vulnerability class and for the levels from V to XII of the local macro-seismic intensity the EMS98 scale gives an incomplete and qualitative matrix defining expected relative frequencies of 6 damage degrees (from D0 to D5),. The scale attaches qualifying adjectives or names (*null, negligible, moderate, heavy, very heavy, destruction*) to the degree of damage; moreover, the scale gives the surveyor fairly precise descriptions (separately for masonry and r.c. buildings) in order to classify degrees of damage and to take into account observed damage to both structural and non-structural elements.

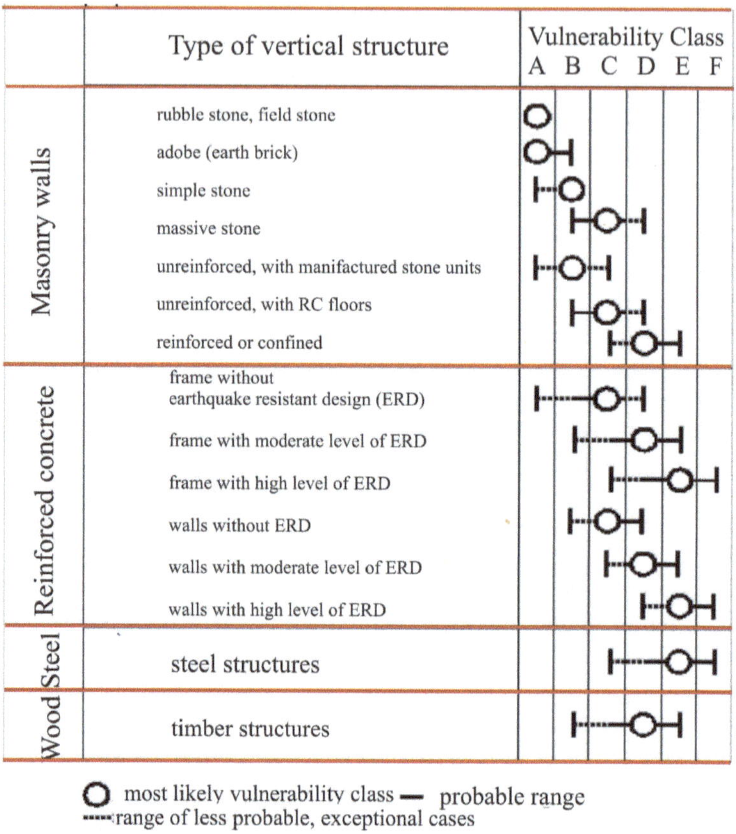

Figure 15. Buildings types according to EMS98 (Grunthal, 1998) and their probabilistic correlation to the vulnerability classes.

For example the matrix in Table 2 summarizes the qualitative linguistic information given by the scale for intensities from VI to IX. Each cell corresponds to a pair of variables (damage degree in rows and intensity level in columns) and provides joint probabilities for some vulnerability classes. Notice both the vagueness and the incompleteness of the information. The measures are not numbers, but adjectives suggesting linguistic frequencies and damage degrees; for example for intensity IX *many* C-class buildings and *few* D-class buildings will experience *heavy* damage D_3. Many cells are empty, in fact all cells in the first row. Each non empty cell specifies a frequency of no more than two classes. Each column (intensity) specifies frequencies of no more than two damage degrees for each class:

for example for intensity IX and class C only frequencies for D_3 and D_4 are suggested.

Table 2. Linguistic frequencies of damage for vulnerability classes (A to F) and macro-seismic intensities (VI to IX) according to the EMS98 scale.

Damage degree	I=VI	I = VII	I=VIII	I= IX
D_0 (null)				
D_1 (negligible)	*Many* A *or* B, *Few* C			
D_2(moderate)	Few A or B	*Many* B, *Few* C	*Many* C, *Few* D	*Many* D, *Few* E
D_3 (heavy)		*Many* A, *Few* B	*Many* B, *Few* C	*Many* C, *Few* D
D_4 (very heavy)		*Few* A	*Many* A, *Few* B	*Many* B, *Few* C
D_5 (destruction)			*Few* A	*Many* A, *Few* B

Considering the frequencies vagueness, the scale suggests some additional quantitative measure (see (Grünthal, 1998)): variable gradations over numerical intervals are attached to the three basic adjectives (*Few*, *Many*, *Most*). Hence modeling of the adjectives by means of fuzzy subsets (or possibility distribution), on the real range of percentages [0, 100], seems reasonable. Figure 16 displays a slightly modified fuzzy definition of the linguistic frequencies used in the scale, including *Nearly None* and *Nearly All*. Of course a singular crisp representation is assumed for the adjectives *None* and *All* (0 and 100 respectively with *membership* equal to 1).

For each value of α, the α-cuts of the fuzzy set attached to the linguistic definitions give a measure of the percentage by means of "*Interval probabilities*" $[{}^{\alpha}l, {}^{\alpha}u]$. Moreover a synthetic measure is the *white expected* percentage of the consonant random set associated to each linguistic frequency; for the *trapezoidal* fuzzy sets shown in Figure 16, it coincides with the mid-point of the interval for $\alpha = 0.5$.

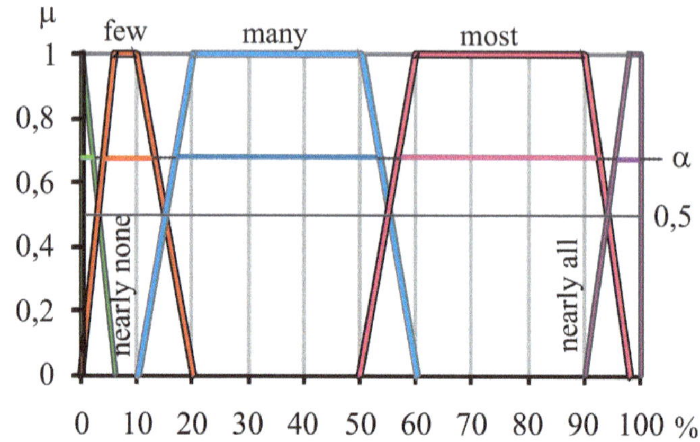

Figure 16. Fuzzy pseudo-partition of the interval [0, 100] by means of 5 fuzzy sets measuring *Nearly None, Few, Many, Most, Nearly All.*

Considering now the scale incompleteness, Bernardini et al. (2007, 2010) have proposed two additional rules:

- The sum of the *white expected* percentages of buildings that evidence various degrees of damage should be equal to 100 for each class and intensity, as required in a precise statistical distribution.
- A unit that increases macro-seismic intensity for a constant class produces a unit increment of the index j of the damage degree D_j. Similarly, for constant intensity, the jump to the next more vulnerable class produces a unit increment of the same index j. This *diagonal shifting* rule is displayed in the shaded cells in Table 2.

The complete linguistic matrix obtained using the rules listed above is shown in Table 3, again for class A and intensities from VI to IX. The linguistic definitions of the EMS98 scale in Table 2 are reported in bold in Table 3, while the proposed ones are in italics. Some combined linguistic definitions appear in the cells; their α-cuts are simply obtained by the rules of interval arithmetic.

In conclusion, a set of "*Interval probabilities*" is derived, for every value of $\alpha \in [0,1]$, in the discrete space of the damage degree indexes $D = \{0,1,2,3,4,5\}$, for vulnerability class $C \in \{\mathrm{A, B, C, D, E, F}\}$ and EMS98 macro-seismic intensity $I \in \{\mathrm{V, VI, VII, VIII, IX, X, XI, XII}\}$:

$$^{\alpha}IP^{C,I} = \left\{ \left[{^{\alpha}l_j^{C,I}}, {^{\alpha}u_j^{C,I}} \right], j = 0,1,\ldots,5 \right\} \qquad (21)$$

The set of the compatible probability distributions is the convex combination of a finite set of extreme distributions that can be easily determined with simple algorithms.

Table 3. Linguistic frequencies of damage for vulnerability class A and macro-seismic intensities VI to IX according to the EMS98 scale (in bold) and the proposes completing definitions (in italics)

Damage degree	I=VI	I = VII	I=VIII	I= IX
D_0 (null)	*Many + 7/3 Few*	*1/3 Few*	*NONE*	*NONE*
D_1 (negligible)	***Many***	*2 Few*	*1/3 Few*	*NONE*
D_2(moderate)	***Few***	*Many*	*2 Few*	*1/3 Few*
D_3 (heavy)	*NONE*	***Many***	*Many*	*3 Few*
D_4 (very heavy)	*NONE*	***Few***	***Many***	*Many*
D_5 (destruction)	*NONE*	*NONE*	***Few***	***Many***

Example. For class B and I = VI only 2 damage levels (1(negligible) and 2 (moderate)) are expected.

The interval probabilities (measured here by percentages of buildings) are specified as follows, for $\alpha = 1$: IP_{D_0} = (many + 7/3 few)= [20+ 7/3 x 6 = 34, 50+ 7/3 x 10 = 73] ; IP_{D_1} = (many)=[20, 50]; IP_{D_2} =(few) = [6, 10] ; $IP_{D_3} = IP_{D_4} = IP_{D_5}$ = [0,0]. Notice that the sum of the white expected values is: 56 + 35 + 9 + 0 + 0 + 0 = 100.

Clearly the lower bound of the interval IP_{D_0} = [34, 73] cannot be reached. In fact no more than 10+50 = 60% of buildings suffer damage > 0, and therefore the percentage of undamaged buildings cannot be lower than 40%. In comparison, at least 26% of buildings suffer damage > 0, so the percentage of undamaged buildings (73%) is reachable because it is less than 100-26=74%. For $\alpha = 1$ the corresponding *lower and upper extreme distributions* are P^1= (0.4, 0.5, 0.1) and P^2 = (0.73, 0.21, 0.06) respectively. However 3 other extreme distributions ((0.5, 0.44, 0.06), (0.73, 0.20, 0.07) and (0.70, 0.2, 0.1)) can be discovered. Figure 17 displays the projection of the convex set of the compatible probability distributions on the 2-dimensional plane ($P(D_1)$, $P(D_2)$), and the larger set evaluated for $\alpha = 0$, corresponding to the convex combination of only 4 extreme points.

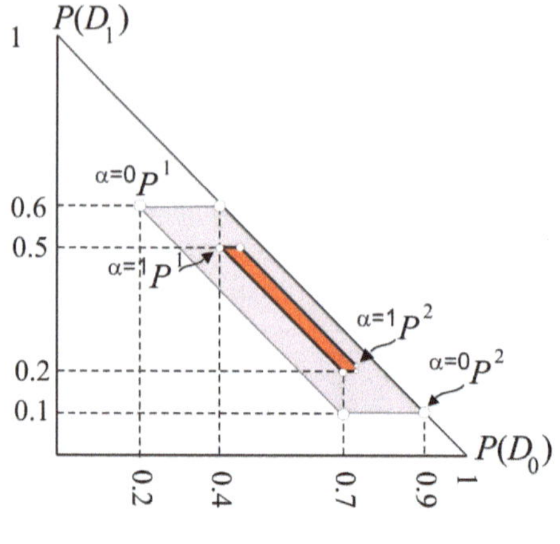

Figure 17. Probability distributions of damage and extreme distributions for $\alpha = 0$ and $\alpha = 1$, Class A, I = VI. In any point $P(D_2) = 1 - P(D_o) - P(D_1)$, $P(D_3) = P(D_4) = P(D_5) = 0$.

Figure 18 gives a synthetic view of the DPMs for each vulnerability class. The upper and lower curves furnish the bounds for the α-dependent relationship between the mean damage degree index j and the macro-seismic intensity I (*vulnerability curves*). Note the intervals for contiguous classes are generally disjointed for $\alpha > 0.5$, partially superposed for $\alpha < 0.5$.

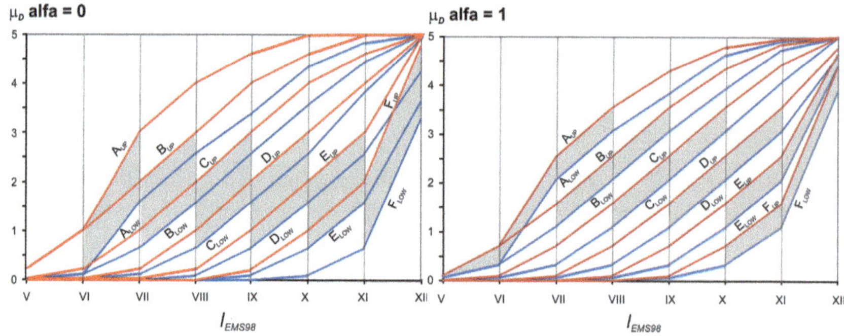

Figure 18. Upper and lower mean damage index of the six vulnerability classes for $\alpha = 0$ and $\alpha = 1$ (vulnerability curves). Shading connects the upper and lower pair of each class.

Damage scenarios from the 1991 INSTAT inventory in the Vittorio Veneto area (Montello Earthquake). The DPMs model allows a damage forecast in urban areas, when some information about the inventory of existing buildings is available. A modest description about structural building types should include, as a minimum, the vertical structural system (masonry or r.c.), or better still simple details, such as the number of stories and the age of the building. In some cases numerical intervals, or qualitative judgments (e.g. well or badly maintained buildings) provide a measure of some simple details. The procedure is based on recognizing the vulnerability class for each type, or at least the probability of membership to the different vulnerability classes. Hence the model furnishes upper/lower bounds for the expectation of functions of the degrees of damage (e.g. interesting functions to forecast number of victims, unusable or collapsed buildings).

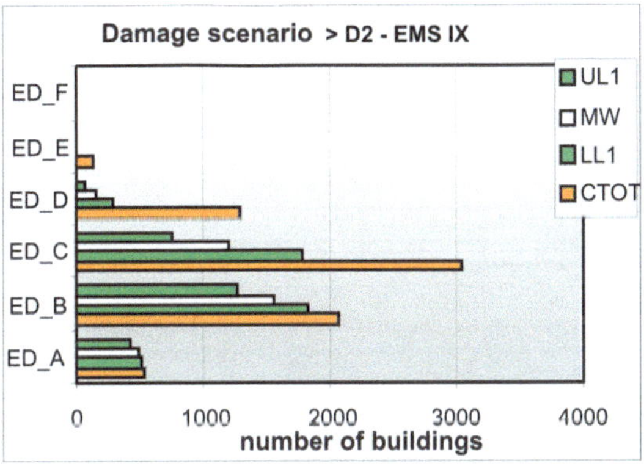

Figure 19. Damage scenario in the Vittorio Veneto municipality for each vulnerability class and local intensity IX: total number of buildings (CTOT), upper/lower bounds for $\alpha = 1$ (UL1 And LL1) and white expected (MW) number of buildings with damage degree over D_2 (after Bernardini, 2008; reprinted with permission)

Below are some numerical results obtained from the inventory of residential buildings ISTAT 1991 in a selected area of the provinces of Treviso, Belluno and Pordenone, around the municipality of Vittorio Veneto. Empirical criteria (Bernardini et al., 2008) have been calibrated to recognize the buildings vulnerability class. We surveyed buildings in a sample of census tracts using the AeDES form (Baggio et al., 2002) and compared their actual vulnerability to the vulnerability assessed by means of ISTAT data. The DPMs presented above allow us to evaluate specific levels of damage in each census tract, conditional to the local macro-seismic intensity or to a correlated parameter of the ground motion (PGA in particular). Evaluation is measured by upper and lower bounds of the expected percentage of buildings.

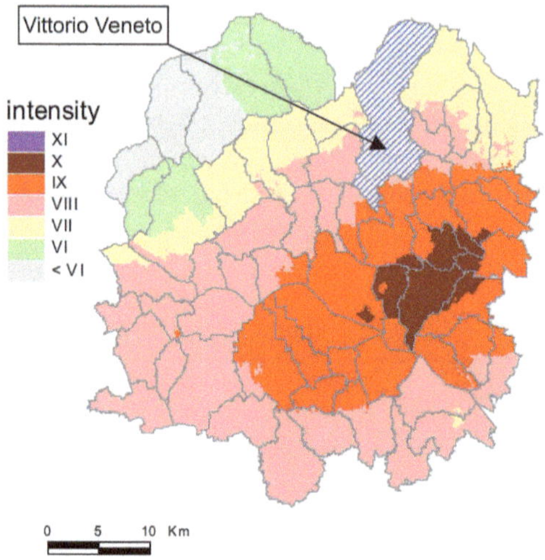

Figure 20. Intensity map for the Montello scenario (after Meroni et al., 2008; reprinted with permission).

Figure 19 summarizes for the municipality of Vittorio Veneto and local intensity equal to IX: 1) the total number of buildings in each vulnerability class; 2) the upper/lower for $\alpha = 1 > D_2$; 3) white expected number of buildings with same level of damage.

Maps of risk or deterministic scenarios can be evaluated through the bounds of monotonic or non monotonic functions of damage degrees. The local intensities in census tracts or municipalities correspond to specified return periods in the first case or, in the second case, are obtained by local ground motion simulation due to specific earthquakes.

Figure 20 shows the EMS98 macro-seismic intensities in a group of municipalities around Vittorio Veneto for a hypothetic M = 6.7 seismic source in the Montello area (Meroni et al., 1998). Average PGA values evaluated by Laurenzano and Priolo (2008) have been converted to EMS98 intensities through a site dependent correlation law proposed by Slejko et al. (2008).

The number of collapsed and unusable buildings can be computed through suggested functions, specifically, conditional probabilities of damage degrees [28]. Collapsed buildings are all buildings with D^5 level damage. Unusable buildings are all buildings with damage D^4 or D^5 plus 40% of buildings with damage D^3. Figure 21 gives white expected numerical results.

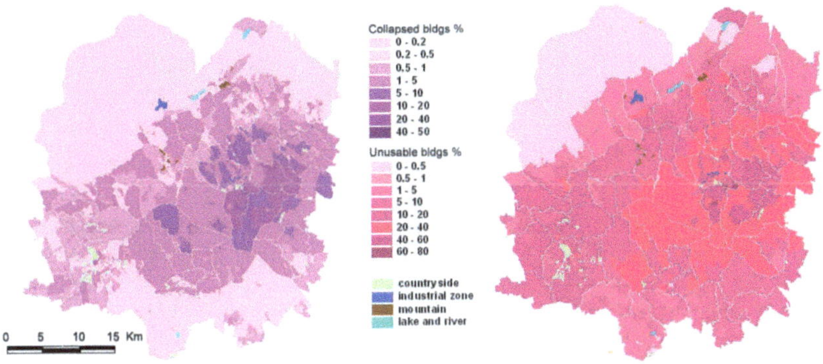

Figure 21. White expected percentages of collapsed and unusable buildings for the Montello earthquake in the Vittorio Veneto area (after Meroni et al., 1998; reprinted with permission).

Damage scenarios from the 2001 INSTAT Inventory in the Abruzzi Region. In 2008 the RELUIS Research Unit at the University of Genova (leader S. Lagomarsino) and CNR-ITC Research Unit in L'Aquila (leader A. Martinelli) carried out an extended application of the macro-seismic methodology presented above to the entire Abruzzo Region (Central Italy). The complete inventory of buildings for this application is the database recorded by the Italian Statistics Agency (ISTAT 2001).

Firstly in each census tract the database furnishes the frequency of types listed in the EMS98 scale and displayed in Figure 15. The selection takes into account for each municipality the seismic protection level imposed by the rules in force in the year of construction of each building. Moreover for masonry buildings available information on masonry units and techniques used in different areas (bricks in the coastal area and stones in mountainous area) has been considered.

Finally the qualitative probabilistic correlation between types and vulnerability classes displayed in Figure 15 allows to evaluate the frequency of vulnerability classes.

On April 6, 2009 a medium intensity earthquake (moment magnitude M_w of 6.3) hit the town of l'Aquila and surrounding area. Total or partial collapse of both masonry and r.c. buildings in the town and in some villages nearby resulted in nearly 300 deaths and 1500 people injured. Nearly one third of the residential buildings were rendered unfit for use.

Independent estimates of the local intensity from a survey of the observed damages (Tertulliani et al., 2009) and analysis of the recorded accelerograms in comparison with Code intensities (Chiauzzi et al., 2009) suggest a local EMS98 between VIII and IX, with a return period close to 500 years. Table 4 shows a comparison of predicted and observed consequences. See further details in (Bernardini et al., 2010). More precisely the local macro-seismic intensity used in the simulation is equal to 8.7. In fact the correlation suggested by Margottini et al. (1992) gives for this value a PGA with a local return period equal to 475 years.

Table 4. Summary of the losses due to earthquake on 6 April 2009 in the L'Aquila municipality and comparison with simulations based on ISTAT 2001 inventory of residential buildings).

	Simulation			**Post earthquake survey 2009**
	$I_{EMS98} = 8.7$ ($T_r = 475$ years)			I_{EMS98} = VIII - IX
	Lower	White	Upper	
Percentage of unusable buildings	8.2	29.5	42.9	33,4
Percentage of victims in the population	0.07	0.47	1.28	1.85

The percentage of unusable buildings resulting from the inspection by means of the AeDES form (33.4%) is slightly greater than the forecasted white value (29.5%) and lower than the upper bound. The percentage of population dead or seriously injured is equal to 1,85%. This value is greater than the expected upper bound (1.28%) and much higher than the white expected value (0.47%). The discrepancy could derive from the time of the earthquake (2 a.m.) and/or from a higher than expected percentage of victims due to the brittle collapse of some r.c. buildings.

8. Conclusions

The concept of random set can be useful to model the actual uncertainty about variables relative to engineering problems. This model not only appears to be a powerful generalization of classical probability theory. but also it is a particular case of a more general theory of measures related to classes of events, through monotone non-additive functions.

The theory of random sets contains, in addition to the probability measures, both the models of uncertainty based on the classical set theory (convex modelling, interval analysis) and the so-called possibility theory. Possibility distributions, on the other hand, can be considered as a particular interpretation of the general fuzzy sets theory.

The applications presented in seismic vulnerability analysis, demonstrate that imprecise probabilities allow the decision maker to preserve the epistemic uncertainty of final numerical results. Similar considerations are valid in many other areas of civil engineering, when the epistemic uncertainties are significant (for example in geotechnical problems).

By knowing these bounds, engineers may ascertain whether what he knows about the system's behaviour is sufficient to make final decisions about the design, safety assessment or management of the system. When the reply is affirmative, any further investigation is not justified. When the reply is negative, new or more precise information is necessary, or more sophisticated models should be employed to narrow the bounds of the final evaluations.

References

Aven, T. (2010). On the need for restricting the probabilistic analysis in risk assessments to variability. *Risk Analysis* 30(3):354–360.

Baggio, C., Bernardini, A., Colozza, R., Corazza, L., Della Bella, M., Di Pasquale, G., Dolce, M., Goretti, A., Martinelli, A., Orsini, G., Papa, F. and Zuccaro, G. (2002). *Manuale per la compilazione della scheda di 1° livello di rilevamento danno, pronto intervento e agibilità per edifici ordinari nell'emergenza post-sismica.* Versione 05/2000, Roma: SSN-GNDT(in Italian)

Ben-Haim, Y. and Elishakoff, I. (1990). *Convex Models of Uncertainty in Applied Mechanics.* Amsterdam et al.: Elsevier Science Publishers.

Bernardini, A. (1995).Valutazioni di frequenze attese di eventi da misure sfuocate di vulnerabilità sismica", In *Atti 7° Convegno ANIdIS L'ingegneria sismica in Italia*, Siena, Collegio degli Ingegneri della Toscana (in Italian).

Bernardini, A. (2008).Vulnerability analyses in a sample of 18 Municipalities in the Veneto-Friuli area. *Bollettino di Geofisica-An Int. Journ. of Earth Sciences* 49 (3-4):447–462,

Bernardini, A., Giovinazzi, S., Lagomarsino, S. and Parodi, S. (2007).The vulnerability assessment of current buildings by a macroseismic approach derived from the EMS-98 scale. In *Proceedings 3° Congreso Nacional de Ingenieria Sismica*, Girona (Espagna).

Bernardini, A., Lagomarsino, S., Mannella, A., Martinelli, A., Milano, L. and Parodi, S. (2010). Forecasting seismic damage scenarios of residential buildings from rough inventories: a case study in the Abruzzo Region (Italy). *Proceedings of the Institution of Mechanical Engineers, Part O: Journal of Risk and Reliability*, Special issue on *Uncertainty in Engineering Risk and Reliability* 224 (4): 279–296.

Bernardini, A., Salmaso, L. and Solari, A. (2008). Statistical evaluation of vulnerability and expected seismic damage of residential buildings in the Veneto – Friuli area. *Bollettino di Geofisica-An Int. Journ. of Earth Sciences* 49 (3-4):427–446,

Bernardini, A. and Tonon, F. (2010). *Bounding uncertainty in Civil Engineering. Theoretical background.* Berlin – Heidelberg: Springer Verlag.

Braga, F., Dolce, M. and Liberatore, M. (1982). A statistical study on damaged buildings and ensuing review of MSK-76 scale. In *Southern Italy November 23, 1980 Earthquake*. Roma: Ed. Scientifiche Associate.

Campos, L. M. D., Huete J. F. and Moral, S. (1994). Probability Intervals: a tool for uncertainty reasoning. *International Journal of Uncertainty, Fuzziness and Knowledge-Based Systems* 2(2):167–196.

Chiauzzi, L., Masi, A., Ditommaso, R., Mucciarelli, M. and Vona, M. (2009). Il terremoto di L'Aquila del 2009: analisi delle registrazioni accelerometriche e confronto con le azioni sismiche previste nelle NTC 2008", In *Proc. 28° Conv. Naz. Gruppo Nazionale di Geofisica della Terra Solida*, Trieste (Italy):439–443(in Italian).

Choquet, G. (1954). Theory of capacities. *Annales de L'Institut Fourier* 5:131–295.

Dempster, A. P. (1967). Upper and lower probabilities induced by a multivalued mapping. *Annals of Mathematical Statistics* 38(2):325–339.

Dubois, D. (2010). Representation, propagation and decision issues in risk analysis under incomplete probabilistic information. *Risk Analysis* 30:361–368.

Dubois, D. and Prade H. (1982).On several representations of an uncertain body of evidence. In Gupta, M.M. and Sanchez E., eds. *Fuzzy Information and Decision Processes*. Amsterdam: North-Holland. 167–181.

Dubois, D. and Prade, H. (1988). *Possibility Theory: an Approach to Computerized Processing of Uncertainty*. New York – London: Plenum Press.

Fetz, T. and Oberguggenberger, M. (2004).Propagation of uncertainty through multivariate functions in the framework of sets of probability measures. *Reliability Engineering & System Safety* 85(1-3):73–87.

Grünthal, G. (1998). European Macroseismic Scale 1998. *Cahiers du Centre Européen de Géodynamique et de Séismologie* 15:1–97.

Laurenzano, G. and Priolo, E. (2008).Numerical modeling of earthquake strong ground motion in the area of Vittorio Veneto (Italy). *Bollettino di Geofisica-An Int. Journ. of Earth Sciences* 49 (3-4):401–425.

Lucantoni, A., Bosi, V., Bramerini, F., De Marco, R., Lo Presti, T., Naso, G. and Sabetta, F. (2001). Il rischio sismico in Italia. *Ingegneria Sismica* XVIII(N1):5–36 (in Italian).

Margottini, C., Molin, D., Narcisi, B. and Serva, L. (1992). Intensity versus ground motion: a new approach using Italian data. *Engineering Geology* **33**:45–48.

Meroni, F., Pessina, V. and Bernardini, A. (2008).Damage risk and scenarios in the Veneto - Friuli area. *Bollettino di Geofisica-An Int. Journ. of Earth Sciences* 49 (3-4):485–503.

Moore, R. E. (1966). *Interval Analysis*. Englewood Cliffs: Prentice-Hall.

Oberkampf, W. L., Trucano, T. G. and Hirsch, C. (2003). *Verification, Validation, and Predictive Capability in Computational Engineering and Physics*. SAND REPORT 2003-3769, Sandia National Laboratories.

Shafer, G. (1976). *A Mathematical Theory of Evidence*. Princeton – London: Princeton University Press.

Slejko, D., Rebez, A. and Santulin, M. (2008). Seismic hazard estimates for the Vittorio Veneto broader area (NE Italy). *Bollettino di Geofisica-An Int. Journ. of Earth Sciences* 49 (3-4):329–356.

Tertulliani, A., Bernardini, F., Camassi, R., Castellano, C., Ercolani, E., Graziani, L., Leschiutta, I., Rossi, A and Vecchi, M. (2009). Il rilievo macrosismico di dettaglio del centro storico de l'Aquila in termini di EMS98. In *Proc. 28° Conv. Naz. Gruppo Nazionale di Geofisica della Terra Solida*, Trieste (Italy). 494–495(in Italian).

Walley, P. (1991). *Statistical Reasoning with Imprecise Probabilities*. London et al.: Chapman and Hall.

Zadeh, L. A. (1965). Fuzzy sets. *Information and Control* 8:338–353.

Zadeh, L. A. (1978).Fuzzy sets as a basis for a theory of possibility. *Fuzzy Sets and Systems* 1(1):3–28.

Part 3
Combined Methods

Combined Methods in Nondeterministic Mechanics

Michael Oberguggenberger [*]

[*] Institute of Basic Sciences in Civil Engineering, University of Innsbruck, Innsbruck, Austria

Abstract The goal of the lectures on *Combined Methods* is to discuss various (mathematical and conceptual) approaches that have been put forth as tools for modeling uncertainty in engineering, among them probability, interval arithmetic, random sets, fuzzy sets, sets of probability measures, and previsions. After recalling the definitions, we stress their interpretations (semantics), axioms, interrelations as well as numerical procedures and demonstrate how the concepts are applied in practice. As an accompanying example we use the dimensioning of an elastically bedded beam. Further applications of combined methods in aerospace engineering, to vibrations of belltowers, in queueing theory, and to tuned massed dampers will be sketched.

1 Introduction

Uncertainty analysis in engineering is a vast domain. *Uncertainty* has come a long way from being ignored (or buried in safety factors) to the idea that it should be actively controlled – as exemplified by the fashionable catch word *risk management.* The reader may have heard of failure probability, stochastic processes, interval arithmetic, fuzzy sets, convex models, info-gap analysis, worst case scenarios, random sets, belief functions, Bayesian reliability theory, interval probability, upper and lower previsions – just to mention a fraction of the methods that have been proposed to serve as a framework for modeling uncertainty in engineering.

It is needless to say that an article of the present length is incapable of addressing all the approaches that can be found in the engineering, mathematical, economic and statistical literature. However, most of the methods can be classified as one or the other combination of the poles *intervals* and *probability.* This point of view yields one of the possible systematic paths through the jungle. The intention of this article is to present some of the major players, to compare them and to show them at work in applications.

The layout plan of the paper is as follows. Section 1 starts with some general considerations on uncertainty in engineering, followed by a critical look at the idea of failure probability, and a short account of the development from purely deterministic to probabilistic models and onwards to more general models of the uncertainty. Section 2 provides a survey of the most widely used models of uncertainty, with a strong emphasis on their *meaning* – semantics – an often neglected, but utterly essential ingredient for their proper application. To emphasize it once more: the semantics provide the road map for the transition from reality to the model and back. Numerics are just as important for propagating the uncertainty through an input-output map (practically all structural engineering models are input-output maps – mapping the system input to the system response).

Section 3 is devoted to demonstrating the discussed methods (and various of their combinations) at work in a simple engineering example, an elastically bedded beam. Section 4 addresses a companion of uncertainty analysis – sensitivity analysis, shown at the hand of a concrete application in aerospace engineering. It turns out that sensitivity analysis can be tackled by the same methods. In Section 5, various further applications are discussed: fuzzy modeling of vibrations of bell towers, a fuzzy-probabilistic queuing model in construction management, and a combination of random sets and stochastic processes in modeling tuned mass dampers in earthquake engineering. The conclusion – Section 6 – contains some remarks on decision making in engineering and further reading. With the extensive bibliography, it is hoped that the readers can follow up the paths which they might have found attractive and enter more deeply in one or the other of the areas of uncertainty analysis. The lectures are based on material from [135, 137, 138, 139, 144, 146, 171].

1.1 Modeling uncertainty in engineering

Scientific modeling in engineering has to deal with three facets. First, there is reality (with materials, soils etc.). Second, there is the model of reality (formulated in mathematical terms and containing physical laws and constitutive equations). Third, correspondence rules (prescribing how to translate one into the other) are needed. The physical model establishes what are the state variables and what are the material constants, the *parameters* to be observed. Once this has been decided, the values of the parameters have to be determined from information extracted from the real world and will serve as input in the physical model. This plus the design of the structure enters in numerical computations that result in the proper dimensions, construction procedures, assessment of safety measures. The actual implementation usually undergoes several transitional phases of construction – each requiring different modeling – until the final structure is completed. Ideally, the model output should provide a design that works, reliable guidelines for action, and aids for decision making (as emphasized, e.g., by Einstein [43]).

Models of the data uncertainty should reflect and incorporate the level of information available on the data and, second, must be able to propagate it through numerical computations and deliver an output whose uncertainty is formulated in the same terms. In addition, the uncertainty models need correspondence rules themselves, that is, well-defined semantics. A further aspect that has to be taken into account is how the information on the uncertainty of different parameters is combined – this refers to modeling the dependence of variables as well as the combination of information from different sources. For example, in a sum $A+B$ of two parameters A and B, will the joint uncertainty be the smallest interval containing all realizations $a+b$, or do we believe that extreme combinations of realizations $a+b$ are less *probable* than those near the *standard values*? The first choice implies adherence to the axiomatics of interval arithmetic, the second to the axiomatics of probability theory. As we will see, many more axiomatic systems exist in between these two extremal viewpoints.

Thus three aspects of the modeling of uncertainty are isolated:

- *Definition and axiomatics:* How is uncertainty described and what are the combination rules?
- *Numerics:* How is uncertainty propagated through the computational model?
- *Semantics:* What is the meaning of the results – what do they say about our conception of reality?

There are two major categories of uncertainty: model uncertainties and parameter uncertainties, the latter encompassing the uncertainty of the data used to determine the parameter values.

Model uncertainties. The choice of the structural model is one of the central engineering decisions to be made. For example, for the description of the soil, there are continuum models and granular models, there are two- and three-dimensional models, there are multi-phase models (solids, liquids, gases), and so on. The next point is the adequate selection of state variables and parameters (constant or not). Failure in engineering is described by the so-called *limit state function* that separates the safe states from the unsafe states. The choice of the limit state function is again an engineering decision. For example in geotechnics, is failure due to bounds exceeded by average values of the total loads or due to localized disturbances? Another issue is the proper incorporation of transitional states during construction.

Parameter uncertainties. Parameter variability can be attributed to a large number of causes. There are random fluctuations, lack of information, random measurement errors, but also systematic measurement errors (deriving, e.g., from uncontrollable changes of the properties of the soil material caused by its extraction in bore holes before being analyzed in the laboratory). There are fluctuations due to spatial inhomogeneity, and errors made by assigning parameter status to state variables. This is one of the essential and often unavoidable errors in engineering, because models typically are valid in certain ranges only. If the state of the structure exits the intended range, constants may turn into variables depending on external forces (for example, the friction coefficient of most materials is approximately constant for small loads, but starts becoming a function of the internal stresses for larger loads). Finally, there is variability arising from the fact that parameters have to carry the burden of model insufficiency. The available information on data uncertainty may range from frequency distributions obtained from large samples, values from small samples or single measurements, interval bounds, to experts' point estimates and educated guesses from experience.

1.2 Failure probability

Traditionally, engineers have dealt with uncertainty by employing *safety factors*. That is, the traditional codes would require that the load carrying capacity of the structure exceeds the design loads by a certain factor > 1, typically 1.35 for permanent loads (such as dead weight) and $1.5 - 2.0$ for temporary loads. These factors have been negotiated in the committees of standards. This state of affairs has been considered as unsatisfactory: no information about the *actual* distance to failure can be extracted from such a procedure. Based on the desire for a more analytical description of the uncertainties, engineering codes have been put on a probabilistic foundation, starting with the pioneering work of Freudenthal [64], Bolotin [23] and others in the 1950ies. Under this point of view, every relevant parameter of the engineering model is a random variable. There is no absolute safety, but rather a probability of failure.

To make it more precise, let the vector R comprise all random variables describing the resistance of a structure, S the loads and denote by $g(R,S)$ the limit state function (that is, $g(R,S) < 0$ means failure, $g(R,S) > 0$ signifies a safe state). Then $p_f = P\big(g(R,S) < 0\big)$ is the *failure probability*; $R = 1 - p_f$ is the *reliability* of the structure. To determine the probability p_f, the types and parameters of the probability distributions of R and S are needed. This multiplies the number of parameter values that have to be provided by the designing engineer: each model parameter comes with a distribution type and set of (uncertain) distribution parameters. Actually, the current codes employ critical values R_k and S_k (certain percentiles of R and S) and partial safety factors γ_R and γ_S, so that the designing engineer has to verify a relation of the type

$$R_k/\gamma_R \geq \gamma_S S_k.$$

In theory, the critical values and the partial safety factors are computed in such a way that this inequality holds if and only if p_f attains a certain required value p_{fr}. In practice, γ_R and γ_S are not computed but rather prescribed in the codes. Starting with the 1980ies and 1990ies, the European codes have been changed into probability based codes. By now, this is the standard in civil engineering (see e.g. EN 1990:2002 [48]). In addition, risk analysis and risk management have become a major ingredient in construction and project management. Especially calls for tender by public organizations ask the contractor to supply a risk analysis together with the proposed design.

The civil engineering codes require that the designed structure obtains an instantaneous probability of failure of $p_f = 10^{-6}$ and a longterm failure probability of $p_f = 10^{-5}$. To credibly estimate tail probabilities of such a small magnitude, *a lot* of information is needed. In addition, if time dependent reliability is to be assessed, failure rates and the additional parameters of the reliability function are required. It is no surprise that simultaneously with the establishment of the probabilistic codes, criticism started in the engineering community, especially from the geotechnical side. The uneasiness about the concept has focused on three points:

- the meaning of the failure probability;
- the need for information which as a rule is not available in engineering practice;
- the artificial separation in resistance R and loads S.

Already the pioneers warned that the failure probability cannot have the meaning of a frequency of failure (Bolotin [23, Section 26]). This is quite contrary to public perception and has been widely disputed from many aspects in the probabilistic and engineering literature. It is true that the probabilistic foundations have played a fruitful role in putting arguments and negotiation of the codes on a rational basis. Accordingly, some of the authors of the codes refer to the failure probability as an *operational* or *normative probability* [48, Annex C4(3)], [95, Section 1.3]. The way the probabilistic concept is applied in practice quite clearly puts it into the realm of *subjective probability*.

Here is an example showing that the magnitude of failure probability strongly depends on the model assumptions. It is a simple geotechnical example, taken from [140, 142, 143]: the bearing capacity of a square footing. In this example, the angle of internal friction φ enters as an important parameter. Data are from twenty direct shear tests [25] with disturbed samples of silt that had been performed with equal initial conditions. In each experiment, the unit weights were nearly identical: $\gamma = 19.8$ kN/m^3 and the cohesion resulted to $c = 0$. The friction angles φ [°] were obtained as shown in Table 1.

Table 1. Sample of friction angles φ[°] obtained from $n = 20$ direct shear tests.

25.6	25.5	24.0	26.0	24.1	24.0	28.5	25.3	23.4	26.5
23.2	25.0	22.0	24.0	24.9	30.0	27.0	24.4	24.3	29.5

The value affecting the limit state is the shear strength of the soil $\tau_f = c + \sigma \tan\varphi$, with the normal stress σ. Accordingly, the uncertainty originating from the friction coefficient $\nu = \tan\varphi$ will be analyzed. From the sample of Table 1, the mean value and standard deviation are obtained as $\bar{\nu} = 0.474$, $s_\nu = 0.0452$. A normal distribution, a two and a three parameter lognormal distribution and a triangular distribution were fitted to the data for ν. The resulting fit can be seen from Figure 1. All fits passed the 95% Kolomogorov-Smirnov test and the χ^2-test, thus can be accepted as valid models of the data uncertainty.

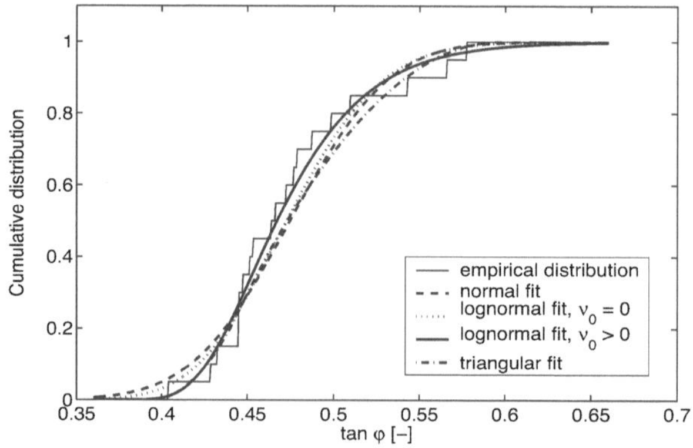

Figure 1. Fits, cumulative distribution functions.

These data will now be used to compute the bearing capacity of a square footing in silt. The footing is centrically loaded and has the dimensions $a = b = 1.0$ m and $t = 0.5$ m (Figure 2).

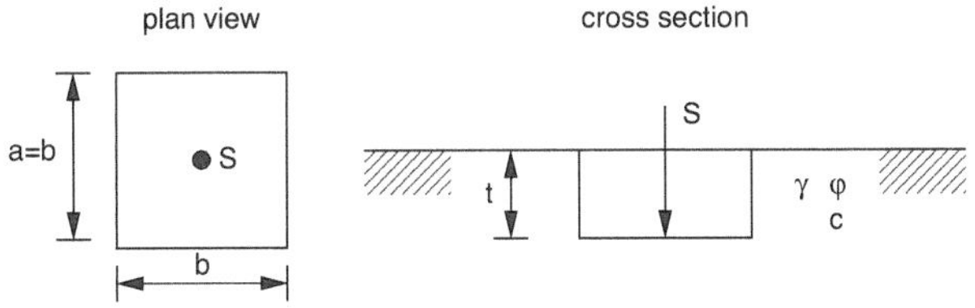

Figure 2. Bearing capacity example.

The resistance $R = Q_f = Q_f(\nu)$ depends on the friction angle $\nu = \tan\varphi$ according to the following formulas

$$R = Q_f = \gamma(N_\gamma s_\gamma + 0.5\, N_q s_q) \tag{1}$$

with

$$N_q = \frac{1+\sin\varphi}{1-\sin\varphi}\, e^{\pi\tan\varphi}, \quad N_\gamma = (N_q - 1)\tan\varphi, \quad s_q = 1 + \sin\varphi, \quad s_\gamma = 0.7.$$

To assess the load S, the procedure for determining the design bearing capacity from the Austrian code of standards [151] was employed, resulting in a mean value of $\mu_S = 87.53$ [kN] and a standard deviation of $\sigma_S = 8.75$ [kN]; for details, see [142]. In this example, the failure probabilities can be analytically expressed as a single integral involving the distribution functions F_S of the load and the probability density f_ν of the friction coefficient. Indeed,

$$\begin{aligned} p_f &= P\big(g(R,S) < 0\big) = P\big(Q_f(\nu) < S\big) \\ &= \iint\limits_{Q_f(\nu)<s} f_S(s) f_\nu(\nu)\, \mathrm{d}s\, \mathrm{d}\nu = \int_{-\infty}^{\infty} \int_{Q_f(\nu)}^{\infty} f_S(s)\, \mathrm{d}s\, f_\nu(\nu)\, \mathrm{d}\nu \\ &= \int_{-\infty}^{\infty} \Big(1 - F_S\big(Q_f(\nu)\big)\Big)\, f_\nu(\nu)\, \mathrm{d}\nu. \end{aligned}$$

Assuming the load to be normally distributed according to $S \sim \mathcal{N}(\mu_S, \sigma_S^2)$ results in

$$p_f = \int_{-\infty}^{\infty} \left(1 - \Phi\left(\frac{Q_f(\nu) - \mu_S}{\sigma_S}\right)\right) f_\nu(\nu)\, \mathrm{d}\nu$$

where Φ denotes the cumulative standard normal distribution function. The function $Q_f(\nu)$ is given by formula (1). The integral can be evaluated numerically. The values computed in this way are listed in Table 2. The

Table 2. Failure probabilities for the footing.

distribution	p_f
normal	0.811×10^{-3}
two parameter lognormal ($\nu_0 = 0$)	1.11×10^{-4}
three parameter lognormal ($\nu_0 > 0$)	1.03×10^{-9}
triangular	0.96×10^{-11}

example shows that the failure probability may vary by several magnitudes, when different (but all acceptable) statistical models of the input data are used. Further critical appraisal of the dependence of p_f on the choice of probability distributions can be found in [46, 140, 141].

1.3 Imprecise probability models

As described above, there are many types of uncertainties in an engineering model. It has been questioned whether a purely probabilistic approach is capable of catching all aspects, for example, ignorance or fluctuations due to systematic model errors. In addition, large samples allowing a frequentist assessment of data are rarely available. Due to high costs, sample sizes in laboratory experiments are in the range of $5 - 10$. The number of soundings in soil investigations depend on the situation, but typically may be in the range of $0 - 3$ at small construction sites. However, what is generally known about a model parameter is a central value and a coefficient or range of fluctuation. In geotechnics, geologists can provide interval estimates of soil parameters; a geological report in tunneling may deliver rock classes and interval probabilities for their occurrence. Thus, in practice, the determination of distribution parameters takes recourse to normative assumptions (types of distributions extracted from the engineering literature) and to expert estimates (sometimes put in a Bayesian framework).

Finally, the requirement of separating resistance and loads brings the geotechnician into a delicate situation: in the stability of a foundation, say, the soil acts both as resistance and as load (this specialized question will not be pursued here).

The uneasiness about the probabilistic safety concept and the desire for models of the data uncertainty that reflect and incorporate the level of available information led to the search for alternative concepts in the engineering community. On the one hand, probabilistic models and probabilistic reasoning were considered as too tight a concept. On the other hand, engineering practice quite blatantly showed that interval estimates should be incorporated in the framework of uncertainty analysis.

Starting in the 1990ies, the usability of fuzzy sets has been explored in the engineering literature. One of the earliest ventures is in Haldar and Reddy [71], who modeled the material properties of existing buildings by fuzzy sets (acknowledging ignorance about the exact values which would not be available in a non-destructive way). In the subsequent years, a vast literature on fuzzy methods in civil engineering appeared, starting from fuzzy finite element methods, fuzzy geotechnical stability, fuzzy rule based decision models, project planning with fuzzy durations and costs, to the proposal of a fuzzy safety concept as a complete replacement of the probabilistic approach (Möller et al. [121]). Later, the capabilities of evidence theory (Ferson, Oberkampf et al. [51, 148]) and interval probability (Blockley, Hall et al. [22, 73]) were explored as well. In the decade after the year 2000, the usefulness of random sets was increasingly acknowledged, starting with the pioneering work of Bernardini et al. [191, 192] and Hall et al. [75]. In a sense, random sets were seen as a framework bridging the gap between probability and interval analysis and admitting easily accessible visualization tools such as probability boxes (Ferson et al. [51]). Lower and upper previsions have found applications in systems reliability (Utkin, Kozine [196, 197]). In addition to these approaches, a number of further alternative frameworks have been proposed in the engineering literature: for example, clouds (Neumaier [130]), info-gap-analysis (Ben-Haim [16]), ellipsoidal modeling (cf. Ben-Haim et al. [17]), anti-optimization (Elishakoff [45, 47]).

To subsume all these different approaches, the term *Imprecise Probability* has been introduced. As pointed out in the beginning of this section, the probabilistic safety concept has become the wide-spread standard in engineering uncertainty analysis. Thus any imprecise probability approach should be formulated in a way that can be communicated to an engineering community trained in probability theory. The probabilistic language comes natural in random sets, sets of probability measures, and previsions. On the other hand, thinking in interval estimates is also common practice in engineering. Here fuzzy sets come in as an extension of interval arithmetic. On the other hand, fuzzy sets can also be interpreted as special types of random sets, and profound work has been done on the relations of fuzzy sets and sets of probability measures [40, 69]. However, the probabilistic interpretation does not fully capture the semantics inherent to fuzzy set theory. For example, it seems there is no probabilistic counterpart of the idea of a fuzzy controller.

An adequate understanding of the uncertainties in an engineering task cannot be reduced to the calculation of a single number – the failure probability. It requires a multitude of steps, some of which may be isolated as follows:

- reflection about the choice of model and the failure mechanisms;
- assessing the variability of input and output variables and model parameters;
- sensitivity analysis;
- assessing the reliability of the structure;
- decision about acceptance or non-acceptance of the design.

Some of the tasks can be aided by quantitative methods, some have to be dealt with in a qualitative way. Thinking in scenarios is essential (possibly aided by fault trees and decision trees), as is the continuous revision of the model in a feedback loop. For example, the results of the sensitivity analysis may lead to changes in design towards increased robustness. Validation of parts of the model and calibration using experimental data is important as well. Care should be taken that the framework used for describing the data uncertainty reflects and incorporates the level of information available. Finally, one should keep in mind that the role of practical engineering is not so much the production of factual knowledge but rather decision making with the help of scientific tools.

2 Models of the uncertainty

2.1 Definitions

In this section we shall focus on describing the main theories of uncertainty in the univariate case of a single parameter. The following convention will be in use throughout the paper: parameters will be denoted by upper case letters, e.g. A, while corresponding lower case letters, such as a, will be reserved for their realizations. The description of the semantics is deferred to the next subsection, except for the straightforward first two cases.

Deterministic values. The simplest approach is what in engineering terminology usually is called *deterministic description*, that is, the parameter A is described by a single value a, see Figure 3. The semantics is simply that a is an expert estimate (or educated guess). Clearly, possible variations are not modeled in this approach, though their influence can be assessed to some extent by a classical sensitivity analysis. This will be elaborated a bit more in the section on *numerics*; a detailed exposition is in [152].

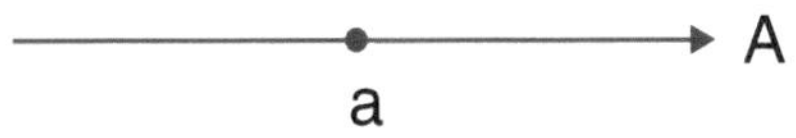

Figure 3. Deterministic value a of parameter A.

Intervals. The next level in modeling uncertainty is interval analysis. The uncertainty of the input A is described by an interval $[a_L, a_R]$, signifying bounds in terms of a worst/best case assumption. In this way the total variability is captured, but no detailed information on the uncertainty is provided (Figure 4).

Figure 4. Interval bounds for parameter A.

Probability. The most informative, but also most stringent description of the uncertainty of a parameter A is by means of *probability*. If the probability distribution is given by a density $p_\lambda(a)$, the probability that the realizations of the parameter A lie in a set S is

$$P(A \in S) = \int_S p_\lambda(a)\, da,$$

see Figure 5. The notation p_λ indicates that, usually, the probability dis-

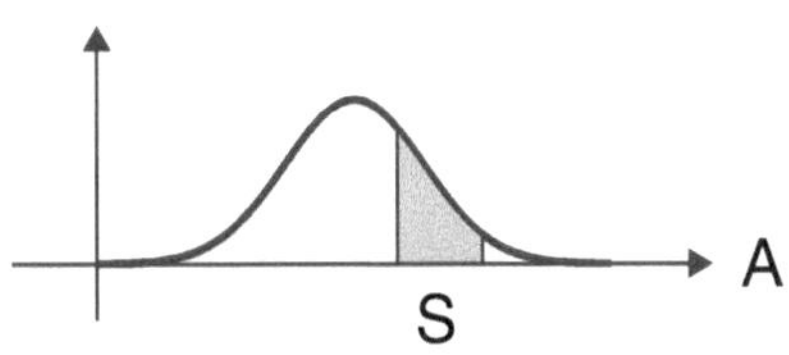

Figure 5. Probability distribution.

tributions arise as members of a class of distributions which in turn are parametrized by parameters λ. For example, the class of Gaussian normal distributions $\mathcal{N}(\mu, \sigma^2)$ is given by the Gaussian densities $p_\lambda(a)$ with parameters $\lambda = (\mu, \sigma)$,

$$p_\lambda(a) = \frac{1}{\sqrt{2\pi}\sigma} e^{-\frac{(a-\mu)^2}{2\sigma^2}}.$$

Thus the complete specification of a probability distribution requires determination of the type it belongs to as well as the values of its parameters.

Sets of probability measures. A central idea in relaxing the precision inherent in a probabilistic model is to replace the single measure by a set of probability measures, say a family

$$\mathcal{M} = \{p_\lambda : \lambda \in \Lambda\}$$

where the parameter $\lambda \in \Lambda$ specifies each participating single measure. A set of probability measures defines lower and upper probabilities according to the rules

$$\begin{aligned} \underline{P}(A \in S) &= \inf\{P(A \in S) : P \in \mathcal{M}\}, \\ \overline{P}(A \in S) &= \sup\{P(A \in S) : P \in \mathcal{M}\}. \end{aligned}$$

The lower probability is the greatest lower bound of all probabilities that are assigned to the event $A \in S$ by choosing one of the probability measures from the set $\mathcal{M}$ (if attained, it is the minimum of those values). Similarly, the upper probability is the least upper bound of such probabilities (if attained, it is the maximum of those values). The idea of sets of probability measures is inherent in Bayesian statistics, where the distribution parameters λ are considered to be random variables themselves. In the frequentist interpretation, there is the notion of sample statistics with its confidence intervals for certain sample parameters.

A set $\mathcal{M}$ of probability measures is called *convex*, if for all probability measures P_1, P_2 belonging to $\mathcal{M}$, also every convex combination $\lambda_1 P_1 + \lambda_2 P_2$ is a member of $\mathcal{M}$ ($\lambda_1 \geq 0$, $\lambda_2 \geq 0$ and $\lambda_1 + \lambda_2 = 1$).

Convex sets of probability measures are often called *credal sets* and play an important role in Choquet's theory of capacities [29], see [35, 84, 202], which in turn has become an important tool in economic decision theory, notably utility theory [132, Chapter 6]. Credal sets can be conveniently described by means of their extremal points [115], which is an important tool for the numerical calculation of the lower and upper probabilities.

An axiomatic theory of interval probability has been developed by [203]. We will say more about the role of sets of probability measures in the section on semantics.

As most of the models of uncertainty, the two subsequent theories – random and fuzzy sets – can be seen as special prescriptions for obtaining *sets of probability measures.*

Random sets. A *random set*, sometimes also referred to as a *Dempster-Shafer structure*, is given by finitely many subsets $A_i, i = 1, \ldots, n$ of a given set $\mathcal{X}$, called the *focal sets*, each of which comes with a *probability weight* $m_i = m(A_i), \sum m(A_i) = 1$. An example of a random set is shown in Figure 6. The general case of an infinite number of focal sets can be treated as well; the concept of a set-valued random variable being the defining notion (see Subsection 2.2).

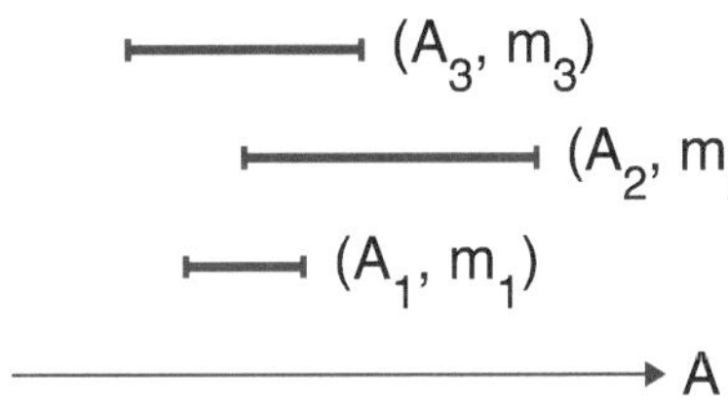

Figure 6. A random set.

In the Dempster-Shafer approach [34, 181], the random set allows one to define a degree of belief $\gamma(S)$ and a degree of plausibility $\eta(S)$, respectively, that the realizations of the parameter A lie in S by

$$\gamma(S) = \sum_{A_i \subset S} m(A_i)\,, \quad \eta(S) = \sum_{A_i \cap S \neq \emptyset} m(A_i). \tag{2}$$

A random set can also be interpreted as a prescription for a set of probability distributions. Denote by $\mathcal{M}(A_i)$ the totality of all probability measures supported by A_i, that is, a probability measure P on the underlying set $\mathcal{X}$ belongs to $\mathcal{M}(A_i)$ iff $P(A_i) = 1$. The set of probability measures induced by the given random set is

$$\mathcal{M} = \{P : P = \sum m(A_i)P_i, P_i \in \mathcal{M}(A_i)\}. \tag{3}$$

One can show [54, 59] that the corresponding lower and upper probabilities coincide with the degrees of belief and plausibility, respectively. That is, for any subset[1] S of $\mathcal{X}$, it holds that

$$\underline{P}(S) = \gamma(S)\,, \quad \overline{P}(S) = \eta(S). \tag{4}$$

[1]When considering sets of probability measures a common framework is needed. Full

Also, it is quite obvious that the set $\mathcal{M}$ is a convex set of probability measures – thus random sets constitute one of the operational methods for constructing credal sets.

Fuzzy sets. Fuzzy sets can be viewed as ordered families of sets or as membership functions. It is simplest to describe the ideas by means of the special case of a fuzzy real number A. From the first point of view, A is a family of parametrized intervals. The parametrization is done in terms of levels α, $0 \le \alpha \le 1$. Each level α has a corresponding interval A^α so that $A^\beta \subset A^\alpha$ if $\alpha \le \beta$. Thus the intervals are stacked and can be depicted by their left/right contour functions, see Figure 7. More generally, one could allow the A^α to be arbitrary, stacked subsets of a given set of objects under investigation (complex numbers, vectors, matrices, functions or the like).

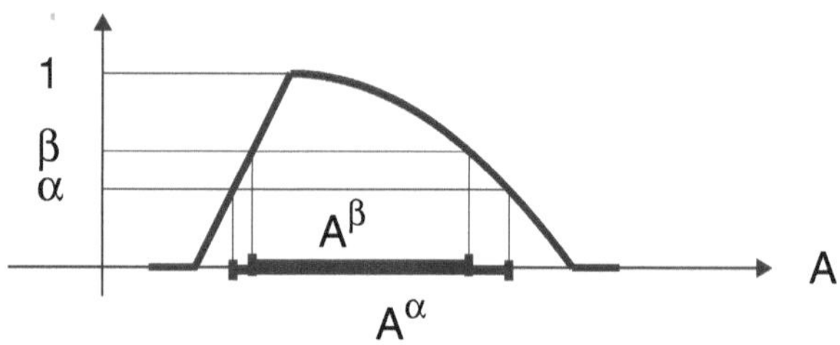

Figure 7. Fuzzy set as family of intervals.

In the second approach, the contour function is taken as the primary object, and a fuzzy set A (over the real numbers) is just a map from the real line to the interval $[0, 1]$, assigning to each real number a a value $\pi_A(a) \in [0, 1]$. This value may be interpreted as the membership degree to which a belongs to the fuzzy set A, or in the language of parameters, as the degree of possibility that the parameter A takes the value a, see Figure 8. In classical set theory, the membership degree is either 0 or 1; fuzzy set theory permits gradual membership as well. The intervals from the first interpretation are now the *α-level sets* $A^\alpha = \{a : \pi_A(a) \ge \alpha\}$. In analogy to the situation in probability theory, one can introduce a *possibility measure* on the underlying set, defining a degree of possibility for each subset by

$$\pi_A(S) = \sup\{\pi_A(a) : a \in S\},$$

mathematical precision can be achieved, for example, by requiring that $\mathcal{X}$ is a topological space, probability measure is interpreted as *regular Borel measure of mass one*, and subset means *Borel-measurable subset*, cf. [76].

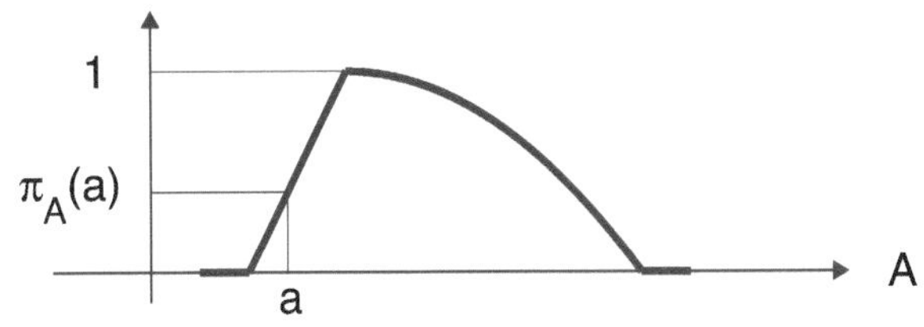

Figure 8. Fuzzy set as degree of possibility.

giving the degree of possibility that the parameter A takes a value in S. The possibility measure is monotone, i.e., $\pi_A(S) \leq \pi_A(T)$ if $S \subset T$. Possibility measures are actually in one-to-one correspondence with fuzzy sets; given a possibility measure π, its evaluation on singletons defines the membership function of a fuzzy set: $\pi_A(a) = \pi(\{a\})$.

Lower and upper previsions. We follow here the exposition in [194]. In the probabilistic setups discussed so far, probability P is the fundamental quantity. The outcomes of a random variable X can be described by their probabilities $P(X \in S)$. The expectation $\mathrm{E}(X)$ and the moments $\mathrm{E}(X^m)$ are derived quantities, e.g., in terms of a probability density $p(x)$:

$$\mathrm{E}(X) = \int xp(x)\,\mathrm{d}x, \qquad \mathrm{E}(X^m) = \int x^m p(x)\,\mathrm{d}x.$$

Conversely, the probability of an event S can be viewed as the expectation of its indicator function[2] $X(x) = \mathbf{1}_S(x)$:

$$P(S) = \mathrm{E}(\mathbf{1}_S).$$

This opens the way to setting up a theory based on expectations of random variables as fundamental quantities. Actually, the approach is more general, as one may admit smaller or larger sets of random variables, now called *gambles*. A *linear prevision*, as introduced by [33] is a linear functional E, assigning to each gamble a value between 0 and 1. The extension to interval-valued previsions, see e.g. [199], can be done as follows. The basic gambles are given by a random variable X and a number of functions $f_1(X), ..., f_m(X)$ of it. *Lower and upper previsions* are functionals on the

[2]The indicator function of a set S is defined by $\mathbf{1}_S(x) = 1$ when $x \in S$ and $\mathbf{1}_S(x) = 0$ when $x \notin S$.

set of gambles with $\underline{\mathrm{E}}(f_i) \leq \overline{\mathrm{E}}(f_i)$. Various types of information can be modeled by means of lower and upper previsions. For example, if $\mathbf{1}_S$ is the indicator function of an event S, then the previsions $\underline{\mathrm{E}}(\mathbf{1}_S)$ and $\overline{\mathrm{E}}(\mathbf{1}_S)$ can be regarded as lower and upper probabilities of the event S. If $f_i(X) = X$, then $\underline{\mathrm{E}}(f_i)$ and $\overline{\mathrm{E}}(f_i)$ are bounds on the mean value of the corresponding random variable.

For computing new previsions $\underline{\mathrm{E}}(g)$ and $\overline{\mathrm{E}}(g)$ of a gamble $g(X)$ from the available information, *natural extension* is used which can be written as the optimization problem

$$\underline{\mathrm{E}}(g) = \min_p \int g(x)p(x)\,\mathrm{d}x, \qquad \overline{\mathrm{E}}(g) = \max_p \int g(x)p(x)\,\mathrm{d}x$$

subject to

$$p(x) \geq 0, \quad \int p(x)\,\mathrm{d}x = 1, \quad \underline{\mathrm{E}}(f_i) \leq \int f_i(x)p(x)\,\mathrm{d}x \leq \overline{\mathrm{E}}(f_i), \ 1 \leq i \leq m. \tag{5}$$

Here the minimum and maximum are taken over the set of all possible probability density functions $p(x)$ satisfying conditions (5).

For additional theoretical background on the concepts discussed here we refer the interested reader to the following sources: interval analysis [128], fuzzy sets [13, 22, 40, 78, 92, 96, 162, 204, 205], possibility theory [40, 207], random sets [51, 70, 118, 131, 132, 181, 204], random sets and possibility theory in engineering [1, 9, 14, 18, 20, 22, 53, 58, 74, 75, 119, 148, 191, 192], probability in geotechnics [7, 43, 142, 161], comparative studies and further concepts [32, 45, 46, 96, 103, 110, 122, 146, 147, 149, 179, 188, 190, 199, 200]; see also the unifying concept of *clouds*, proposed by [130].

2.2 Semantics

As outlined in the introduction, the interpretation of a theory is an essential ingredient for achieving an adequate translation from model into reality and back. Needless to say that the assertions made by a model become meaningful only in the context of the underlying semantics. Different semantics imply different meanings. One has to be aware of the interpretations used when comparing assertions made by different authors, all the more so as often the same vocabulary is employed for notions that differ in the various interpretations.

Probability. The interpretation of probability has been the subject of scientific dispute for centuries; see [63] as an encompassing reference. The most prevalent and important semantics in engineering practice are:

1. *Classical probability*, based on principles like the *principle of non-sufficient reason* would, in colloquial terms, determine the probability of an event S as the fraction of favorable cases among the possible cases.
2. *Frequentist probability*, based on the idea of random occurrence of an event in a sequence of independent trials, would approximate the probability of an event S by its relative frequency.
3. *Subjective probability* is meant to be a measure of personal confidence. It can be assessed by introspection and/or elicitation through experts.

Example 2.1. As an illustration, let us assess the probability p of throwing a six with a dice. If nothing is known about the dice, there is no sufficient reason to assume that it is biased. Thus the classical probabilist would assess the probability as 1 (favorable outcome) over 6 (possible outcomes), hence $p = 1/6$. A frequentist person would repeatedly roll the dice a large number of times and use the fraction of sixes among all results as an estimate for the probability. (If the dice is indeed unbiased and N is large, say $N = 10000$, the fraction will be close to $1/6$.) The subjectivist would guess the probability as $p = 1/6$, if he/she has enough trust into the unbiasedness of the dice. If the subjectivist is willing to bet 1 monetary unit for a gamble that rewards him/her with 6 units when a six is thrown, an observer could infer that the subjectivist believes the probability p to be at least $1/6$.

In its applications, classical probability often takes the form of *combinatorial* probability. Aside from the obvious application in computing the chances in a lottery, it is often the means by which the standard probability distributions are derived, like the binomial or geometric distributions. Another example would be the exponential distribution for the survival time of a radioactive particle which is an immediate combinatorial consequence of the law of radioactive decay.

The central idea of frequentist probability is the sample with its statistical parameters. It is viewed or designed as a sequence of independent realizations of the random variable whose distribution parameters have to be determined – keeping the boundary conditions constant. The relative frequencies of the realizations of an event are taken as estimates for the probability of the event. The sample parameters like sample mean or sample variance correspond to moments of the random variable – expectation value and variance in this case. From there, the parameters of the distribution of the random variable can be estimated. This is one of the wide-spread procedures for fitting models based on the frequentist interpretation.

Going one step further and allowing the notion of multiple samples, criteria for decision making about the quality of a frequentist estimate are obtained: confidence intervals or tests of fit, and so on. For example, a single sample of size N delivers the sample mean as an estimate for the expectation value of a random variable. If we imagine collecting many samples of size N, say 100 samples, we get 100 sample means. The 95%-confidence interval of the expectation value is constructed in such a way that it contains, on average, 95 of those 100 sample means.

From the viewpoint of the philosophy of science, the frequentist interpretation carries a number of problems, among them the question *whose* probability is realized in the sample (of the random variable, or of the experiment which was designed to measure it – a clear answer to this question has been given by Popper with his notion of *propensity* [157]). A pragmatic approach with a cautious and critical attitude has proven to provide a successful basis for probabilistic models in science and engineering.

A further issue of debate has been the fact that the decision aids mentioned above provide meaningful evaluations only if the sample size is *sufficiently large*, a condition which remains vague, and in civil engineering with sample sizes of 0, 1, 2 or 3 is often lacking. This is the point where subjective probability enters engineering. When such a switch of interpretation is undertaken, we believe that it is the responsibility of the engineer to put it in the open. Otherwise the meaning of probability in the final result is lost or at least obscured.

Turning to subjective probability, we first mention that schemes have been developed that allow one to deduce it from decision theoretic principles, assuming *rational behavior* of the agent. This has been done to obtain operational ways of extracting the personal probability assessment of an agent/decision maker quantitatively. One way promoted by [167] has been the notion of *indifference price.* To determine a probability of an event S, the decision maker is required to imagine a gamble which pays one monetary unit if event S occurs and zero otherwise. The decision maker surely would buy the gamble at the price of zero units, but surely not for more than one unit. Raising the lower bound for the price and lowering the upper bound should eventually lead to a price at which the decision maker is indifferent to buying the gamble or not. This indifference price is the probability $P(S)$ of the event. It is argued that the indifference price also equals the minimal price $\underline{P}(S)$ at which the decision maker is willing to sell the gamble, as well as the maximal buying price $\overline{P}(S)$.

This has been a point of critique, because real world persons do not behave strictly rational in this sense and often lack the information to decide about the minimal selling and maximal buying price, see e.g. the discussions in [37, 44]. Thus an interval $[\underline{P}(S), \overline{P}(S)]$ appears to be a more accurate description of a decision maker's information. This line of argument directly leads to *imprecise probability*, probability intervals, and lower and upper probabilities [32, 199, 203].

In practical engineering applications, *elicitation* of probabilities from *experts* is the paradigm for obtaining subjective probabilities quantitatively. We refer e.g. to [44, 113, 163]. For the purpose of demonstration, we exemplify an elicitation procedure for a continuous probability distribution, following [150].

Example 2.2. To determine the distribution of a random variable X, experts are first asked to determine a lower bound L, an upper bound U and a modal value M. Next, they are required to assign probabilities to the intervals listed in Figure 9. The procedure is intended to unveil more details

(a) $[L, M]$
(b) $[L, (L+M)/2]$
(c) $[(M+U)/2, U]$
(d) $[L, (L+3M)/4]$
(e) $[(3M+U)/4, U]$

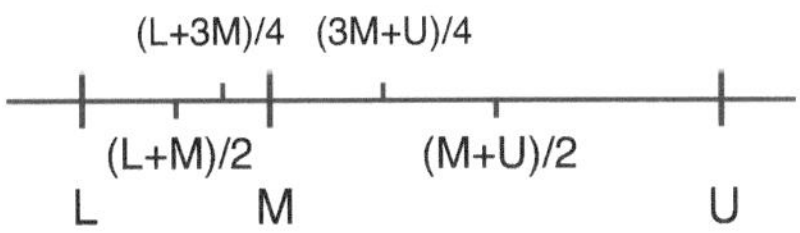

Figure 9. Elicitation intervals, according to [150].

near the modal value M. As an example, we use here data that were elicited from the participants of the CISM Advanced School on Nondeterministic Mechanics in 2011. Participants were asked to estimate the price (in Euro) of the dinner they intended to consume on the same evening. The following bounds were given:

$$L = 5, \ M = 15, \ U = 35.$$

The following probabilities were estimated from their answers:

(a) $[5, 15] : P = 0.5$,
(b) $[5, 10] : P = 0.15$,
(c) $[25, 35] : P = 0.1$,
(d) $[5, 12.5] : P = 0.15$,
(e) $[20, 35] : P = 0.25$.

This implies the probabilities for the non-overlapping subintervals displayed in Table 3; to plot them in a histogram, the heights have to be adjusted so that the probabilities correspond to the areas of the bins.

Table 3. Elicited probabilities and calculation of histogram.

bin	bin center z_j	probability p_j	height of bin
$[5, 10]$	7.5	0.15	0.03
$[10, 12.5]$	11.25	0.15	0.06
$[12.5, 15]$	13.75	0.2	0.08
$[15, 20]$	17.5	0.25	0.05
$[20, 25]$	22.5	0.15	0.03
$[25, 35]$	30	0.1	0.01

The next step consists in fitting a probability distribution. For the sake of simplicity, we demonstrate the procedure with a lognormal distribution $\mathcal{L}og\mathcal{N}(\mu, \sigma^2)$ which is easily fitted using the mean and variance of the logarithms of the bin centers: $\mu \approx \sum_j \log z_j\, p_j \approx 2.7122$, $\sigma^2 \approx \sum_j (\log z_j - \mu)^2\, p_j$, $\sigma \approx 0.4057$. Displaying the histogram together with the fitted lognormal distribution reveals a satisfactory fit (Figure 10). Normally, the elicitation procedure would include a discussion with the participants (experts) and the possibility to incorporate a second turn, *stepping back* with the intention to get a distribution which finds the approval of the experts.

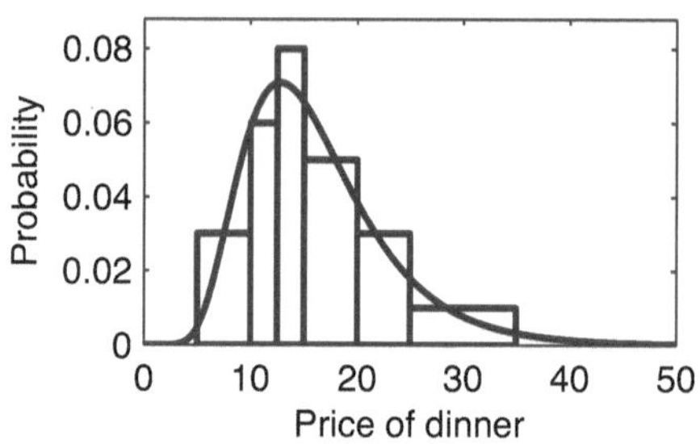

Figure 10. Histogram and lognormal fit.

Finally, we should not fail to mention the *Bayesian* approach to assessing probability distributions. From the Bayesian viewpoint, everything is a random variable, including the parameters, say Θ, of the distributions of the original variables, say X, to be assessed. The Bayesian approach has interpretations both in the frequentist as well as the subjective setting. In the civil engineering literature, it has been found useful for combining expert knowledge with sample data [38, 77, 112, 161, 172]. The expert knowledge may be based on known frequencies or on subjective estimates and is encoded in the *prior distribution* of the parameter θ. Sample data x (or again expert estimates obtained in situ) are then used to produce a *posterior distribution* of the distribution parameter θ according to Bayes' rule, loosely stated as

$$p(\theta|x) = \frac{p(x|\theta)p(\theta)}{p(x)}.$$

The data vector x may have length 1, or may consist in a large sample of size n, so that the Bayesian procedure may accommodate single estimates up to sample sizes satisfying frequentist requirements. The assessment of the distribution of the original variable X is completed by the *Bayes estimate* of its distribution parameter

$$\hat{\theta} = \int_{-\infty}^{\infty} \theta\, p(\theta|x)\, \mathrm{d}\theta.$$

Further powerful concepts start out from here, like odds ratios, high posterior density regions, posterior predictive distributions, and so on. We refer e.g. to [15, 66, 112, 198] for further information on Bayesian methods.

Sets of probability measures. Sets of probability measures can arise both in a frequentist as well as in a subjectivist approach. In a frequentist setting, sets of probability measures arise as sets of fitted distributions: in fact, confidence regions for the distribution parameters imply that parametrized *families* of distributions are employed. Further, *robust statistics* is based on distributions neighboring a given distribution, see e.g. [84].

As we have just seen above, sets of probability distributions are inherent in Bayesian statistics (each parameter value θ defines a distribution of the original variable). In order to avoid implausible determinations due to the choice of a prior distribution, families of prior distributions have been employed (robust Bayesian methods), as well as fuzzy prior distributions [198].

Further, referring to the decision theoretic foundation of probability, replacing the indifference price by an interval $[\underline{P}(S), \overline{P}(S)]$ leads to interval-valued probabilities, which again imply that a set of probabilities is considered.

Multicriteria decision making and extensions of expected utility theory have already been mentioned in Section 2.1. In this respect, we refer to [88, 126]. For decisions under uncertainty in a fuzzy framework, see [209].

Random sets. In a frequentist interpretation, this might correspond to a sample of size n of interval data for a parameter A, the probability weight being approximated by the relative frequency. The difference to a histogram then is that the focal sets A_i may overlap. In geotechnical engineering, for example, interval data arise as ranges of rock parameters associated with certain rock classes. These rock classes in turn may be the outcomes, obtained with a certain frequency from in situ measurements. This approach has been proposed and elaborated by [191, 192]. In a subjectivist interpretation, the focal sets A_i may be (possibly conflicting) estimates given by different experts and the weights might correspond to each expert's relative credibility.

Random sets have turned out to be useful for bracketing probability estimates given by different sources [100, 188] as well as for combining information of different type, due to the observation that every histogram, every interval and every fuzzy set can be viewed as a random set, without a need for artificial transformations [69, 99].

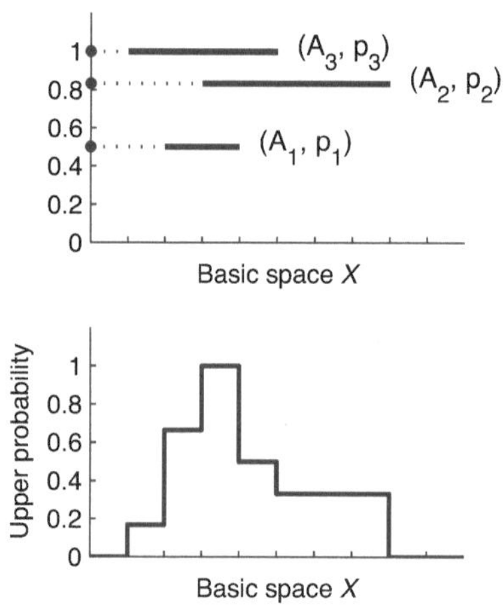

Figure 11. A random set (top) and its contour function (bottom). The three dots on the vertical axis symbolize the underlying 3-point probability space.

A good visualization of a random set can be given through its *contour function* on the basic space $\mathcal{X}$, assigning each singleton x its plausibility:

$$x \to \overline{P}(\{x\}).$$

It is simply obtained by adding the probability weights p_i of those focal sets A_i to which x belongs. Figure 11 shows a random set and the resulting contour function where weights have been chosen as $p_1 = 1/2, p_2 = 1/3, p_3 = 1/6$. For later purposes it is useful to plot the focal sets A_1, A_2, A_3 in a coordinate system at heights $p_1, p_1 + p_2, p_1 + p_2 + p_3 = 1$.

When $\mathcal{X}$ is the set of real numbers, the random set defines a so-called *probability box* which is bounded by the upper and lower distribution functions

$$\overline{F}(x) = \overline{P}(-\infty, x], \quad \underline{F}(x) = \underline{P}(-\infty, x], \tag{6}$$

see [51]. Any distribution function $F(x)$ that arises from one of the probability measures from the set $\mathcal{M}$ (cf. Equation (3)) is necessarily bounded by the probability box: $\underline{F}(x) \leq F(x) \leq \overline{F}(x)$. Thus the probability box is a good representation of the variability of a quantity described by a random set, see Figure 12 (left).

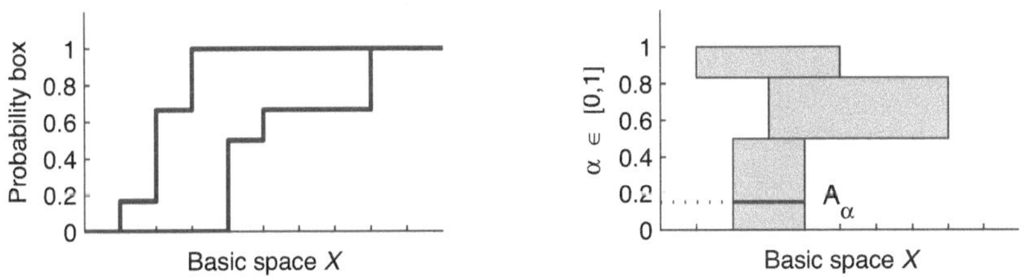

Figure 12. Probability box (left) and representation as an infinite random set (right).

We now turn to the more general case of infinite random sets. As shown in [2], for most applications in reliability theory it suffices to consider *closed random sets* defined on the probability space $\Omega = [0, 1]$ with the uniform probability distribution. Recall that the uniform probability P on the interval $[0, 1]$ assigns to each subinterval of $[0, 1]$ its length, that is, $P[\alpha, \beta] = \beta - \alpha$ for $0 \leq \alpha \leq \beta \leq 1$.

Thus – in our narrower sense – a random set is a function

$$[0, 1] \to \mathcal{X} : \alpha \to A_\alpha$$

where each A_α is a closed subset of the real line $\mathcal{X}$ as basic space.[3] Often, the *focal sets* A_α are just closed intervals. In the multivariate case, we would take the d-dimensional unit hypercube $\Omega = [0, 1]^d$ as probability space and the focal sets as closed subsets of d-dimensional coordinate space.

[3]The rigorous technical requirement is that the sets $\{\alpha \in [0, 1] : A_\alpha \cap S \neq \emptyset\}$ are measurable for every measurable subset $S \subset \mathcal{X}$.

In generalization of formulas (2) and (4) we define the *plausibility* of an event S as

$$\overline{P}(S) = P\big(\alpha \in [0,1] : A_\alpha \cap S \neq \emptyset\big) = \int_{\{\alpha\in[0,1]:A_\alpha\cap S\neq\emptyset\}} \mathrm{d}\alpha \qquad (7)$$

and the *belief*

$$\underline{P}(S) = P\big(\alpha \in [0,1] : A_\alpha \subset S\big) = \int_{\{\alpha\in[0,1]:A_\alpha\subset S\}} \mathrm{d}\alpha. \qquad (8)$$

Again, *the contour function* is defined by

$$x \to \overline{P}(\{x\}) = \int_{\{\alpha\in[0,1]:x\in A_\alpha\}} \mathrm{d}\alpha$$

and the upper and lower distribution functions are given by formula (6). An example of the contour function of an infinite random set can be seen in Figure 13 (left) with the corresponding probability box (right). The probability box may be viewed as the envelope of all cumulative distribution functions encoded in the random set.[4]

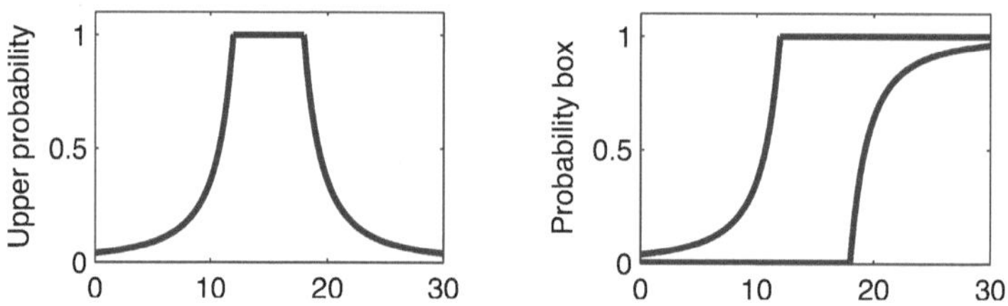

Figure 13. Random set for a real-valued parameter: contour function and probability box (values from the dinner price example below).

Example 2.3. (a) Every Dempster-Shafer structure can be viewed as an infinite random set, in which the focal sets S_i arise as certain A_α with repetition. For this purpose, we subdivide the interval $[0,1]$ into n subintervals of length p_i and put

$$A_\alpha = A_1,\ 0 \le \alpha \le p_1, \quad A_\alpha = A_2,\ p_1 < \alpha \le p_1 + p_2,$$

[4]Technically, these distribution functions are those coming from the set of all probability measures of the form $P(S) = \int_0^1 P_\alpha(S)\,\mathrm{d}\alpha$, $P_\alpha \in \mathcal{M}(A_\alpha)$, cf. (3).

and so on until $A_\alpha = A_n$, $p_1+p_2+\ldots+p_{n-1} < \alpha \leq 1$ (see Figure 12, right). It is not difficult to see that the two expressions for belief and plausibility in formulas (2), (7) and (8) coincide.
(b) Fuzzy sets: every normalized fuzzy number[5] can be seen as a random set; the sets A_α are just the α-level sets. It is not difficult to prove that the possibility measure of a subset S of the real line coincides with its plausibility: $\overline{P}(S) = \sup\{\pi(x) : x \in S\}$.
(c) Random variables. A random variable X can be reconstructed as a random set on $[0,1]$ by putting $A_\alpha = F^{-1}(\alpha)$ where $F(x)$ is the distribution function of the random variable and F^{-1} its inverse or, more generally, pseudo-inverse. The plausibility and belief of an event B coincide with the probability of the outcome $P(X \in B)$.

Example 2.4. Frequentist elicitation of a random set. During the CISM Advanced School on Nondeterministic Mechanics in 2011, the author asked the participants to give an interval estimate (minimum – maximum) of the price of a meal at a wellknown international fastfood chain. The results of the survey (sample size 33), the proposed intervals, are depicted in Figure 14. For easier visualization, they have been ordered according to their left endpoints; differences in height correspond to relative frequency.

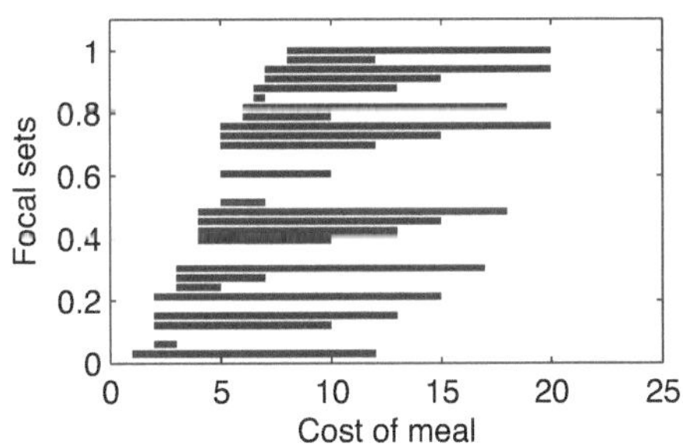

Figure 14. Elicited interval estimates for the price of a fastfood meal.

Interpreting these intervals as focal sets of a finite random set, one obtains the probability box and contour function, respectively, depicted in Figure 15.

The p-box gives estimates of the upper and lower cumulative distribution function. For example, the upper probability that the price is at most

[5] A normalized fuzzy number is a fuzzy subset of the real line with upper semi-continuous membership function $\pi(x)$ that attains the value 1.

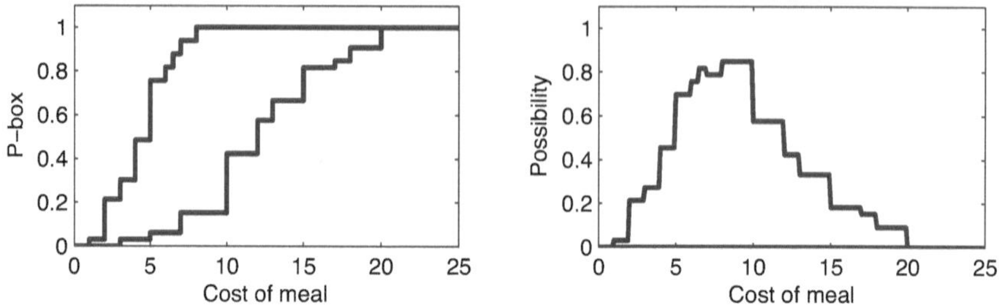

Figure 15. Price of fastfood meal: p-box and contour function.

5 Euros is estimated at 0.76, the lower probability at 0.06, i.e., the probability of a price of at most 5 Euros is enclosed by the interval $[0.06, 0.76]$ (according to the collective opinion of the 33 participants of the Advanced School). Using the relation $\overline{P}(S) = 1 - \underline{P}(S^c)$, we can also read off that the probability of a price above 17 Euros is supposed to lie in the interval $[0, 0.15]$. The contour function gives a visual impression of the upper probabilities of individual prices. The fact that the contour function remains below the value 1 indicates the lack of a commonly agreed upon core of possible prices.

Example 2.5. Random sets constructed from Tchebycheff's inequality. A major application of the tools outlined above is the construction of random sets from minimal information on a random variable X. Let $\mu = \mathrm{E}(X)$ be its expectation and $\sigma^2 = \mathrm{V}(X)$ its variance. Tchebycheff's inequality asserts that

$$P\big(|X - \mu| > d_\alpha\big) \le \alpha \quad \text{with} \quad d_\alpha = \sigma/\sqrt{\alpha} \tag{9}$$

for $\alpha \in (0, 1]$. Let

$$A_\alpha = [\mu - d_\alpha, \mu + d_\alpha]. \tag{10}$$

By Tchebycheff's inequality, the probability of A_α is greater or equal to $1 - \alpha$, while the probability of its complement A_α^c is less or equal to α. Thus A_α will contain approximately a fraction of $1 - \alpha$ of the realizations of the random variable X, e.g., $A_{0.05}$ contains about 95%, $A_{0.10}$ contains about 90%, etc. This is a conservative, non-parametric estimate valid for whatever distribution of the random variable X. It encodes the minimal information that can be extracted from the expectation and the variance of a random variable without further parametric assumptions.

We formalize this information as an infinite random set (actually a random interval) $\alpha \to A_\alpha$ on the space $\Omega = (0, 1]$, equipped with the uniform probability distribution. By construction, the following formulas for the belief in the set A_α and the plausibility of its complement A_α^{c} hold:

$$\begin{aligned}\underline{P}(A_\alpha) &= \int_{\{\beta\in[0,1]:A_\beta\subset A_\alpha\}} \mathrm{d}\beta = 1-\alpha \leq P(A_\alpha),\\ \overline{P}(A_\alpha^{\mathrm{c}}) &= \int_{\{\beta\in[0,1]:A_\beta\cap A_\alpha^{\mathrm{c}}\neq\emptyset\}} \mathrm{d}\beta = \alpha \geq P(A_\alpha^{\mathrm{c}}).\end{aligned}$$

Thus we may say that our degree of belief in $A_{0.05}$ is 0.95, while the complement of $A_{0.05}$ has plausibility degree 0.05. Further, the probability box, see (6), can be calculated explicitly:

$$\overline{F}(x) = \begin{cases} \frac{\sigma^2}{(x-\mu)^2}, & x < \mu - \sigma,\\ 1, & x \geq \mu - \sigma, \end{cases} \qquad \underline{F}(x) = \begin{cases} 0, & x < \mu + \sigma,\\ 1 - \frac{\sigma^2}{(x-\mu)^2}, & x \geq \mu + \sigma. \end{cases}$$

Returning to the example of the estimated price of a dinner, suppose the information (e.g., elicited from one participant) is a mean value of $\mu = 15$ Euros and a coefficient of variation of 20% (i.e., a standard deviation of $\sigma = 3$. Encoding this type of information in a Tchebycheff random set yields the contour function and the p-box depicted in Figure 13. If needed, one may truncate the random set to avoid impossible values (e.g., values below zero in the dinner price example).

We remark that the use of Tchebycheff's inequality combined with *finite random sets* was initiated in [141]; more detailed interpretations can be found in [60, 56, 143]. Of course, Tchebycheff's inequality can be replaced by narrower estimates if more is known about the distribution of the given random variable, for example, symmetry or unimodularity (cf. [159]).

Fuzzy sets. In engineering and in risk analysis applications, probabilistic models have been criticized as requiring more input from the designing engineer or the decision maker than could be plausibly provided – or that would be reasonably required for a rough estimate. In particular, the requirement that probabilities have to add up to 1 causes the problem that probabilities of events change when additional events are taken into consideration. Further, probabilities have to be set up in a consistent way, e.g. satisfying the rule $p(S\cup T) = p(S)+p(T)-p(S\cap T)$ and thus do not admit incorporating conflicting information (in the sense that the intersection of an event S with its complement can never have positive probability). Fuzzy set theory appears to provide a resolution of these difficulties in as much as it admits much more freedom in modeling. Further, fuzzy sets may be used to model vagueness and ambiguity.

The notion of *possibility* provides an interpretation for a fuzzy set and an operational method of constructing it, as we wish to argue here. We imagine that a fuzzy set describing the uncertainty of a real-valued parameter has to be designed. The procedure proposed below is based on a scale $0 \leq \alpha \leq 1$. The linguistic meaning of the α-values is specified verbally by the designing engineer or the decision maker in advance, but then remains fixed during the whole modeling process (for example, $\alpha = 1$ signifies the standard value of the parameter, $\alpha = 2/3, 1/3, 0$ might indicate high, medium, and low *degree of possibility*).

One would start by specifying the standard value a_S of the parameter, in engineering terminology often referred to as the *deterministic approximation* and assign degree of possibility $\alpha = 1$ to it. Next, possible deviations of the parameter from the standard value are taken into account, corresponding to decreasing degree of possibility, until the minimal and maximal values, which are assumed with very small degree of possibility, are reached at level $\alpha = 0$, see Figure 16.

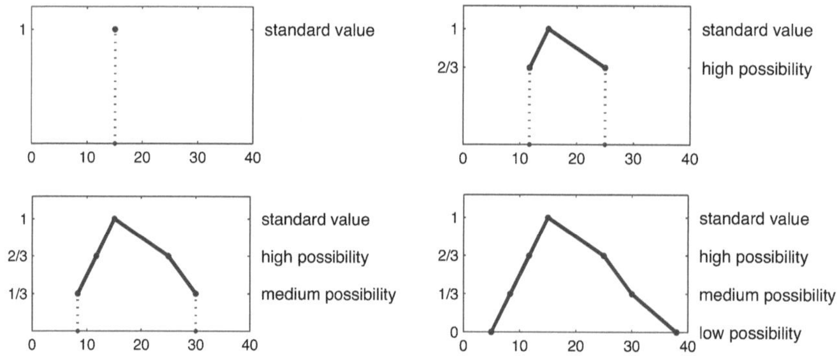

Figure 16. Construction of a fuzzy set (values from dinner price example below).

The notion of *possibility* can be given an operational meaning, using the language of subjective risk assessments. Thereby, it is envisaged that the risks leading to parameter fluctuations at the corresponding possibility level are established in an analysis of scenarios. The level sets correspond to the bounds a parameter attains under a certain risk level, see Figure 17.

Example 2.6. We return to modeling the uncertainty of the estimated dinner price by means of a fuzzy set. The participants of the CISM Advanced School agreed on the standard value of 15 Euros. Certain risks/chances for a higher or lower price have been identified with their consequences as follows: ending up in a tourist trap type restaurant (high risk, increase of price up to 10 Euros); having an intentionally cheating waiter (medium risk, additional

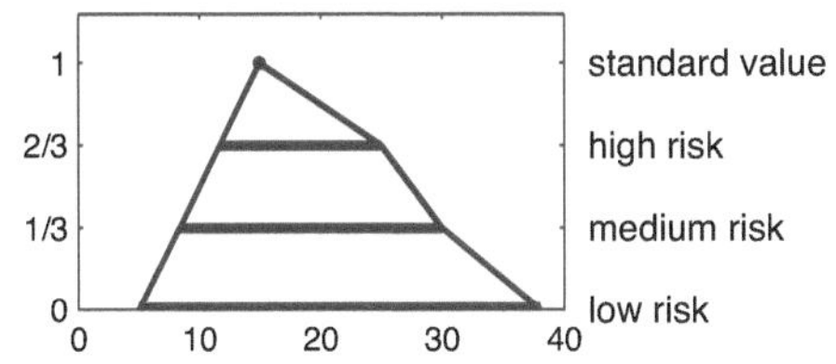

Figure 17. Fuzzy set describing interval bounds at given risk levels.

price increase of 5 Euros); encountering an unintentionally miscalculating waiter (low risk, additional price increase up to 8 Euros). There are also chances for a lower price: ending up in a good, low budget restaurant for locals (reduces price up to 3.5 Euros); meeting a friendly owner who offers the drinks on the house (further reduction of 3.5 Euros); unintentionally miscalculating waiter (possibly a further 3 Euros less). The data are summarized in Table 4. One may the construct the membership function of a fuzzy set starting at the standard value of 15 Euros with degree of possibility 1, and successively plot the maximal increase at the estimated risk level (degree of possibility) and join the points by straight lines. This procedure is depicted in Figure 16. Variations in the setup are possible (e.g., assigning

Table 4. Variations of dinner price: risks and chances with consequences and degrees of possibility.

risk	assessment	possibility	price increase
tourist trap	high risk	2/3	up to +10
cheating waiter	medium risk	1/3	up to +5
miscalculation	low risk	0	up to +8
chance	**assessment**	**possibility**	**price decrease**
low budget	high chance	2/3	up to −3.5
generous owner	medium chance	1/3	up to −3.5
miscalculation	low chance	0	up to −3

other degrees of possibility to the linguistically assessed risks, constructing a fuzzy set with horizontal lines and jumps). To be operational, it just must be made clear from the start how the visualization has to be translated.

Of course, in an engineering risk analysis, more effort has to be invested in the identification and quantification of the risks. We refer to [105] where the method has been demonstrated in the case of network analysis of a civil engineering project. Similar procedures have been proposed by [11]. There is also a normative approach which uses pre-shaped, parametrized membership functions as well as the suggestion to use elicitation procedures; see [12, 40, 163, 209, 210] for further details.

An alternative way of establishing the semantics of *possibility* is to start from the notion of *potential surprise* and to define possibility as its complementary notion [180] or as a transformed quantity [129] thereof. For a decision-theoretic foundation, see [41], for possibility as a transformation of probability, see [42].

Lower and upper previsions. A behavioral interpretation of lower and upper previsions has been elaborated by Walley [199], who also set up an axiomatic system of lower and upper previsions, derived from certain principles, such as *avoiding sure loss.* Previsions are functionals defined on gambles, thus the betting analogy (already used in the operational definition of subjective probability) comes in naturally. The lower prevision $\underline{\mathrm{E}}(g)$ of a gamble $g(X)$ is the supremum buying price (the largest sum the decision maker is willing to pay for the gamble), while the upper prevision $\overline{\mathrm{E}}(g)$ is the infimum selling price. Lower and upper probabilities of events are obtained as lower and upper previsions of their indicator functions. In contrast to subjective probability (which would translate into linear previsions), the decision maker is not obliged to end up with a single number (the indifference price), but has the freedom to remain undecided about the probability of an event, within the bounds given by its lower and upper probability. For details on interpretations and relations with other concepts of uncertainty, see [199].

2.3 Axiomatics

Referring to probability theory, it has been emphasized by Popper [158, Section 71] that a formal mathematical system of axioms and postulates is required in order to approach the problem of relations between the different interpretations of probability. All the more so, this applies to the problem of comparing the different theories of uncertainty. By now, all these theories can be based on appropriate axioms, the most well known being the Kolmogorov [98] axioms of probability (though criticized in [158]) and the Choquet [29] axioms of capacities.

We do not wish to give a detailed exposition of the axioms here, but just highlight a few of them, showing that they lead to different combination rules in the various models of uncertainty. This may serve as a means of distinguishing the models as well as a guiding criterion to decide which model should be selected for what purpose.

The axioms fix the algebraic properties of the corresponding set functions. For example, probability measures p define additive set functions, that is,

$$p(S \cup T) = p(S) + p(T) - p(S \cap T)$$

for sets S, T. This is not true of possibility measures π, which in turn satisfy

$$\pi(S \cup T) = \max\{\pi(S), \pi(T)\}.$$

Both probability measures and possibility measures are special cases of plausibility measures η which enjoy the more general property

$$\eta(S \cup T) \leq \eta(S) + \eta(T) - \eta(S \cap T).$$

All these set functions are contained in the largest class of monotone set functions μ, also called fuzzy measures, characterized by the property

$$\mu(S \cup T) \geq \max\{\mu(S), \mu(T)\}.$$

Suitably completing these algebraic properties to full systems of axioms, it is actually possible to *characterize* all these type of measures (and many more). The following hierarchical relations obtain:

$$\begin{array}{lcl} \text{probability} & \rightarrow & \text{plausibility} \\ \text{possibility} & \rightarrow & \text{plausibility} \\ \text{plausibility} & \rightarrow & \text{monotonicity.} \end{array}$$

This means that every probability measure is a plausibility measure, and so on (thus plausibility theory is more general than probability theory, in the sense that it admits a larger class of measures). Probability measures and possibility measures are in no implication relation in either direction.

Alternatively, if one starts from the notion of random sets, probability would correspond to singletons as focal sets, possibility to nested focal sets, and plausibility to arbitrary focal sets. These short indications may suffice; a discussion of the complete axiomatic systems as well as much more elaborated hierarchies can be found in [35, 96, 200, 201].

2.4 Numerics

Practically all engineering models are input-output systems. Given certain input values (model parameters, initial conditions, dimensions, etc.) the model produces output values (displacements, stresses, costs, etc.). In other words, the model is a function F that assigns to the input data A

certain output values $F(A)$. Both A and $F(A)$ may be multidimensional (for simplicity, we shall consider $F(A)$ as one-dimensional in the sequel – corresponding to a single component of a multidimensional output). Often, the function F is a computer code, in which case the output is obtained as a numerical approximation. If the input data consist of a single, *deterministic* data set, then the model produces a uniquely determined output. If the input data fluctuate, so does the output. If the fluctuation of the input is described by one or the other theories of uncertainty discussed so far, the fluctuation of the output should be captured on the same terms. This is the issue of this section: how is data uncertainty propagated through an input-output system?

Deterministic values. If a is an expert estimate of some input parameter A, the output is just the value $F(a)$ of the function F at a. In this framework, the effects of the uncertainty of the input parameters can still be assessed by performing a *sensitivity analysis*. In its classical form, sensitivity analysis means the study of the derivatives of the function F at the fixed value a of interest, that is, the linear approximation to the change in output when the input parameter a is changed to a nearby value b:

$$F(b) \approx F(a) + \frac{\mathrm{d}F}{\mathrm{d}a}(a)(b-a)$$

or in case the parameter $a = (a_1, \ldots, a_n)$ is n-dimensional,

$$F(b) \approx F(a) + \sum_{i=1}^{n} \frac{\partial F}{\partial a_i}(a)(b_i - a_i)$$

where the approximation error is of order $|b-a|^2$. The sign and size of the partial derivative $\frac{\partial F}{\partial a_i}(a)$ is an indicator of the influence of the i-th component of the parameter a on the output. For more details on sensitivity analysis, see Section 4.

Intervals. If A is an interval, the functional evaluation $F(A)$ results in a set of values (an interval, if F is continuous and one-dimensional). In general, both A and $F(A)$ could be sets of arbitrary geometry. In interval arithmetic, one would bound these sets by the smallest multidimensional intervals (boxes) that contain them (see [128]). In any case this approach represents the full range of the possible output values without further fine structure.

Probability measures and random sets. Given a single probability measure and a (measurable) map F, the output probabilities are determined through the induced image measure, that is, $P(F(A) \in S) = P(A \in F^{-1}(S))$. Though the distribution of the random variable $F(A)$ can be computed by this prescription in principle, this is practically impossible as soon as F attains a rather moderate complexity. *The* numerical method for approximating the output distribution by means of an artificially created sample is *Monte Carlo simulation.* We refer e.g. to [15] as well as the survey articles [80, 175].

In case the uncertainty of the input is modeled by a set $\mathcal{M}$ of probability measures, the map F induces a set of probability measures as well, namely the collection of all image measures, obtained from $\mathcal{M}$ under this map. The computation of lower and upper probabilities turns into an optimization problem.

If the input is described by a random set with focal sets $A_i, i = 1, \ldots, n$ and probability weights $m(A_i)$, the output is again a random set which consists of the focal sets $F(A_i), i = 1, \ldots, n$, supplied with the original weights $m(A_i)$ (or sums of weights if case some of the image sets $F(A_i)$ coincide). In case the sets $F(A_i)$ are intervals, their boundaries can be found by optimization (minimizing/maximizing the function F on A_i). The determination of lower and upper probabilities is then a combinatorial task involving the formulas for belief and plausibility. For further details on computational methods with random sets, see [54, 59].

Fuzzy sets. The question of propagating the uncertainty of a fuzzy input A through a function F needs some more explication. If the input is a fuzzy set with membership function $\pi_A(a)$, say, the output will also be a fuzzy set, described by a membership function $\pi_{F(A)}(b)$. According to the *Zadeh extension principle* [206], it is given by

$$\pi_{F(A)}(b) = \sup \big\{ \pi_A(a) : F(a) = b \big\}.$$

In case the input consists of a vector of parameters $A = (A_1, \ldots, A_m)$, the extension principle takes the form

$$\pi_{F(A)}(b) = \sup \big\{ \min \big(\pi_{A_1}(a_1), \ldots, \pi_{A_m}(a_m) \big) : F(a_1, \ldots, a_m) = b \big\}.$$

Note that this comes naturally from the possibility-theoretic interpretation: to compute the degree of possibility that $F(A)$ takes the value b, one has to look for all combinations $a_1, \ldots, a_m$ producing the value b; each single combination gets the smallest possibility among its participants, while b gets the supremum of all possibility degrees that can be obtained in this way. In case F is continuous and the α-level sets of $A_1, \ldots, A_m$ are compact ($0 < \alpha \leq 1$), this corresponds exactly to computing the range of the function

F on each α-level set,

$$F(A)^\alpha = F(A^\alpha), \quad \text{respectively}, \quad F(A_1, \ldots, A_m)^\alpha = F(A_1^\alpha, \ldots, A_m^\alpha).$$

When the A_j^α are intervals, in addition, the set $F(A_1, \ldots, A_m)^\alpha$ is an interval as well. The computation of its boundaries is then a task of global optimization: finding the minimum and maximum value of F on the set A^α. In any case, the procedure is consistent: if the input data are structured as stacked intervals, so is the output.

2.5 The multivariate case

In principle, the multivariate case, that is, the case when the input A has several components $A_1, \ldots, A_n$, has been covered by what has been said above – all applies to multidimensional intervals, random sets, fuzzy sets and multivariate distribution functions. However, the issue is how to model mutual dependence, correlation, interaction, influence of the different components. In addition, the task remains to model infinitely many components, as arising in spatial fields or temporal processes, when the parameters are functions of space and/or time.

There is a rather well established notion of independence in probability theory: two random variables are *independent* when their joint distribution function is the product of the individual (marginal) distribution functions. The situation is also clear in interval analysis: two parameters taking interval values are *non-interactive* when their joint behavior is described by the product of the two intervals (a rectangle), and *interactive* when their joint range is a proper subset of the product interval. This idea extends to fuzzy sets: Two fuzzy sets are non-interactive, when all their α-level sets are rectangles.

More precisely, given d univariate fuzzy sets $A^1, \ldots, A^d$, the non-interactive joint fuzzy set has the α-level sets

$$A_\alpha = A_\alpha^1 \times \cdots \times A_\alpha^d, \quad \alpha \in (0, 1].$$

Interactivity can be modeled by certain parametric restrictions on the α-level sets. To avoid combinatorial complications, consider interactivity of at most two out of the d variables. Since an α-level set of the form $A_\alpha^i \times A_\alpha^j$ is a homothetic image of the unit square, it suffices to give the definitions for $A_\alpha^1 = A_\alpha^2 = [0, 1]$. Following [62, 193], interactivity can be modeled by replacing the unit square by a diamond-shaped region, symmetric around one of the diagonals. Let $0 \le \rho \le 1$ and define the points $P_1, \ldots, P_4$ by

$$\begin{aligned} P_1 &= (\rho/2, \rho/2), & P_2 &= (1 - \rho/2, \rho/2), \\ P_3 &= (1 - \rho/2, 1 - \rho/2), & P_4 &= (\rho/2, 1 - \rho/2). \end{aligned}$$

Interactivity of *positive degree* ρ is modeled by taking the rhombus with corners $(0,0)$, P_2, $(1,1)$, P_4 as joint level set, while interactivity of *negative degree* $-\rho$ is modeled by the rhombus with corners $(0,1)$, P_1, $(1,0)$, P_3 as joint level set (Figure 18). This is a straightforward way of introducing

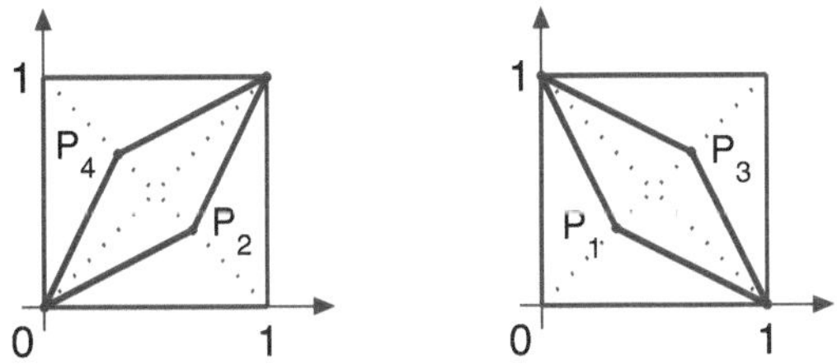

Figure 18. Positive/negative interactivity.

parametric interactivity in fuzzy sets. Of course, one can imagine replacing the diamond shapes by other geometric shapes.

Switching to random sets, the situation becomes more complicated. The concept of independence splits into a number of different concepts, all coinciding when the focal sets are singletons (i.e., when the random set is actually a scalar random variable). *Random set independence* is characterized by two properties: the joint focal sets are product intervals, and the joint weights are the products of the corresponding individual (marginal) weights.[6] *Strong independence* is obtained when the underlying set of joint probability measures is required to consist of product measures only. Going deeper into the structure of the underlying set of probability measures, many more notions of independence can be considered, one of the more prominent being *epistemic independence*. For further details on this, see [54, 59].

To model *dependence*, the notion of *copulas* has gained increased attention in the past decade. An n-dimensional copula is a multivariate probability distribution function on the n-dimensional unit hypercube whose marginals are uniform distributions [127]. Thus in the two-dimensional case, it is a function $C(u,v)$ of two variables $u, v \in [0,1]$ which satisfies

$$C(u,0) = 0, \quad C(u,1) = u, \quad C(0,v) = 0, \quad C(1,v) = v$$

[6] Note that when forming the joint random set of two marginal random sets A and B, all possible combinations of the focal sets of A with those of B have to be taken into account. In the infinite case, the joint random set is defined on $[0,1] \times [0,1]$ with the two-dimensional uniform distribution and focal sets $(\alpha, \beta) \to A_\alpha \times B_\beta$, $\alpha, \beta \in [0,1]$.

and which is two-monotone, i.e.,

$$C(u+h, v+k) + C(u,v) - C(u+h, v) - C(u, v+k) \geq 0.$$

The joint distribution $F_{XY}(x,y)$ of two random variables X, Y can be reconstructed from the marginal distributions $F_X(x)$, $F_Y(y)$ by means of a copula:

$$F_{XY}(x,y) = C(F_X(x), F_Y(y)).$$

The point is that given the marginals $F_X(x)$, $F_Y(y)$, parametric copluas can be found that produce a joint distribution with a prescribed correlation structure. In the situation of random sets, copulas can be used to introduce correlations on the basic probability weights, cf. e.g. [3, 4].

What concerns spatial fields or temporal processes, there is the fully developed probabilistic theory of stochastic processes, see e.g. [111]. Processes involving random sets have been considered in [74]. A theory of stochastic differential equations with random set parameters has been developed recently [170] with engineering applications [171]. A different approach based on set-valued integrands is in [89, 107, 109]. A fully intrinsic theory of processes of fuzzy sets (with interactivity replacing dependence) has not yet been developed. A useful substitute has been the notion of a fuzzy random process, either obtained as a stochastic process with fuzzy parameters or as a process of fuzzy numbers whose α-level sets are made up by *imbedded* stochastic processes. For more references on both approaches, see [21, 28, 108, 138]; a theory of fuzzy random processes with engineering applications has been developed in [120].

3 Application to an example from geotechnics

In this section, some of the methods will be shown at work in an example from geotechnics: an infinite beam on a linear elastic bedding. We begin by discussing the setup and a straightforward probabilistic model, which will be refined subsequently.

3.1 Deterministic/probabilistic dimensioning

The simplest, one-dimensional model is the so-called Winkler beam, extending along the real line with coordinate $x \in \mathbb{R}$. The displacement $u(x)$ is described by the bending equation

$$EI\, u^{IV}(x) + bc\, u(x) = q(x), \quad -\infty < x < \infty,$$

see e. g. [23, Section 61]. Here EI is the flexural rigidity of the beam, b its effective width, c the bearing coefficient of the foundation and $q(x)$ the loading. One may imagine that the beam describes a buried pipeline, the loading $q(x)$ resulting from the covering soil. The parameters EI and b of the beam may be considered as precisely known, whereas the soil properties c and q vary in an imprecisely known fashion. We will study the singular boundary value problem for the standardized equation

$$u^{IV}(x) + 4k^4 u(x) = p(x), \ -\infty < x < \infty \tag{11}$$

with $bc/EI = 4k^4$, $p(x) = q(x)/EI$, requiring that the solution should remain bounded at $\pm\infty$. In case k is a constant and $p(x)$ is an integrable function, both deterministic, its unique deterministic solution is given by

$$u(x) = \int_{-\infty}^{\infty} G(x,y)\, p(y)\, dy$$

in terms of its Green function

$$G(x,y) = \frac{1}{8k^3} e^{-k|x-y|} \big(\sin k|x-y| + \cos k|x-y| \big).$$

In case the load $q(x) \equiv q$ (and hence $p(x) \equiv p$) is constant, the displacement is constant as well and simply given by

$$u(x) \equiv \frac{p}{4k^4} = \frac{q}{bc}.$$

For the computational examples to follow we let the parameters vary around central moduli of $k = 10^{-2}$, $p = 10^{-8}$. Approximately, this corresponds to the case of a buried cast-iron pipeline with an effective diameter of 6 [cm], covered by about 100 [cm] of top soil ($q = 10$ [N/cm]) and bedded in loosely packed sand ($c \approx 6.7$ [N/cm^3]). The resulting overall displacement would amount to $u(x) \equiv 0.25$ [cm] in the deterministic case.

In a probabilistic design, one would assume that the input parameters are random variables. In a standard engineering approach, one would argue that their mean values are given by the deterministic design values; further, a coefficient of variation is assumed (for material properties, usually around 5%, for soil parameters up to 15%, cf. [161, 173]). In the simple example, the most uncertain parameters are q (mean $\mu_q = 10$) and bc (mean $\mu_{bc} = 40$); we take a coefficient of variation of 10%. Next, a type of distribution for the parameters has to be assumed. Both q and c are soil- and bedding related parameters, whose values strongly depend on what actually happens at construction site. Thus little evidence about the type of distribution is available. For the sake of presentation, we make the assumption that both parameters are normally distributed, that is, $q \sim \mathcal{N}(10, 1)$, $bc \sim \mathcal{N}(40, 16)$.

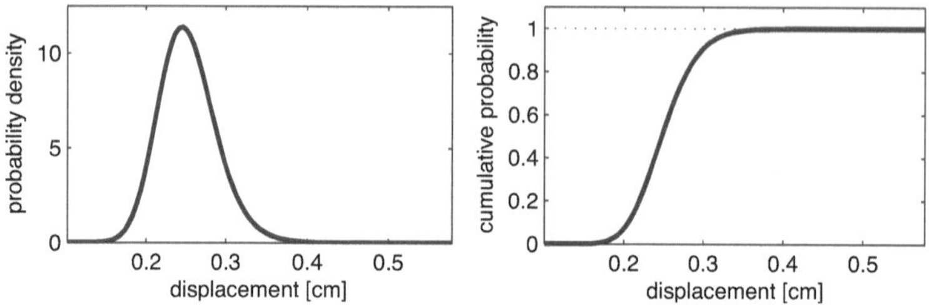

Figure 19. Probability density (left) and distribution function (right) of displacement under constant, but random load.

Under this assumption, we can compute the probability density of the displacement $u = q/bc$, see Figure 19, and read off the quantiles. For example, the probability that the displacement is larger than 0.5 [cm] is $\approx 3.9 \cdot 10^{-6}$.

It is quite clear that the model is too simple to be credible. In particular, the load will certainly not be given by a single, albeit random, value along the whole beam, but will rather vary from point to point.

This suggests describing the load as a random field $q(x), x \in \mathbb{R}$, and thus brings us to a second important aspect of probabilistic modeling in engineering. At each point x in space, the load $q(x)$ is assumed to be a random variable. To define the field, the joint distributions of the loads at any finite number of points $q(x_1), \ldots q(x_n)$ should be specified. The standard assumption in soil engineering is that the random field is stationary (i.e. the finite dimensional distributions are translation invariant) and Gaussian. In this case, the field is completely specified by the mean value μ_q and the second moments, i.e., the covariance $\mathrm{COV}(q(x), q(y))$ for any two points x, y. Due to stationarity, the covariance depends only on the distance $\rho = |x - y|$ of the points and is of the form

$$\mathrm{COV}(q(x), q(y)) = \sigma^2 C(\rho)$$

with the variance σ^2 and the so-called autocorrelation function $C(\rho)$. A typical autocorrelation function is of the form

$$C(\rho) = \exp\big(-|\rho|/\ell\big),$$

where ℓ is the so-called correlation length (available in the literature, e.g. [161]).

Thus for modeling the load as a stationary Gaussian field, we need to provide the mean value, the variance and the autocorrelation function. As above, the mean value is assumed to be $\mu_q = 10$; for the field variance we took $\sigma_q^2 = 4$. In the following, we assumed a moderate correlation length of $\ell = 100$ [cm] and took bc fixed at its deterministic design value 40 [N/cm^2]. A realization of the load and the corresponding displacement is shown in Figure 20; Figure 21 (left) shows the corresponding realization of the bending moment.

At this stage, we can demonstrate how the failure probability can be assessed. This requires a failure criterion. In this situation, failure is assumed to occur when the maximal stress in the beam exceeds the admissible stress (corresponding to the 0.1% yield strength, that is, the stress after which 0.1% plastic deformations remain). Thus the limit state is defined by $\sigma_{\max} \le \sigma_{\text{crit}}$ where $\sigma_{\max} = M_{\max}/W$. Here $M_{\max} = \max(EIu''(x))$ is the maximal bending moment in the beam. The quantity W is the section modulus of the cross section of the beam; for the purpose of demonstration, we assume a thin-walled circular pipe of diameter 6 [cm] and take $W \approx 5$ [cm^3]. The value of σ_{crit} can be taken from the literature; for cast iron, it is between 9800 – 16500 [N/cm^2]. Note that $\sigma_{\max}$ is a random variable resulting from the random field model of the beam. In order to demonstrate how $M_{\max}$ would be assessed probabilistically, a Monte Carlo simulation of $N = 500$ trajectories has been undertaken, resulting in an estimate for the distribution of $M_{\max}$. Figure 21 (right) shows the result, together with a fitted Gaussian distribution ($\mu \approx 2509.2$, $\sigma \approx 562.5$). The so-called safety distance is $Z = \sigma_{\max} - \sigma_{\text{crit}}$, the failure probability is $p_f = P(Z \le 0)$. Usually, the resistance is assumed to be normally distributed. The values mentioned above suggest a mean of 10000 [N/cm^2] for σ_{crit}. Assuming a coefficient of variation of 15%, we would conclude that $\sigma_{\text{crit}} \sim \mathcal{N}(10000, 1500^2)$ and $\sigma_{\max} \sim \mathcal{N}(2509.2/5, (562.5/5)^2)$. Thus Z is distributed according to $\mathcal{N}(10000 - 2509.2/5, 1500^2 + (562.5/5)^2)$. This results in a failure probability of $p_f \approx 0.14 \cdot 10^{-9}$. Of course, fitting other types of distribution would lead to different results.

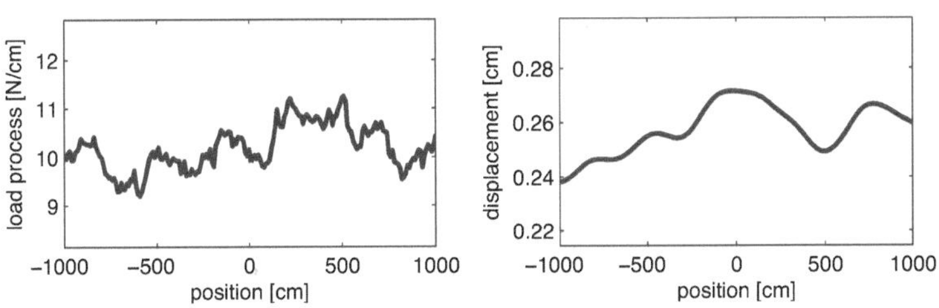

Figure 20. Random field model: trajectories of load process (left) and corresponding displacement (right).

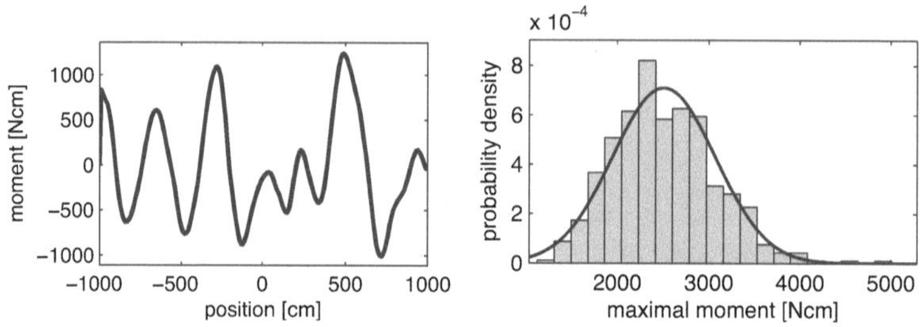

Figure 21. Random field model: trajectory of bending moment (left) and simulation of maximal bending moment (right).

The possibilities of more refined statistical methods, like confidence intervals for quantiles or robust estimates, are left aside here, and we close our brief survey of classical probability in this example.

3.2 Fuzzy set modeling

In order to present a fuzzy model of the elastically bedded beam, we assume that the parameters k and p are described by fuzzy numbers K and P. More generally, one could admit, e.g., loads of the form $p(x) = \sum_{i=1}^{n} a_i p_i(x)$ as a combination of fixed shape functions $p_i(x)$ with parameters a_i, which in turn can be taken as fuzzy numbers $A_i, i = 1, \dots, n$. We adopt here the simple case that both parameters are fuzzy constants and model the data as a non-interactive fuzzy vector with two components (K, P). This signifies that the joint membership function is given by the formula $\pi_{(K,P)}(k,p) = \min\{\pi_K(k), \pi_P(p)\}$, hence the level sets are two-dimensional intervals. We shall compute the fuzzy point values of the fuzzy response $u(x)$ by applying the Zadeh extension principle to the solution operator

$$(k, p) \to u(x) = L_x(k, p) = \frac{p}{4k^4},$$

see Subsection 3.1. By the discussion above, an α-level set of $L_x(K, P)$ is computed as the collection of the values of the solutions attained when the parameters vary in the respective level sets K^α, P^α. In fact, the Zadeh extension principle allows one to derive a fuzzy solution concept for which this is the unique fuzzy solution. For the details we refer to [136, 145].

For the sake of exposition in the example to follow, we take both K and P as triangular fuzzy numbers, centered around the values indicated in Subsection 3.1, namely

$$K = \langle \tfrac{1}{2}, 1, 2 \rangle \cdot 10^{-2}, \qquad P = \langle 0, 1, 2 \rangle \cdot 10^{-8},$$

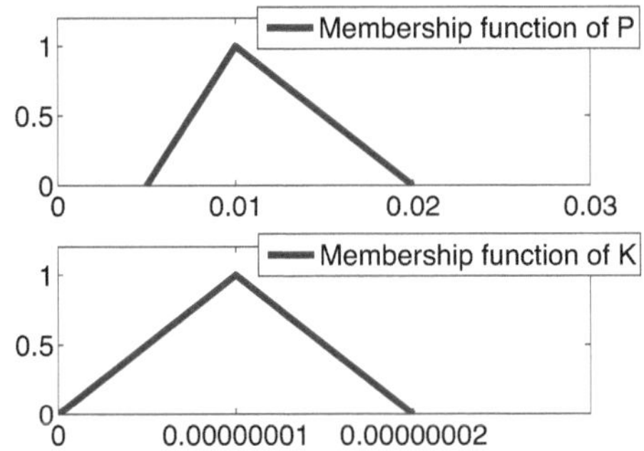

Figure 22. Fuzzy parameters K and P.

see Figure 22. Then the (constant) fuzzy solution is simply given by the fuzzy number $L_x(K,P) = P/4K^4$. It is depicted in Figure 23.

We observe that the fuzzy model consistently describes the fluctuations of the response in dependence on the data variability. In addition, the α-level structure provides a good picture of the sensitivity of the result.

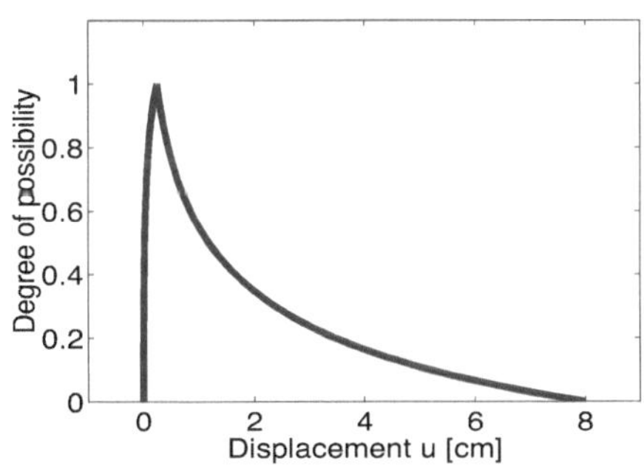

Figure 23. Fuzzy displacement u.

We now discuss a certain limitation of the fuzzy model. The semantics of fuzzy sets allows the parameter p to vary in the range given by its fuzzy description, in the example its support $[0, 2 \cdot 10^{-8}]$. If one permits non-constant realizations of the parameter p, the bounds predicted by the fuzzy output may be exceeded. Indeed, assume that p jumps from 0 to $2 \cdot 10^{-8}$ at the point $x = 0$. This means that we have to solve Equation (11) with a load

$$p(x) = \begin{cases} 0, & x < 0, \\ p = 2 \cdot 10^{-8}, & x > 0. \end{cases}$$

The corresponding displacement is

$$v(x) = \begin{cases} \frac{p}{8k^4}\, e^{kx} \cos kx, & x < 0, \\ \frac{p}{8k^4}\big(2 - e^{-kx} \cos kx\big), & x > 0. \end{cases}$$

Taking the admissible value $k = \frac{1}{2} \cdot 10^{-2}$, it is seen that the the graph of v exceeds the band described by the fuzzy displacement with constant parameters.

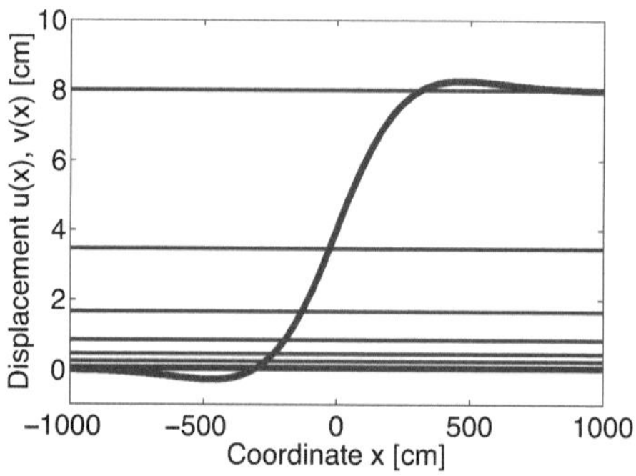

Figure 24. Fuzzy solution versus effect of localization.

This is an example of the effect, observed in other circumstances as well, that a localized parameter fluctuation may produce a response not predicted by the fuzzy model. In Figure 24, the 0-level set of the fuzzy displacement is indicated by the horizontals $u = 0$ and $u = 8$, the curve is the graph of v, and the other horizontals indicate level sets of the fuzzy displacement u for $\alpha = 0.2, 0.4, 0.6, 0.8$. Degree of possibility equal to one occurs at $u = 0.25$.

3.3 Random set modeling

As mentioned in Subsection 3.1, assuming that q and bc are normally distributed random variables is rather artificial. The available information consists of a nominal value and a coefficient of variation. One way of organizing this information is by means of a Tchebycheff random set as described in Subsection 2.2. If the loading q, say, is preliminarily viewed as a random variable with unknown probability distribution, but with expectation value

μ_q and variance σ_q^2, Tchebycheff's inequality asserts that the probability of the event

$$\left\{|q-\mu_q| > \frac{\sigma_q}{\sqrt{\alpha}}\right\}$$

is less or equal to α, where $\alpha \in (0,1]$. Let

$$Q(\alpha) = \left[\mu_q - \sigma_q/\sqrt{\alpha}, \mu_q + \sigma_q/\sqrt{\alpha}\right].$$

As outlined in Example 2.5, we may use the $Q(\alpha)$ to define a random set on the space $\Omega = (0,1]$, equipped with the uniform probability distribution, the Tchebycheff random set arising from μ_q and σ_q^2.

We take up the example of Subsection 3.1 and apply this construction to the loading q with $\mu_q = 10$ and $\sigma_q = 1$ (from a coefficient of variation of 10%). This results in a random set Q, whose contour function is depicted in Figure 25 (left).

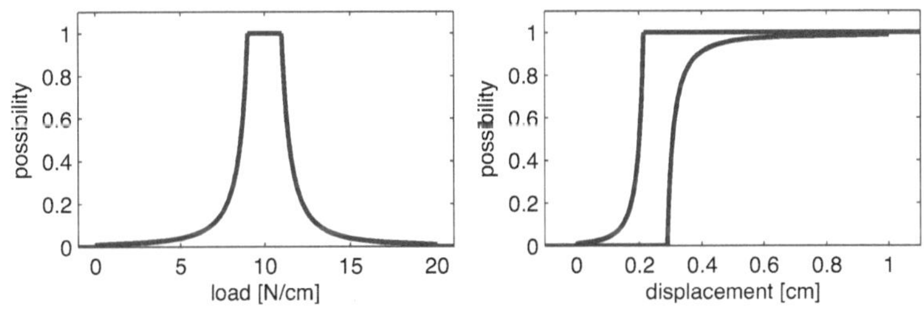

Figure 25. Tchebycheff random set for load (left) and probability box for resulting displacement (right).

In a similar way, we construct a Tchebycheff random set BC for the variable bc, using $\mu_{bc} = 40$ and $\sigma_{bc} = 4$. To form the joint random set (Q, BC), the dependence of Q and BC has to be modeled. To make computations easy, we settle for the so-called *fuzzy set independence*; that is, the joint random set is also defined on $\Omega = (0,1]$ with focal elements $Q(\alpha) \times BC(\alpha)$, $\alpha \in \Omega$ (and thus only focal elements corresponding to the same index α are combined). The random set data can be propagated through the mapping that gives the displacement $u(q, bc) = q/bc$, resulting in a random set U with focal sets $U(\alpha) = u(Q(\alpha) \times BC(\alpha))$, the set of values attained when (q, bc) range in $Q(\alpha) \times BC(\alpha)$.

The evaluation of the interval bounds for $U(\alpha)$ requires a global optimization. It is useful to describe the output random set as a probability box, which is bounded by the lower and the upper distribution functions

$$\underline{F}(x) = \underline{P}(-\infty, x], \qquad \overline{F}(x) = \overline{P}(-\infty, x].$$

The resulting probability box for the displacement is shown in Figure 25 (right).

The probability box immediately gives information on quantile ranges. For the event $A = \{U \geq 0.5\}$, for example, we get the probability interval $[\underline{P}(A), \overline{P}(A)] = [0, 0.04]$, which is more credible than the point estimate $P(A) \approx 3.9 \cdot 10^{-6}$ from the Subsection 3.1 (obtained under the stringent assumption that q and bc were normally distributed).

The even more realistic combination of the random field model with random set parameters will be described in the following subsection.

3.4 A hybrid model

A combination of stochastic differential equations with random set parameters has been recently worked out in [170]. This can be used in the dynamics of structures. Earthquake induced vibrations can be modeled by stochastic processes, like colored noise, whereas uncertain material parameters can be modeled by random intervals; see Subsection 5.3. We shall demonstrate a more modest hybrid model for the elastically bedded beam. The load q will be modeled as a random field as in Section 3.1, while the bedding parameter bc will be modeled as a random set. Of course, the model can be generalized to higher levels by also taking the field parameters σ_q and ℓ as random sets, etc. For the sake of simplicity, we shall take bc as an interval and the field parameters as in Section 3.1, i.e., $\mu_q = 10$, $\sigma_q = 2$, $\ell = 100$. For bc we choose the interval $[20, 40]$, which has the previously assumed mean value for bc as its upper boundary.

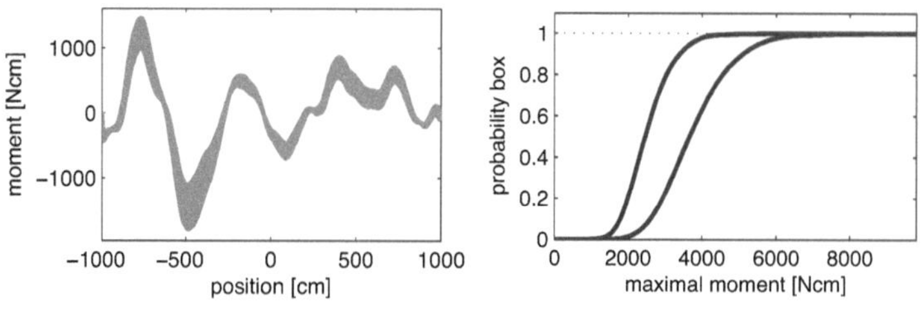

Figure 26. Hybrid model: Single interval trajectory of bending moment (left) and p-box for maximal bending moment (right).

The resulting output w will be a set-valued stochastic process; more precisely, each trajectory will be interval-valued. Figure 26 (left) shows a single trajectory of the bending moment. In order to assess the statistical properties of the output, a sample of trajectories has to be generated. From there, one can compute, e.g., the upper and lower distribution functions of the maximal bending moments in the beam. This is a critical quantity on which the failure criterion from Section 3.1 is based. A p-box of the maximal bending moment is shown in Figure 26 (right), based on $N = 500$ trajectories of the field. From the list of computed values (interpolated using a kernel smoother) one may obtain upper and lower probabilities that given limits are being exceeded, e.g.,

$$\underline{P}(M_{\max} > 6000) \approx 6.51 \cdot 10^{-5}, \qquad \overline{P}(M_{\max} > 6000) \approx 2.05 \cdot 10^{-2},$$
$$\underline{P}(M_{\max} > 8000) \approx 0, \qquad \overline{P}(M_{\max} > 8000) \approx 1.21 \cdot 10^{-3}.$$

Taking interval uncertainty into account typically leads to wider ranges for the input parameters. Consequently, the resulting output probability intervals are often large and possibly of a different magnitude than the sharp purely probabilistic estimates. If the decision maker is not satisfied with the relatively wide range of probabilities, more information about the distribution of the parameters has to be gathered. If the imprecise probability model properly reflects the available data, the results have to be accepted and should not be distorted by artificial probabilistic assumptions.

4 Sensitivity Analysis

An important issue in assessing the behavior of a structural model is sensitivity analysis. It aims at determining the input parameters that have the largest influence on critical output. As such it is already a core ingredient in the understanding of the behavior of a structure. In addition, it can be used as a first step in reliability analysis or in optimizing structural properties.

Sensitivity analysis does not necessarily require knowledge of the probabilistic properties of the input and thus is a nonparametric method. If the input-output function is explicitly given, one may use partial derivatives to assess the sensitivity, as briefly indicated in Subsection 2.4 and further elaborated, e.g., in [50, 152].

Otherwise, the influence of the individual variables can be studied by methods from probability theory or from imprecise probability theory, among them:

- Pinching: freezing individual variables at their central value and studying the change of variability in the output gives indications on the influence of the pinched variable. This can be done in fuzzy set models, random set models, or sampling based probabilistic models.
- Variance decomposition: This is a sampling based method that aims at classifying the importance of variables according to their contribution

to the overall variance of the output.

- Measures of correlation: Various measures of correlation between input variables and output variables can be used, such as the Pearson correlation coefficient, the partial correlation coefficient, the standardized regression coefficients as well as their ranked counterparts.
- Regression in metamodels: A metamodel, surrogate model or response surface is a function interpolating or approximating the actual input-output map at discrete points. This can, e.g., be obtained by linear or nonlinear regression. Methods of regression analysis can serve for assessing the importance of individual regressors in the regression function, for example, by means of partial coefficients of determination.

Variance decomposition can be briefly described as follows. Suppose $A = (A_1, \dots, A_n)$ is a vector of n independent random variables and a is its expectation value. The variance σ^2 of $F(A)$ can be approximated by

$$\sigma^2 \approx \sum_{i=1}^{n} \Big(\frac{\partial F}{\partial a_i}(a)\Big)^2 \sigma_i^2$$

where σ_i^2 is the variance of A_i. Then $|\frac{\partial F}{\partial a_i}(a)|\sigma_i/\sigma$ is an indicator of the fractional contribution of A_i to the total standard deviation σ, and thus an indicator of the influence of the uncertainty of the i-th component on the output. The method can be refined and extended by admitting correlations between the input variables or by assembling the input variables into groups (Sobol variance decomposition [183]).

What concerns the metamodel approach, assessing the importance of individual regressor variables is standard in regression analysis; we refer to [101, 123].

In the following subsections, we shall discuss the pinching strategies and the approach using correlation measures in more detail. For a survey of sampling based methods in sensitivity analysis, see [82].

4.1 An example from aerospace engineering

This example is taken from a research project in aerospace engineering[7] which involved the determination of the buckling load of the frontskirt of the ARIANE 5 launcher under various loading and flight scenarios as well as the development of tools for an encompassing sensitivity analysis. The presentation is based on the article [144].

[7] ICONA-projects 2006 – 2008, Intales GmbH Engineering Solutions and University of Innsbruck, supported by TransIT Innsbruck and EADS Astrium ST.

The frontskirt is a reinforced light weight shell structure. The computation of the decisive parameter indicating failure, the load proportionality factor (LPF), is based on a finite element model.[8] The full finite element model is composed of shell elements and solid elements, altogether with two million degrees of freedom. The load proportionality factor is computed by means of a path following procedure that follows bifurcations as long as possible until failure of the structure is reached, indicated by numerical breakdown of the program. The calculation of the output variable LPF – under a given single set of input parameters – takes about 32 hours on a high performance computer. In addition to the extremely high computational cost, the LPF may depend in a non-differentiable manner on some of the input parameters, especially variations in the geometry. A classical sensitivity analysis of the complete structure thus is not feasible.

We are going to demonstrate sensitivity analysis by means of

- random sets and pinching;
- fuzzy sets, pinching, and Hartley-like measures;
- Monte Carlo simulation, correlation measures and resampling.

In order to test the sensitivity analysis methods, a simplified finite element model keeping the global structure (Figure 27) with about ninety thousand degrees of freedom is used. The computational cost for the simplified model was one hour per call of the program.[9]

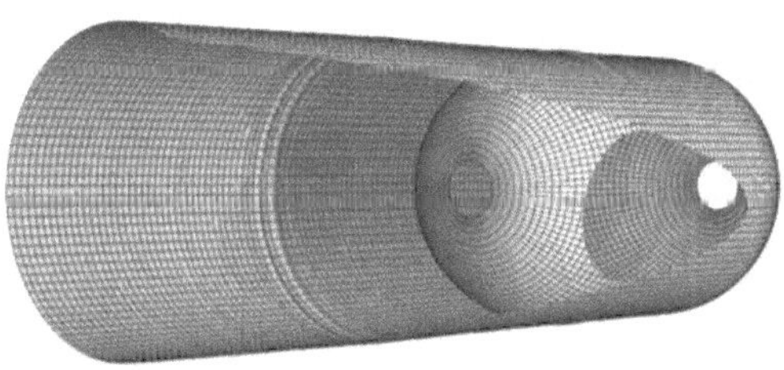

Figure 27. Simplified finite element model of frontskirt.

In the sensitivity analysis, up to 17 input parameters were taken into account. A terse description of the meaning of the parameters as well as

[8]The load proportionality factor is defined as the limiting value in an incremental procedure in which the dynamic loads during a flight scenario are increased step by step until breakdown of the structure is reached.

[9]An extended Monte Carlo based sensitivity analysis of large models was undertaken in the follow-up project ACOSTA, 2008 – 2010, supported by The Austrian Science Fund.

their nominal values can be read off from Table 5. The coefficients of variation of the input variables were estimated – after discussions with various engineering experts and consulting the literature – at 15%.

Table 5. Description of input parameters no. 1 – 17.

i	Parameter X_i	Mean μ_i
1	Initial temperature	293 K
2	Step1 thermal loading cylinder1	450 K
3	Step1 thermal loading cylinder2	350 K
4	Step1 thermal loading cylinder3	150 K
5	Step1 thermal loading sphere1	150 K
6	Step1 thermal loading sphere2	110 K
7	Step2 hydrostatic pressure cylinder3	0.4 MPa
8	Step2 hydrostatic pressure sphere1	0.4 MPa
9	Step2 hydrostatic pressure sphere2	0.4 MPa
10	Step3 aerodynamic pressure	−0.05 MPa
11	Step4 booster loads y-direction node1	40000 N
12	Step4 booster loads y-direction node2	20000 N
13	Step4 booster loads z-direction node1	3.e6 N
14	Step4 booster loads z-direction node2	1.e6 N
15	Step4 mechanical loads x-direction	100 N
16	Step4 mechanical loads y-direction	50 N
17	Step4 mechanical loads z-direction	300 N

Random sets and pinching. In this analysis, input parameters are modeled as Tchebycheff random sets as described in Subsection 2.2, with the multivariate combinations in Subsection 2.5 and the propagation through a function in Subsection 2.4. In the numerical evaluation, the joint random set is approximated by a finite random set with focal sets

$$A^1_{\alpha_1} \times \cdots \times A^d_{\alpha_d}, \qquad \alpha_j \in \{\tfrac{1}{n}, \tfrac{2}{n}, \ldots, 1\} \quad \text{for} \quad j = 1, \ldots, d,$$

each with probability weight n^{-d}. The input-output function is evaluated as follows: First, an interval $Q \subset \mathbb{R}^d$ is determined that bounds the relevant range of the input variables X. Next, the values of the function g are computed at the m^d nodes of a uniform grid on Q. The output $g(Q)$ is approximated by a response surface $\hat{g}(Q)$ obtained by multilinear splines. The interval $g(A_\alpha)$ is approximated by the minimum and maximum value thus obtained. Finally, the probability box is calculated by adding the weights when appropriate. The essential computational effort thus amounts to m^d calls of the finite element program.

The thin outer curves in Figure 28 show the result of the calculation of the load proportionality factor (LPF) where the three input parameters X_3, X_{13}, X_{14} (temperature cylinder 2, booster load node 1 in z-direction, booster load node 2 in z-direction) were taken up as variable in the analysis (probability box on the left, contour function on the right). The variance σ^2 for the Tchebycheff model was adjusted such that the base intervals $[x_{\min}, x_{\max}]$ for each of the parameters was symmetric around the corresponding mean μ with spread $\pm 0.15\mu$. In this case, $d = 3$ and we chose $m = 5$ so that 125 calls of the FE-program were required.

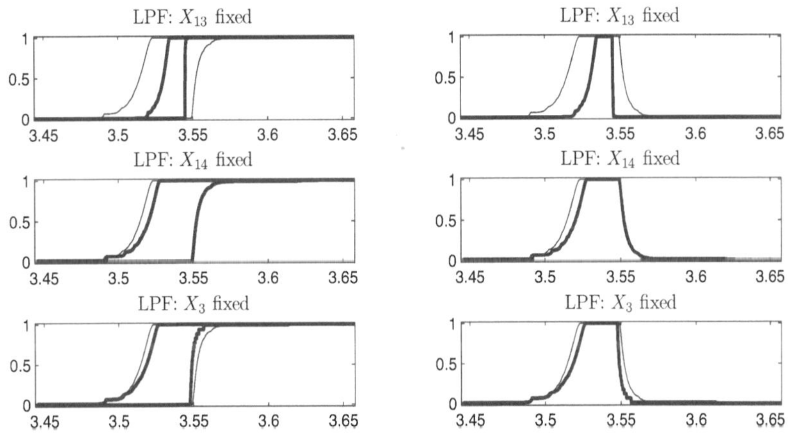

Figure 28. LPF, total variability (thin lines) and LPF, successively pinched variables (solid lines); probability boxes (left) and contour functions (right).

To assess the sensitivity of the load proportionality factor LPF with respect to the parameters X_3, X_{13}, X_{14} we use the pinching strategy, that is, we successively set one of the resulting standard deviations $\sigma_3, \sigma_{13}, \sigma_{14}$ equal to zero (while keeping the others at their given value), go through the calculation indicated above and plot the resulting probability boxes and contour functions (solid lines – the thin lines indicate the unperturbed result). This is displayed in Figure 28 and shows that setting $\sigma_{13} = 0$ produces the biggest reduction of the width of the probability box, while setting $\sigma_{14} = 0$ has little effect. We infer that the parameter X_{14} has the least influence on the variability of the response, while X_{13} exerts the biggest influence. For a way of quantifying this effect, see the discussion below and Table 6. The pinching strategy in the case of probability boxes is further explicated in [52] and applied in [143].

Fuzzy sets, pinching, and Hartley-like measures. In this paragraph, one-dimensional input variables will be modeled as *normalized fuzzy numbers*. In the assessment of the sensitivity of the load proportionality factor LPF with respect to the input parameters X_3, X_{13}, X_{14}, these parameters were modeled as symmetric triangular fuzzy numbers,[10] with central values μ_i from Table 6 and spread $\pm 0.15\mu_i$ as before. The numerical calculation is based on the response surface method explained above. The images of the α-level sets are again computed by piecewise multilinear combination. To handle possible lack of monotonicity of the function g, we start with level $\alpha = 1$ and go the way down to $\alpha = 0$, insuring at each step that the approximations satisfy $g(A_\beta) \subset g(A_\alpha)$ for $\alpha < \beta$.

In the non-interactive case, the procedure for determining the sensitivity of the output with respect to the input variables is the same as in the random set case. The initial calculation is performed with proportional spreads $\pm 0.15\mu_i$. Then we successively replace one of the triangular fuzzy numbers by its crisp central value μ_i, and compute the output as a fuzzy number. The result gives a good visual representation of the change of variability. This can be quantified using, e.g., the Hartley-like measure

$$\mathrm{HL}(B) = \int_0^1 \log_2\left(1 + \lambda(B_\alpha)\right)\, \mathrm{d}\alpha$$

of fuzzy sets B as proposed by [96], where λ denotes Lebesgue measure (see also [3] for further implementations). The result is depicted in Figure 29 (left), where the outer contour is the membership function of the fuzzy LPF with all input parameters fuzzy, while the shaded region is bounded by the membership function of the fuzzy LPF with successively pinched input parameters. It confirms the observations obtained by the random set method: X_{13} is the most influential parameter, followed by X_3 and then X_{14}. This can be explained by the model setup: X_{13} refers to a large booster load on one side of the frontskirt, while X_{14} signifies a much smaller booster load on the opposite side. The Hartley-like measures displayed in Table 6 (left hand side, central column), though, show that some, albeit small, influence of parameter X_{14} is detectable. For reasons of comparison we also display the Hartley-like measure of the random set output in Table 6 (right hand side). The latter Hartley-like measures were computed from the level sets of the probability box as in Figure 28.

[10]Triangular fuzzy numbers were employed for two reasons: first, they do not encode information other than a central value and a spread; second, they appear to be ideally suited for sensitivity analysis. In fact, they are built up linearly from nested intervals – this allows to read off the sensitivity of the output level by level and gives information on how the sensitivity changes when the input intervals increase.

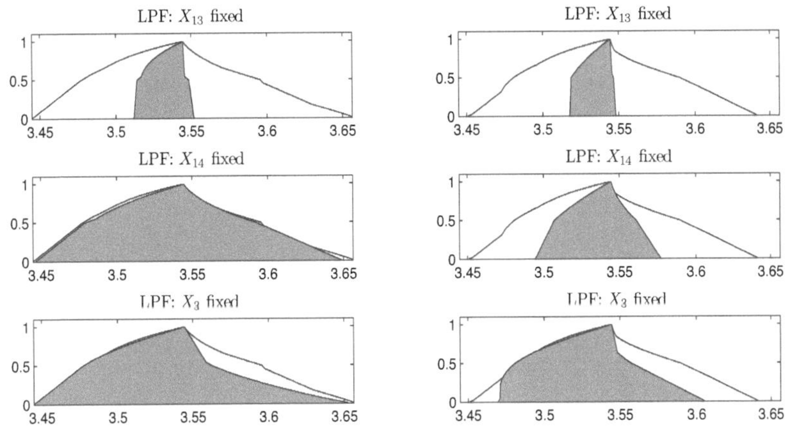

Figure 29. Fuzzy sets: LPF, pinched variables, noninteractive case (left) and interactive case (right).

Table 6. Hartley-like measures of fuzzy outputs, non-interactive vs. interactive input (left); Hartley-like measures of random set output (right).

Fuzzy set	HL-measure non-interactive	HL-measure interactive
no fixing	0.1481	0.1359
X_{13} fixed	0.0398	0.0315
X_{14} fixed	0.1430	0.0666
X_3 fixed	0.1268	0.1020

Random set	HL-measure of p-box
no fixing	0.0603
X_{13} fixed	0.0233
X_{14} fixed	0.0534
X_3 fixed	0.0504

We now turn to showing how the effect of possible correlations between two of the input parameters on the sensitivity can be assessed. *Correlation* will be interpreted here as degree of interactivity as described in Subsection 2.5. In this example, we assume a degree of interactivity $\rho = 0.8$ between parameters X_{13} and X_{14}. The remaining parameters are treated as non-interactive. The α-level sets are of cylindrical shape with a rhombic base R_α, say. Their images are again computed by piecewise multilinear combination. Otherwise, the procedure of successively pinching variables is similar: For example, when X_{13} is frozen at its central value μ_{13}, the interactivity restricts X_{14} to vary along the intersection of R_α with the line through μ_{13} parallel to the x_{14}-axis, while X_3 varies in its original α-level interval.

The result is shown in Figure 29 (right). The outcome confirms the prominence of parameter X_{13}; as a consequence of the correlation, parameter X_{14} is seen to exert a comparable influence. The result also demonstrates that the correlation changes the sensitivity of the output with respect to parameter X_3. Table 6 (left hand side, right column) again shows the Hartley-like measures of the fuzzy output under successive pinching of input variables. One may note that the study of the influence of correlations can be implemented in the fuzzy approach with ease.

Monte Carlo simulation, correlation measures and resampling. To complete the analysis, we discuss the merits of direct Monte Carlo simulation in sensitivity analysis. For the sake of comparison, we place ourselves in the standard probabilistic setting. In the Monte Carlo simulation to follow, we will use uniformly distributed input variables. This choice is motivated by the purpose of our study: modeling the variability of the input with the smallest amount of information necessary so as not to introduce information that might distort the simulated sensitivities. Symmetric triangular distributions offer another option; for the sake of exposition, we remain with uniformly distributed input here. For further information on sampling based sensitivity analysis, see [82, 166].

An explorative analysis usually starts with checking the scatterplots of individual input variables vs. the output, obtained by Monte Carlo simulation. However, scatterplots cannot help in detecting hidden influences of co-variates on the correlation between a given input variable X_i and the output variable Y, nor do they give a quantification. It is therefore advisable to use further correlation measure, like the partial correlation coefficients and standardized regression coefficients. To detect monotonic (rather than linear) dependencies, the ranked counterparts of these indicators can be used.

We recall that partial correlation between two random variables X_i and Y given a set of co-variates $X_{\smallsetminus i} = \{X_1, \dots, X_{i-1}, X_{i+1}, \dots, X_d\}$ is defined as the correlation between the two residuals $e_{X_i \cdot X_{\smallsetminus i}}$ and $e_{Y \cdot X_{\smallsetminus i}}$ obtained by regressing X_i on $X_{\smallsetminus i}$ and Y on $X_{\smallsetminus i}$, respectively. More precisely, one first constructs the two regression models

$$\widehat{X}_i = \alpha_0 + \sum_{j \neq i} \alpha_j X_j, \quad \widehat{Y} = \beta_0 + \sum_{j \neq i} \beta_j X_j,$$

obtaining the residuals

$$e_{X_i \cdot X_{\smallsetminus i}} = X_i - \widehat{X}_i, \quad e_{Y \cdot X_{\smallsetminus i}} = Y - \widehat{Y}.$$

Since $e_{X_i \cdot X_{\smallsetminus i}}$ and $e_{Y \cdot X_{\smallsetminus i}}$ are those parts of X_i and Y that remain after subtraction of the best linear estimates in terms of $X_{\smallsetminus i}$, the partial correlation coefficient (PCC)

$$\rho_{X_i,Y\cdot X_{\smallsetminus i}} = \rho(e_{X_i\cdot X_{\smallsetminus i}}, e_{Y\cdot X_{\smallsetminus i}})$$

quantifies the linear relationship between X_i and Y after removal of any part of the variation due to the linear influence of $X_{\smallsetminus i}$. Applying a rank transformation to the variables X_i and Y leads to the partial rank correlation coefficient (PRCC). The advantage of the partial correlation coefficients over the usual correlation coefficients is that they are more discriminating. In fact, if the input-output map is a truly linear function, the PCC of an input variable that enters with a non-zero coefficient is equal to one. On the other hand, if two input variables are multiples of each other, their PCCs with the output are zero. In reality, input-output maps are not ideally linear functions and so the effect is somewhat moderated. Still the PCCs are an accentuating measure of influence. For further background on PCCs and PRCCs, see [79, 85, 165].

To estimate the influence of each of the 17 input parameters from Table 5 on the output LPF, we performed a Monte Carlo simulation of size $n = 100$ with uniformly distributed input variables (on the intervals defined by the mean values and a spread of $\pm 15\%$), using Latin hypercube sampling. For additional accuracy in view of the rather small sample size we subjected the simulated variables to correlation control (see [86, 87]). This procedure consists in a rearrangement of the originally simulated values such that the resulting empirical rank correlation matrix is close to diagonal.

The scatterplot of Figure 30 gives a first impression of the influence of the input variables on the output. However, as noted above, the scatterplot alone delivers questionable information. Accordingly, we computed the various correlation coefficients that discount certain possible interactions between the input variables.

The results are displayed in Table 7; the meaning of the various coefficients is as follows. CC: Pearson correlation coefficient (quantifying the strength of a linear relationship between X_i and LPF); PCC: partial correlation coefficient (quantifying the strength of a linear relationship between X_i and LPF after removal of linear interactions with $X_{\smallsetminus i}$); SRC: standardized regression coefficient between X_i and LPF (quantifying the effect of deviations from μ_i); RCC: Spearman rank correlation coefficient (quantifying the strength of a monotonic relationship between X_i and LPF); PRCC: partial rank correlation coefficient (quantifying the strength of a monotonic relationship between X_i and LPF after removal of monotonic interactions with $X_{\smallsetminus i}$); SRRC: standardized rank regression coefficient between X_i and LPF (quantifying the effect of deviations from μ_i within a rank regression model).

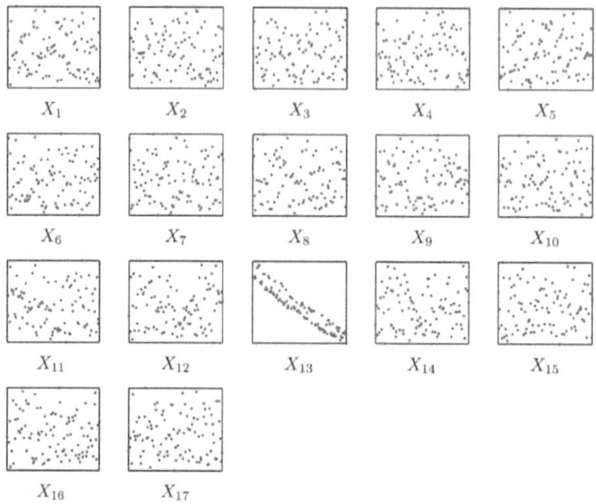

Figure 30. Scatterplots of 17 input variables vs. output.

Table 7. Various measures of correlation input – output; superscripts indicate the respective ranks.

#	*Raw data*			*Rank data*		
	CC	PCC	SRC	RCC	PRCC	SRRC
1	−0.0553	**−0.2631** ^{3}	**−0.0662** ^{3}	−0.0566	**−0.2788** ^{4}	**−0.0659** ^{4}
2	−0.0781	−0.0888	−0.0217	−0.0913	−0.1420	−0.0326
3	0.0954	**0.4680** ^{2}	**0.1285** ^{2}	0.0629	**0.3957** ^{2}	**0.0978** ^{2}
4	0.0553	0.1038	0.0254	0.0574	0.1173	0.0268
5	0.0525	0.0612	0.0149	0.0653	0.1315	0.0301
6	0.0418	0.0210	0.0051	0.0521	0.0720	0.0164
7	−0.0043	−0.0223	−0.0054	0.0062	0.0237	0.0054
8	−0.0397	−0.1230	−0.0301	−0.0233	−0.0629	−0.0143
9	0.0479	0.1787	0.0442	0.0684	**0.2753** ^{5}	**0.0652** ^{5}
10	0.0216	0.1803	0.0446	0.0237	0.1942	0.0450
11	−0.0181	0.0893	0.0219	−0.0332	0.0340	0.0077
12	0.0019	0.0203	0.0049	0.0152	0.0644	0.0147
13	**−0.9534** ^{1}	**−0.9690** ^{1}	**−0.9574** ^{1}	**−0.9593** ^{1}	**−0.9728** ^{1}	**−0.9602** ^{1}
14	0.0407	**0.2596** ^{4}	**0.0653** ^{4}	0.0380	**0.2810** ^{3}	**0.0665** ^{3}
15	0.0403	0.0352	0.0086	0.0498	0.0667	0.0152
16	−0.0844	−0.0880	−0.0216	−0.0777	−0.0805	−0.0184
17	−0.0056	0.0744	0.0181	−0.0050	0.0791	0.0180

The linear regression model constructed in order to obtain standardized regression coefficients (including all variables) has a corresponding coefficient of determination R^2 of 0.9414, thus indicating a fairly good fit. Changing to rank transformed data only slightly improves the quality of approximation (yielding $R^2 = 0.9488$). The resulting sensitivity indices induce a ranking of the input parameters according to their influence on the output (denoted by superscripts in Table 7). It is seen that the evaluations based on RCCs, PRCCs and SRRCs do not differ drastically from those obtained from raw data (CCs, PCCs and SRCs).

The rankings of the input parameters were based on the absolute value of the index under study. Small index values may or may not be considered as significant. For this reason, we constructed bootstrap confidence intervals (see e.g. [182]) with nominal level 0.95. In an overall assessment of the ranking, only those sensitivity estimates with a resulting confidence interval not including 0 (indicated by bold values) should be regarded as assertive.

As an example, we visualize the computed PRCCs and the corresponding bootstrap confidence intervals in Figure 31. Accordingly, only the PRCCs of the parameters X_1, X_3, X_9, X_{13} and X_{14} test to be nonzero.

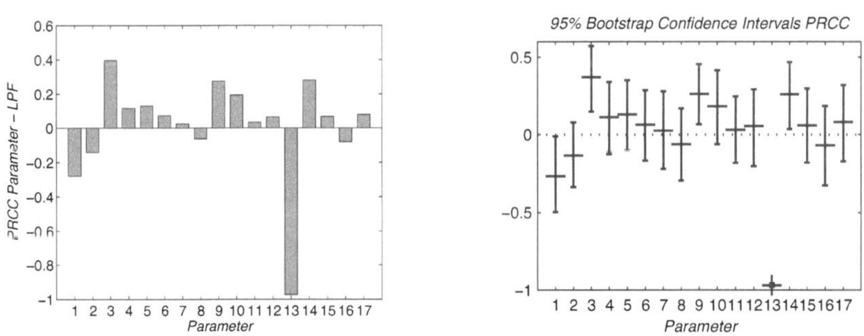

Figure 31. Partial rank correlation coefficients (left) and confidence intervals (right).

The outcome confirms the results of the previous sensitivity analysis: Among the parameters X_3, X_{13} and X_{14}, the one with the biggest influence is X_{13}, followed by X_3 and X_{14}. In addition, it shows the influence of parameters that would not have been visible in the scatterplots of Figure 30.

4.2 The elastically bedded beam

We take up the simple model $u = q/bc$ from Section 3, using the random set approach described in Subsection 3.3. Recall that the probability box in Figure 25 (right) shows the total variability of the output. We successively freeze the variables q and bc at their nominal value, compute the displacement u and observe whether the probability box gets narrower or not. The narrower it gets, the more influence the pinched variable has on the variability of the output, as mentioned earlier. The result of the pinching analysis can be seen in Figure 32. In the left picture, bc was pinched at its nominal value; the pinching of q in the right picture shows a more pronounced reduction of the variability, thus q is seen to be more influential on the displacement u than bc.

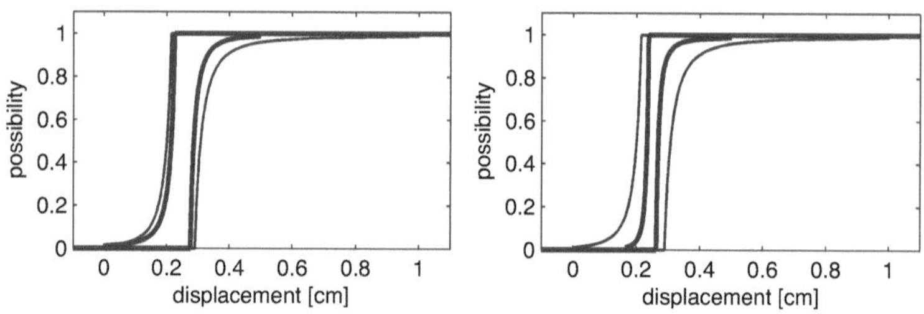

Figure 32. Sensitivity analysis: probability box with bc pinched (left) and with q pinched (right). The outer thin lines indicate the p-box without pinching.

To round off the picture, we have a look at Monte Carlo based sensitivity analysis in the random field model described in Subsection 3.1. As explained, a random sample of the parameters of interest is drawn, and the corresponding sample of output values is computed. Then the various correlation measures of the previous subsection can be used to asses the input-output relationship. The peculiarity of this example is that it involves a random field, and so the field parameters can be viewed as input variables whose influence is of interest.

Thus the input parameters under scrutiny are q, bc, the flexural rigidity EI of the beam and the field parameters σ and ℓ. To assess the sensitivities, we again started with uniformly distributed input, each in an interval around its nominal value with spread $\pm 15\%$. Then an independent random sample of size $N = 500$ was generated and put through the input-output function. The correlation coefficients of the five input parameters with the maximal displacement $|u|_{max}$ turned out to be

$$R(|u|_{max}, bc) \approx -0.67, \qquad R(|u|_{max}, q) \approx 0.71$$

with $R(|u|_{max}, EI)$, $R(|u|_{max}, \sigma_q)$, $R(|u|_{max}, \ell)$ all around 0.035. The values are depicted in Figure 33. Thus the simple probabilistic sensitivity analysis confirms that q and bc are the most influential parameters. We defer from repeating the extended methods presented in Subsection 4.1.

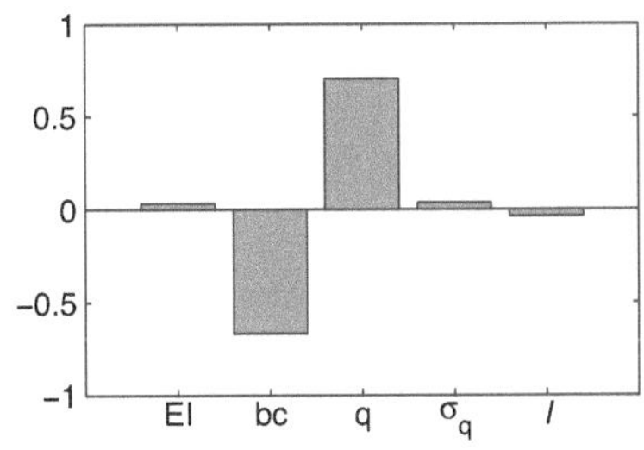

Figure 33. Correlation coefficients of input parameters with maximal displacement of the beam.

4.3 Comments on computational aspects

Summarizing the observations of this section, one may say that Monte Carlo simulation methods are computationally least expensive in sensitivity analysis. In the sensitivity study in Subsection 4.1, a sample size of $n = 100$ appeared sufficient. In fact, the mean error of the Monte Carlo simulation is well known to be of order $1/\sqrt{n}$ times the standard deviation of the simulated variable. In addition, the sample size required for a given accuracy does not depend on the dimension of the input (apart from the effort for correlation control). In the present application, the required accuracy of the respective correlation coefficients is in the range of 5–10 %. In view of the coefficient of variation of the input variables, this accuracy could be obtained with the mentioned sample size of $n = 100$.

In conclusion, these methods are clearly applicable to large scale problems. A disadvantage is that pinching of variables requires repetition of the $n = 100$ simulations, as would be studying input-output sensitivities under additionally imposed correlations of the input variables. Thus computing PRCCs plus resampling is possible irrespective of the problem scale, but variance decomposition by pinching variables may become rather expensive.

We point out that the results of a Monte Carlo simulation are amenable to resampling, which introduces little additional computational effort (no further evaluations of the costly input-output map are needed). In this way, bootstrap confidence intervals can be obtained that may serve as statistical estimates of the accuracy of the results.

Both in the fuzzy set and random set methods, the output α-level sets and focal sets, respectively, are computed by searching for the maximum and minimum of the corresponding output range. Sufficient accuracy can only be obtained by a larger number of calls of the output function, evaluated on a grid of input data. In addition, the grid size increases exponentially with the number of input variables. These methods appear feasible only in the case of medium size problems and a small number of input variables. Monotonicity or partial monotonicity of the output function increases accuracy and helps reducing the number of computations required.

However, the computational effort for the pinching of random or fuzzy sets is low, once the computation with the overall input has been performed: the values needed for the calculation with pinched variables are already among the computed values. The fuzzy model in combination with the response surface technique has an additional advantage: it allows the a posteriori introduction of interactivity between the input variables without the need for new calls of the output function. The effect of interactive input can simply be evaluated by interpolation in the response surface.

5 Further applications

This section demonstrates further applications of the discussed methods of uncertainty analysis. We begin by vibrations of belltowers, modeled by differential equations with fuzzy parameters. Next, a combination of a probabilistic queueing model with state variables satisfying fuzzy differential equations is exhibited. The section ends with a model from earthquake engineering combining stochastic differential equations with random sets.

5.1 Fuzzy modeling of vibrations of belltowers

Our first example concerns a system of differential equations with fuzzy coefficients. It arises in analyzing bell-induced vibrations of bell-towers. In a simplified model [133], the situation is described by two coupled pendulums obeying the system of ordinary differential equations

$$\begin{aligned}\psi'' L_1^2(m_1+m_2) + m_1 L_1 L_2[\varphi''(\cos\varphi + \psi\sin\varphi) - \varphi'^2(\sin\varphi - \psi\cos\varphi)]\\ + g\psi(m_1L_1 + m_2L_1) + k\psi = 0,\\ \varphi''(m_2L_2^2 + \theta_2) + m_2L_1L_2[\psi''(\cos\varphi + \psi\sin\varphi) + \psi'^2(\sin\varphi - \psi\cos\varphi)]\\ + m_2L_2 g\sin\varphi = 0,\end{aligned}$$

where ψ_j denote the angles of inclination of the tower and the bell, L_j their lengths, m_j their masses, g the acceleration due to gravity and k a stiffness constant describing the coupling of the soil/tower (assumed elastic), see Figure 34.

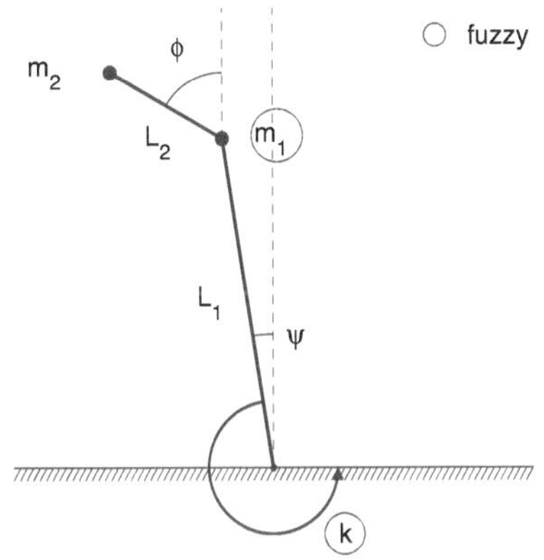

Figure 34. Bell-tower and bell as double pendulum.

If the investigation concerns an existing tower, all data can be measured sufficiently precisely except for the mass m_1 of the tower and the stiffness constant k. We model these parameters by fuzzy numbers; for the purpose of our example, we choose them as depicted in Figure 35.

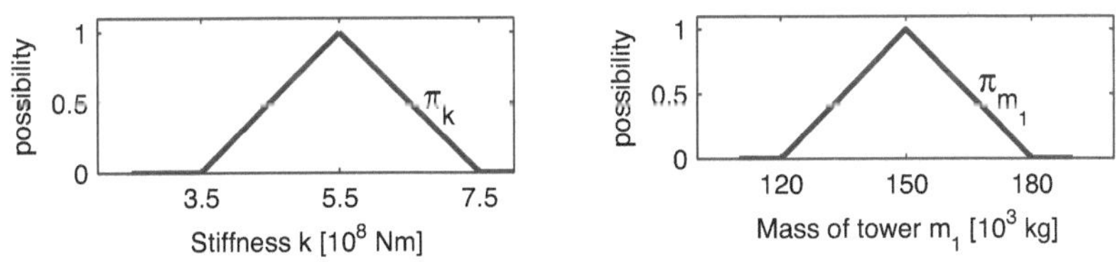

Figure 35. Input data for bell-tower problem.

The remaining parameters have been assigned deterministic values. The fuzzy solution concept has been elaborated in detail in [145]. As in Subsection 3.2, we solve the system of equations by applying the extension principle to the solution operator. In particular, this may be used to compute the fuzzy values of the solution at fixed time t. Here we perform this calculation with the angles of inclination $\varphi(t), \psi(t)$, depending on the fuzzy parameters m_1, k. We may depict the results as fuzzy numbers depending on time t either by showing the level sets (Figure 36) or by plotting the time-depending membership degree as a three-dimensional plot (Figure 37).

In Figure 36, the time evolution of the trajectory with membership degree 1

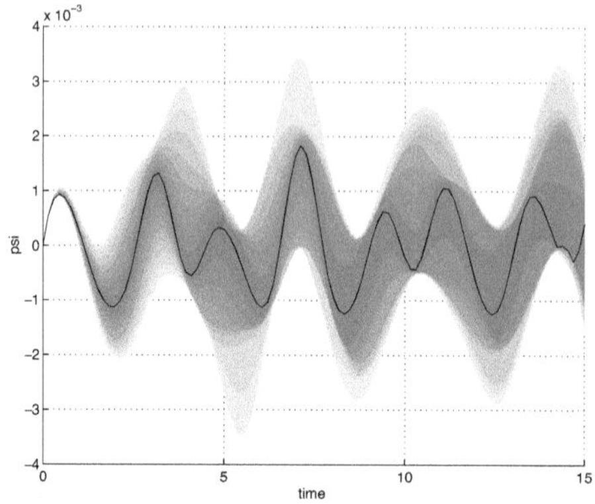

Figure 36. Level sets for angle of inclination $\psi(t)$.

is clearly visible. Trajectories with lower membership degree are somewhere hidden in the shaded regions. In Figure 37, cutting the three-dimensional graph at fixed t gives a fuzzy number, the membership function of $\psi(t)$. Further functions of the output of interest, like the dominant frequency, can be computed from there, making use of the extension principle again.

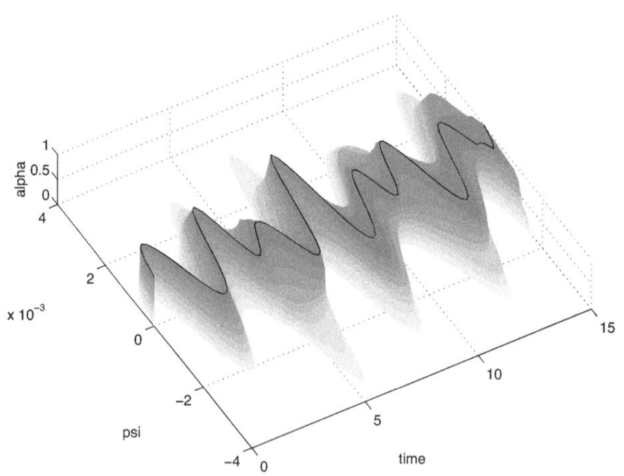

Figure 37. Fuzzy angle of inclination $\psi(t)$.

5.2 A random-fuzzy queueing model

The purpose of this part is to demonstrate a combination of random and fuzzy ingredients. We use a simple model from queueing theory with exponentially distributed interarrival and service times with fuzzy mean values. In civil engineering, queueing problems typically arise in earth work at larger construction sites, see e. g. [102].

To set the stage, we briefly describe this situation formally. The problem is a closed loop transport system, consisting of a single excavator (server) and N transport vehicles. After loading by the server, the vehicles transport and unload the material and return to the excavator. Due to variations in the service and return times, a waiting queue will build up in front of the server, see Figure 38. The engineering problem is to design the system in a cost efficient way. Having chosen a certain excavator, this will chiefly be decided by the number of transport vehicles employed: Too few vehicles will incur costs due to idle time of the server, while too many will incur costs due to waiting time spent in the queue. The transportation and excavation costs per unit time given, the essential performance index is the average time needed by each transport vehicle to complete a full run, or the average number of completed runs in a given period.

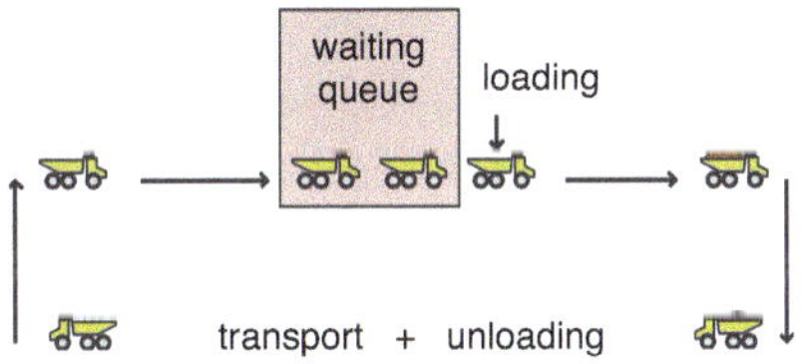

Figure 38. Scheme of the queueing system.

The input parameters in the system are the service time (loading of a single transport vehicle by the excavator) and the travel or return time of the transport vehicles. In the project planning phase, the designing engineer has to determine these parameters in order to calculate the required capacity of the equipment. The loading time of the excavator depends on a large number of uncontrollable conditions: soil parameters, like grain structure, angle of internal friction, loosening; accessibility of construction site; effective slewing angle of the excavator; weather conditions, and so on. Similarly, the return time of the transport vehicles depends on road and traffic conditions, conditions at the unloading site, and so on. In the planning phase, these data are only vaguely known. Statistical data – if collected from previous projects at all – are not assertive under the always different and only partially predictable circumstances arising in new building projects.

On the other hand, the standard probabilistic queueing models, as the Markov (M/M/1):N model, assume certain probability distributions on the service and return times and require precise estimates for the average service rate and average return rate as input. Such a model is incapable of dealing with the data uncertainty as described above, unless augmented by a second level model for the data uncertainty. It is the aim of this subsection to use fuzzy sets for describing the parameter uncertainty in a standard probabilistic queueing model. This will result in fuzzy state probabilities from which the required performance measures can be estimated. As the method of computation, we use fuzzy differential equations.

Setting up the probabilistic model. As outlined in the beginning, we consider a closed loop queueing system, consisting of a single server (excavator) and N transport vehicles. We adopt a first-come-first-served queueing discipline. The required input parameters are the average service time $1/\mu$ (service rate μ) and the average return time $1/\nu$ (driving rate ν) of each vehicle. We note that the arrival rate λ is not an input variable in a closed system; it rather establishes itself during the process.

What concerns the theoretical analysis of the behavior of queues, we adopt the standard Markov (M/M/1):N queueing model (see e.g. [94]). The model assumes that the service time and the return time are exponentially distributed with expectation values $1/\mu, 1/\nu$. The decisive state variables are the probabilities $p_k(t), k = 0, \dots, N$ that k transport vehicles are present in the queueing system at time t. These probabilities give preliminary information on the initial behavior of the system, but will chiefly be used to compute the stationary state with limiting probabilities $q_k = \lim_{t\to\infty} p_k(t)$. The stationary state will serve as a good approximation to the behavior of the system (simulations with actual data from construction management indicate that it is usually reached within one to two hours).

Following standard arguments (e.g. [94]) one can deduce the system of ordinary differential equations

$$\begin{aligned} p_0'(t) &= -N\nu p_0(t) + \mu p_1(t), \\ p_k'(t) &= (N-k+1)\nu p_{k-1}(t) - (\mu + (N-k)\nu)p_k(t) + \mu p_{k+1}(t) \\ &\quad \text{for } \ k = 1, \cdots, N-1, \qquad\qquad (12) \\ p_N'(t) &= \nu p_{N-1}(t) - \mu p_N(t) \end{aligned}$$

Due to the fact that the components of the system are probabilities, we have to adjoin the constraint

$$\sum_{k=0}^{N} p_k(t) = 1. \tag{13}$$

We normally will assume deterministic initial data $p_j(0) = 1$ for some j, $p_i(0) = 0$ for $i \neq j$. We note that $(\sum p_k(t))' = 0$ so that the constraint (13) is automatically satisfied for all times if it is satisfied initially. There is a unique equilibrium state given by

$$q_0 = \left(\sum_{n=0}^{N} \frac{N!}{(N-n)!} \left(\frac{\nu}{\mu} \right)^n \right)^{-1} \tag{14}$$

$$q_k = q_0 \frac{N!}{(N-n)!} \left(\frac{\nu}{\mu} \right)^n \quad \text{for } k = 1, \ldots, N. \tag{15}$$

The convergence of the probabilities $p_k(t)$ to the equilibrium probabilities q_k is a well known result in the theory of Markov processes. What we need later will be the stronger result of uniform convergence with respect to the parameters μ and ν (when they vary in bounded intervals). This stronger result is proved, e.g., in [61].

Once the equilibrium probabilities are found, other performance indicators can be computed by well known arguments from queueing theory. For instance, the average number of vehicles in the queue (in the equilibrium state) is $L = N - (\mu/\nu)(1 - q_0)$. By equating the arrival with the departure rate one gets for the arrival rate $\lambda = \nu(N - L) = \mu(1 - q_0)$. Finally, the average total run time (travel plus waiting time in the queue plus service time) of each vehicle equals $T = N/\lambda$.

The probabilistic model with fuzzy input and fuzzy output. In this paragraph, we adhere to the probabilistic (M/M/1):N queueing model constructed above, but assume that the distribution parameters μ and ν are fuzzy. In the terminology and spirit of [108], this is an (FM/FM/1):N queueing system. Our approach is based on the system of differential equations for the probabilities $p_k(t)$ that k transport vehicles are present in the queueing system at time t. We apply the extension principle to the solution operator $(\mu, \nu) \to L_t(\mu, \nu)$ of system (12) with constraint (13). When μ and ν are given by fuzzy numbers, the extension principle yields the fuzzy solution

$$p(t) = L_t(\mu, \nu), \tag{16}$$

consisting of the fuzzy vector of time-dependent state probabilities. The k-th component of the solution operator produces the fuzzy probabilities that k vehicles are in the system at time t:

$$p_k(t) = L_t^k(\mu, \nu). \tag{17}$$

We note that the fuzzy equilibrium probabilities q_k could be computed from formula (14) and (15) with the aid of the extension principle. However, having our machinery at hand, it is simpler to pass to the limit in the expressions (17) for the time-dependent fuzzy marginals as $t \to \infty$. As has been proven in [61], the convergence holds in the sense of the distance

$$d(A, B) = \sup_{\alpha \in (0,1]} d_H(A^\alpha, B^\alpha)$$

between fuzzy numbers A, B where d_H denotes the Hausdorff distance (see e.g. [36]).

The performance indicators mentioned above can now easily be computed, again by the extension principle, from the fuzzy output probability q_0 and the fuzzy service rate. The total time needed by each vehicle on average to complete a full run is given as the fuzzy number

$$T = N/\mu(1 - q_0); \tag{18}$$

due to the fact that μ and q_0 are interactive, this is actually an upper estimate. Various further information can be extracted from there. First, the total cost per transported unit mass is proportional to T and $C_V + C_S/N$, where C_V and C_S denote the cost per unit time of a single vehicle and the server, respectively. Thus estimates (with their degrees of possibility) of how the total cost emanates depending on the number N of vehicles can be obtained. The average number of completed runs by all vehicles together in a given period T_0 is $T_0 N/T = T_0 \mu(1 - q_0)$. Thus the degree of possibility of achieving a certain performance in a given period of time can be computed as well, giving the basis for assessing the risk of not achieving a required threshold.

Example of a random-fuzzy queueing system. To illustrate the analysis above we present a computational example with three transport vehicles ($N = 3$).

The fuzzy mean serving time is assumed to be given by a triangular fuzzy number with supporting interval $[2, 6]$ and center at $z = 4$, while the average return time is modeled as a triangular fuzzy number with support $[9, 11]$ and center at $z = 10$ (see Figure 39). The initial state is taken deterministic as described above with $p_3(0) = 1$ (all three transport vehicles present at start). Figure 40 shows the α-level sets of the fuzzy time-dependent probability

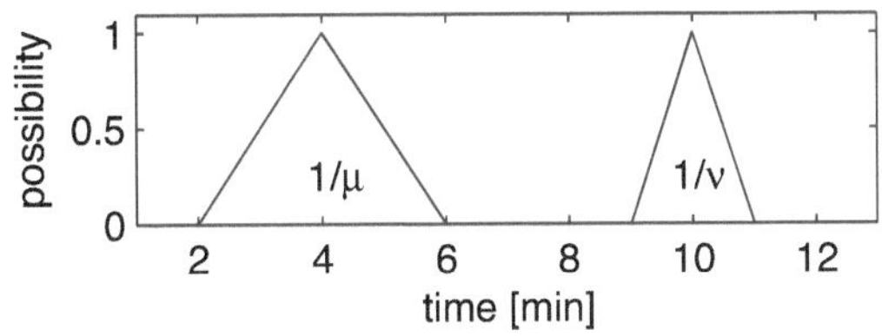

Figure 39. Fuzzy input parameters.

$p_0(t)$ for $0 \leq t \leq 30$; Figure 41 displays the contour of its membership function as a three-dimensional graph with the degree of possibility as third axis.

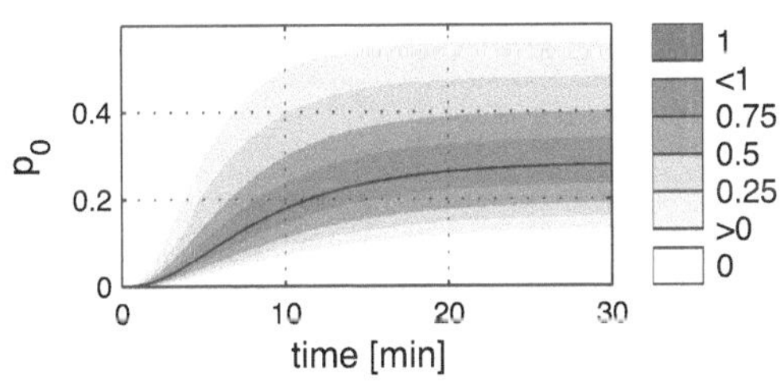

Figure 40. α-level sets of $p_0(t)$.

One can read off that at time $t = 30$ the equilibrium state is almost reached; the probability q_0 can be approximated by a triangular fuzzy number with supporting interval $[0.13, 0.56]$ and center at 0.28.

The average total time to complete a full run is given by the fuzzy number depicted in Figure 42; the approximate bounds for its support are $[6.9, 40.9]$, while its central value lies approximately at 16.7. Similarly, the average number of completed runs in unit time, $\mu(1 - q_0)$, is roughly computed as the triangular fuzzy number with supporting interval $[0.88, 5.22]$ and central value at 2.88. Thus the vagueness of the input data leads to possibly large fluctuations in the performance of the system, in the range of $\pm 70\%$ of the central value. Such a range of variation is quite realistic, as is seen e. g. from the collected project data in [102].

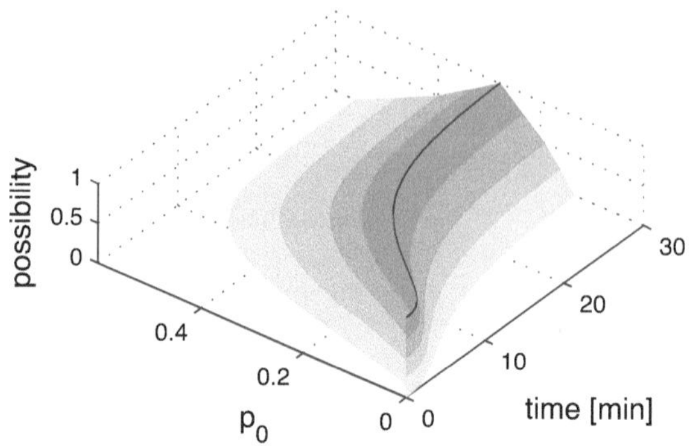

Figure 41. Contour of fuzzy set $p_0(t)$.

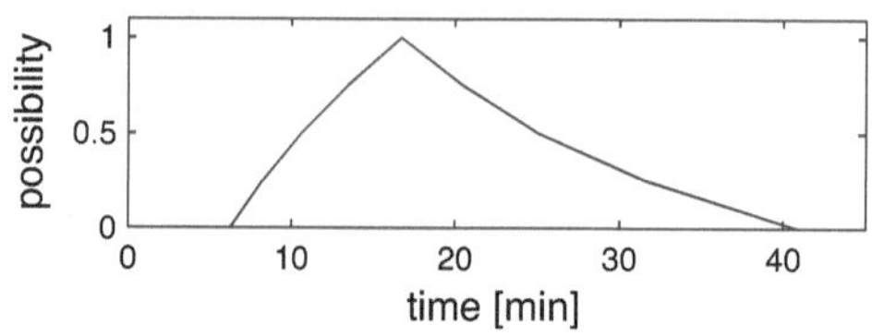

Figure 42. Fuzzy average total runtime T.

5.3 Random set/stochastic modeling of TMDs

In this subsection we address the combination of stochastic differential equations with random sets. It is based on recent work published in [171]. The new framework combining stochastic differential equations with random set coefficients, resulting in set-valued stochastic processes, is due to [169, 170]. A prototypical problem arises in earthquake engineering in the construction of damping devices. In an earthquake, buildings and structures are subjected to stochastic excitations due to ground accelerations. Tuned Mass Dampers (TMDs) are designed to reduce the resulting vibrations of the structure.

On the one hand, there is the earthquake record which can be modeled as a *stochastic process*. On the other hand, various design parameters of the damping device, like the damping coefficient, cannot be determined with certainty, although interval bounds usually are known. The theory of *random sets* lends itself as a framework for describing the parameter uncertainty. Thus the variations of the earthquake record will be described stochastically, while the parameter uncertainty is accounted for by means of random sets.

More specifically, the combined structure-damper-system can be modeled by a system of first order ordinary differential equations in phase space. If the excitation is modeled as a continuous time white noise, this leads to a system of linear Itô differential equations (see e.g. [5]) of the form

$$dx(t) = F(a)\, x(t)\, dt + G\, dw(t)$$

where $F(a)$ is a matrix containing the uncertain parameters a and G is a constant vector distributing the white noise excitation to the appropriate components of the system. Earthquake excitations can be modeled more realistically by colored noise, for example by the output of an Ornstein-Uhlenbeck process

$$dy(t) = \alpha y(t) + \beta dw(t)$$

with coefficients adjusted by means of the spectrum of earthquake records. Introducing colored noise in the present framework is easily possible – it would amount to adding an additional line in the system of Itô differential equations.

The parameters a in the system matrix are uncertain and will be modeled by random sets (often by a collection of random intervals). The response process x will be set-valued (its components actually interval-valued). In this framework, output variables of interest, like the reduction coefficient, are also interval-valued. Another output of interest may be the probability that the response exceeds a certain limit in a given time interval. This concept splits into first entrance and first inclusion times in this setting, reflecting properly the parameter uncertainty

Description of the mechanical model. In the last decades various passive, active and semi-active control mechanism have been developed to suppress earthquake-induced vibrations of structures [30, 83, 186]. An example of a passive control device is the TMD, which reduces the dynamic response when attached to a vibration prone structure [30]. We shall look at TMDs consisting of a mass-spring-dashpot system, which has single-degree-of-freedom (SDOF). Its natural frequency is calibrated to the dominant mode of vibration of the structure to be protected. In a dynamic event the kinetic energy is transferred from the structure to the TMD and subsequently dissipated by its viscous element.

Formally, we study a linear SDOF oscillator of mass m_s, stiffness k_s and viscous damping parameter c_s representing a vibration-prone structure. The base of the oscillator is subjected to a given acceleration $\ddot{x}_g$ leading to structural vibrations. Another SDOF oscillator of mass m_d ($\ll m_s$), stiffness

k_d and viscous damping parameter c_d serves as TMD, and is connected in series to the structure, see Figure 43.

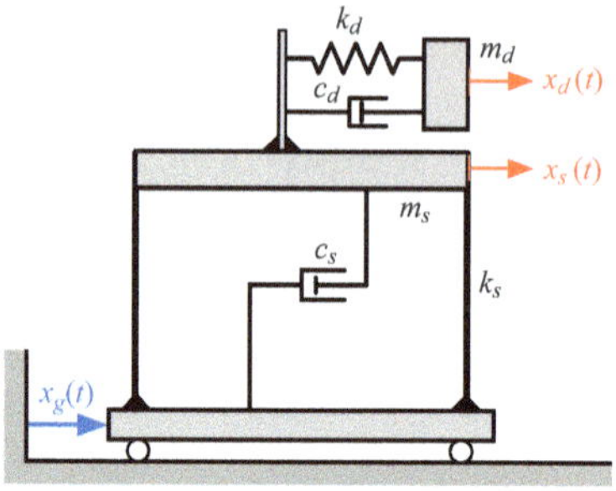

Figure 43. Mechanical model of a SDOF vibration-prone structure equipped with a TMD.

Thus the combined structure-TMD system exhibits two-degrees-of-freedom. Both the displacement x_s of the structure and of the TMD x_d are measured relative to the base displacement x_g. The coupled equations of motion read as follows:

$$M\begin{bmatrix}\ddot{x}_\text{s}\\ \ddot{x}_\text{d}\end{bmatrix} + C\begin{bmatrix}\dot{x}_\text{s}\\ \dot{x}_\text{d}\end{bmatrix} + K\begin{bmatrix}x_\text{s}\\ x_\text{d}\end{bmatrix} = \begin{bmatrix}-1\\ -\mu\end{bmatrix}\ddot{x}_\text{g} \tag{19}$$

where

$$M = \begin{bmatrix}1 & 0\\ 0 & \mu\end{bmatrix},\ C = \begin{bmatrix}2\zeta_\text{s}\omega_\text{s} + 2\zeta_\text{d}\omega_\text{d}\mu & -2\zeta_\text{d}\omega_\text{d}\mu\\ -2\zeta_\text{d}\omega_\text{d}\mu & 2\zeta_\text{d}\omega_\text{d}\mu\end{bmatrix},\ K = \begin{bmatrix}\omega_\text{s}^2 + \omega_\text{d}^2\mu & -\omega_\text{d}^2\mu\\ -\omega_\text{d}^2\mu & \omega_\text{d}^2\mu\end{bmatrix}.$$

The variable μ stands for the mass ratio,

$$\mu = \frac{m_\text{d}}{m_\text{s}} \tag{20}$$

and ω_s and ω_d denote the natural circular frequencies of the stand-alone structure and the detuned TMD, respectively,

$$\omega_\text{s} = \sqrt{\frac{k_\text{s}}{m_\text{s}}},\quad \omega_\text{d} = \sqrt{\frac{k_\text{d}}{m_\text{d}}}$$

while the non-dimensional damping coefficients ζ_s and ζ_d of the structure and the TMD are derived from

$$\zeta_\text{s} = \frac{c_\text{s}}{2\omega_\text{s} m_\text{s}},\quad \zeta_\text{d} = \frac{c_\text{d}}{2\omega_\text{d} m_\text{d}}.$$

For an effective reduction of the dynamic response x_s the parameters of the TMD, i.e., the damping coefficient ζ_d and the frequency ratio

$$\delta = \frac{\omega_d}{\omega_s} \tag{21}$$

must be tuned.

Stochastic differential equations with random set coefficients. As mentioned at the beginning, the earthquake excitation will be modeled as a white noise. We recall some required notions from stochastic analysis.

An $\mathbb{R}^d$-valued *stochastic process* x on a time interval $[t_0, \bar{t}]$ assigns to each point of time t a random variable $x(t)$. The process is specified if the finite dimensional joint distributions of all random variables $x(t)$, $t \in [t_0, \bar{t}]$ are known. More formally, a stochastic process is a map $(t, \xi) \to x(t, \xi)$ for t in a time interval and ξ from a probability space Ξ with probability measure P_Ξ. The distribution of the process can be rephrased in terms of the underlying probability space, e.g., the distribution of $x(t)$ is

$$P(x(t) \in B) = P_\Xi\big(\xi \in \Xi : x(t, \xi) \in B\big)$$

where B is a d-dimensional interval, or more generally, a Borel subset of $\mathbb{R}^d$. We shall need this level of abstraction in the sequel, when we have to distinguish between different probabilities.

A one-dimensional Wiener process (Brownian motion) w is defined for $t \in [0, \infty)$ as follows: each $w(t)$ is a Gaussian variable with mean zero and variance t. Further, the covariance of $w(t_1)$ and $w(t_2)$ equals $\min(t_1, t_2)$ and $w_0 = 0$. For our purposes the specification of the underlying Wiener probability space (Ξ, P_Ξ) is not important. For various constructive approaches to the Wiener process we refer to [93]. A Wiener process can be shifted to start at time t_0; the m-dimensional versions are obtained by taking m independent Wiener processes as components. Continuous time white noise $\dot{w}$ is the weak derivative of Wiener process. It is a generalized process with mean zero, infinite variance and zero covariance. It will be formalized here by means of the Itô integral. For deterministic integrands (and this is all what we need here), the Itô integral can be obtained as square mean limit (with respect to the probability P_Ξ) of Riemann-Stieltjes sums (for details, we refer to the literature, e.g. [5, 68]). For example, one has the relation

$$w(t) = \int_0^t dw(s).$$

A discretized version with time step Δt of white noise is obtained by simulating a sequence of independent Gaussian variables with mean zero and variance $1/\Delta t$.

Systems of ordinary differential equations with white noise excitation will be handled here as Itô stochastic differential equations (SDEs):

$$dx(t) = f\big(t, x(t)\big)dt + G\big(t, x(t)\big)dw(t), \tag{22}$$

interpreted as the integral equation

$$x(t) = x_0 + \int_{t_0}^{t} f\big(s, x(s)\big)ds + \int_{t_0}^{t} G\big(s, x(s)\big)dw(s) \tag{23}$$

where time t ranges in some finite time interval $[t_0, \bar{t}]$, x_0 is a random variable representing the initial value, f a vector-valued, G a matrix-valued function and w an m-dimensional Wiener process. Their solutions are stochastic processes with continuous trajectories.

Suppose the coefficients f, G and possibly the initial value x_0 are depending on parameters that are not known with certainty. If these parameters are represented by a vector a taking values in $\mathbb{R}^d$, Equation (22) can be written in the form

$$dx_a(t) = f\big(t, a, x_a(t)\big)dt + G\big(t, a, x_a(t)\big)dw(t)$$

where f and G are vector- and matrix-valued functions as above and the initial value x_0 may be a random variable on Wiener probability space possibly depending on the parameter a as well.

For each $a \in \mathbb{R}^d$ it is assumed that the conditions from the theory of SDEs are satisfied that guarantee the existence and uniqueness of a solution process $x_a = \{x_a(t) : t \in [t_0, \bar{t}]\}$. This leads to a family of stochastic processes depending on t, a, and the Wiener random element ξ:

$$x : [t_0, \bar{t}] \times \mathbb{R}^d \times \Xi \to \mathbb{R}^d, (t, a, \xi) \to x_a(t, \xi). \tag{24}$$

Under certain conditions (see [169, 170]) that are fulfilled for linear systems of SDEs almost all of the trajectories of this process are continuous with respect to (t, a).

The uncertainty of a will be modeled by a random compact set $\alpha \to A_\alpha$ with α from a probability space (H, P_H); i.e., each A_α is assumed to be a compact subset of $\mathbb{R}^d$. The aim is now to combine both kinds of uncertainty in the model, that is, the parameter uncertainty and the uncertainty implied by the stochastic excitation of the system. To this end we introduce the product probability space

$$(\Omega, P_\Omega) = (H \times \Xi, P_H \otimes P_\Xi),$$

whose elements $\omega = (\alpha, \xi)$ stand for the combined probabilities of the random set A and the Wiener process w. We introduce the set-valued function

$$X : (t, \omega) \mapsto \{x_a(t, \xi) : a \in A_\alpha\} \tag{25}$$

where the process x is defined by Equation (24) and $(t, \omega) \in [t_0, \bar{t}] \times \Omega$. Thus for each time t and each random element $\omega = (\alpha, \xi)$ the corresponding values of all solution processes x_a are merged to one set $X(t, \omega)$, which is a random realization of the set-valued process $X(t)$.

Under the assumption that all sample functions of x depend continuously on t and a it can be inferred that the values of the function X are compact subsets of $\mathbb{R}^d$. In the one-dimensional case one actually gets compact intervals. Furthermore X is a set-valued stochastic process, i.e. $X(t)$ is a random compact set for each time t.

Set-valued assessment of TMD performance. Under the assumption of white noise base excitation, that is $\ddot{x}_{\mathrm{g}}(t) = \dot{w}(t)$, the system of second order SDEs (19) can be written as a four-dimensional linear system of first order SDEs:

$$dx(t) = F\,x(t)\,dt + G\,dw(t) \tag{26}$$

with

$$x = \begin{bmatrix} x_{\mathrm{s}} \\ x_{\mathrm{d}} \\ \dot{x}_{\mathrm{s}} \\ \dot{x}_{\mathrm{d}} \end{bmatrix}, \qquad G = \begin{bmatrix} 0 \\ 0 \\ -1 \\ -1 \end{bmatrix}$$

and the four-dimensional matrix

$$F = \begin{bmatrix} 0 & I \\ -M^{-1}K & -M^{-1}C \end{bmatrix},$$

where I denotes the 2×2 identity matrix. The response of the system depends on the five parameters μ, ζ_{s}, ω_{s}, ζ_{d} and ω_{d}. The parameters of the main structure ζ_{s} and ω_{s} are assumed to be precisely known, whereas the parameters ζ_{d} and ω_{d} of the TMD are uncertain. Since the masses of the structure and of the TMD can be determined comparatively accurately and are not subject to change during varying conditions of the environment (such as temperature), also the mass ratio μ is assumed to be precisely known.

To investigate the effectiveness of the TMD the displacement x_{s} is compared to the displacement $\tilde{x}_{\mathrm{s}}$ of the structure when no TMD is attached. The motion of the corresponding SDOF system is then described by the equation

$$\ddot{\tilde{x}}_{\mathrm{s}} + 2\zeta_{\mathrm{s}}\omega_{\mathrm{s}}\dot{\tilde{x}}_{\mathrm{s}} + \omega_{\mathrm{s}}^2\tilde{x}_{\mathrm{s}} = -\ddot{x}_{\mathrm{g}}$$

or the equivalent first order SDE system

$$d\tilde{x}(t) = \begin{bmatrix} 0 & 1 \\ -\omega_s^2 & -2\omega_s\zeta_s \end{bmatrix} \tilde{x}(t)dt + \begin{bmatrix} 0 \\ -1 \end{bmatrix} dw(t) \tag{27}$$

in the case of white noise base excitation where $\tilde{x} = (\tilde{x}_s, \dot{\tilde{x}}_s)^\mathsf{T}$. In both systems (26) and (27) the initial values are assumed to be zero.

The uncertain parameters ζ_d and ω_d will be modeled as Tchebycheff random sets. The nominal values $\overline{\zeta_d}$ and $\overline{\omega_d}$ are obtained from (21) and (20) using proposed optimal values for μ from the literature [8]. The variability of these parameters is described as follows: it is assumed that the actual values lie in a range of $\pm 40\%$ of their nominal values with 99% certainty.

This information will be encoded in Tchebycheff random sets. Let us temporarily denote the random set to be constructed for the parameter ζ_d by $\alpha_1 \to A^{(1)}_{\alpha_1}$. At level $\alpha_1 = 0.01 = 1 - 0.99$, the parameter ζ_d varies at most by $0.4\,\overline{\zeta_d}$ around $\overline{\zeta_d}$. Thus we may say that

$$A^{(1)}_{0.01} = [\overline{\zeta_d} - 0.4\,\overline{\zeta_d},\ \overline{\zeta_d} + 0.4\,\overline{\zeta_d}]$$

and compute the corresponding variance σ_1 using Formula (10) with $d_\alpha = \sigma/\sqrt{\alpha}$. It follows that $d_{0.01} = 0.4\,\overline{\zeta_d}$. Using $\sqrt{0.01} = 0.1$ this leads to $\sigma_1 = 0.04\,\overline{\zeta_d}$. The random set for the parameter ζ_d has the focal sets

$$\begin{aligned} A^{(1)}_{\alpha_1} &= [\overline{\zeta_d} - \tfrac{\sigma_1}{\sqrt{\alpha_1}},\ \overline{\zeta_d} + \tfrac{\sigma_1}{\sqrt{\alpha_1}}] \\ &= [\overline{\zeta_d}(1 - \tfrac{0.04}{\sqrt{\alpha_1}}),\ \overline{\zeta_d}(1 + \tfrac{0.04}{\sqrt{\alpha_1}})] \end{aligned}$$

for $\alpha_1 \in (0, 1]$. Similarly, one gets a random set $\alpha_2 \to A^{(2)}_{\alpha_2}$ for the parameter ω_d with $\sigma_2 = 0.04\,\overline{\omega_d}$.

The joint random set $A = A^{(1)} \times A^{(2)}$ is obtained by introducing the probability space $H = (0,1] \times (0,1]$, equipped with the bivariate uniform distribution, and taking as focal sets the two-dimensional intervals

$$A_\alpha = A^{(1)}_{\alpha_1} \times A^{(2)}_{\alpha_2}$$

for $\alpha = (\alpha_1, \alpha_2) \in H$. This way of forming the joint random set assumes random set independence. For numerical purposes the Tchebycheff random set has to be truncated to a finite range; for this the focal set obtained by letting $\alpha_1 = \alpha_2 = 0.01$ is chosen. Thus the largest focal set is the rectangle $[0.6\,\overline{\zeta_d}, 1.4\,\overline{\zeta_d}] \times [0.6\,\overline{\omega_d}, 1.4\,\overline{\omega_d}]$.

Computation of relevant output. First of all, one is interested in the relative displacement x_s of the structure, i.e., the displacement scaled by the largest structural displacement $\tilde{x}_{\mathrm{s}}$ when no TMD is attached. This leads to the process

$$y_a(t,\xi) = \frac{x_{\mathrm{s},a}(t,\xi)}{\sup_{t\in[t_0,\bar{t}]} |\tilde{x}_{\mathrm{s}}(t,\xi)|}.$$

According to Equation (25) the set-valued non-dimensional displacement is defined by

$$Y(t,\omega) = \{y_a(t,\xi) : a \in A_\alpha\}. \tag{28}$$

Its values are compact intervals. Thus Y can be represented by the boundary processes L and U defined by

$$L(t,\omega) = \inf Y(t,\omega), \qquad U(t,\omega) = \sup Y(t,\omega),$$

i.e., $Y(t,\omega) = [L(t,\omega), U(t,\omega)]$. For reasons of comparison the non-dimensional displacement $\tilde{y}$ of the structure without TMD

$$\tilde{y}(t,\xi) = \frac{\tilde{x}_{\mathrm{s}}(t,\xi)}{\sup_{t\in[t_0,\bar{t}]} |\tilde{x}_{\mathrm{s}}(t,\xi)|} \tag{29}$$

will be considered, too.

Further information is contained in the mean value functions of the absolute values of the non-dimensional displacements. At fixed t, the absolute value of the non-dimensional displacement $|Y(t)|$ is a random set. Its expectation $\mathrm{E}|Y(t)|$ is an interval, thus $\mathrm{E}|Y|$ is an interval-valued function of time.

Another relevant output is constituted by the first time when the absolute value of the non-dimensional displacement exceeds a certain level $b > 0$. In the single-valued case its value is

$$\tau_a^B(\xi) = \inf\{t : |y_a(t,\xi)| > b\} \tag{30}$$

which is called the first exceedance time of b or the first entrance time of $|y_a|$ into $B = (b,\infty)$. For the set-valued output $|Y|$ the corresponding quantities are

$$\underline{\tau}^B(\omega) = \inf\{t : |Y|(t,\omega) \cap B \neq \emptyset\}, \qquad \overline{\tau}^B(\omega) = \inf\{t : |Y|(t,\omega) \subseteq B\}$$

where $\underline{\tau}^B$ is called the first entrance time of X into B, and $\overline{\tau}^B$ is called the first inclusion time of X in B. Furthermore, $\underline{\tau}^B$ corresponds to the first time when $|U|$ exceeds level b and $\overline{\tau}^B$ corresponds to the first time when $|L|$ exceeds level b. Further theoretical properties of first entrance and inclusion times for set-valued stochastic process can be found in [169, 170].

In addition, one can also consider for each time $t \in [t_0, \bar{t}]$ the lower and the upper probability of the set B, namely

$$\underline{P}(t, B) = P(\{\nu : |Y|(t, \nu) \subseteq B\}), \tag{31}$$
$$\overline{P}(t, B) = P(\{\nu : |Y|(t, \nu) \cap B \neq \emptyset\}). \tag{32}$$

For a study of the effectiveness of the TMD by means of set-valued reduction coefficients we refer to [171].

Numerical simulation and results. We are going to present some results of numerical simulations. In all computations the mass ratio is chosen as $\mu = 0.05$ and the non-dimensional damping coefficient of the main structure is set to $\zeta_s = 0.005$. We exhibit here only results for a structural period of $T_s = 2\pi/\omega_s = 2$ (seconds). In order to compute the dependence of the reduction coefficients on T_s, a parametric study has been undertaken as well. For this topic we refer to [171].

In the numerical approximation, each of the two one-dimensional random sets $A^{(i)}$, $i = 1, 2$, is replaced by a finite random set consisting of 10 intervals $A^{(i)}_{\alpha_{j_i}}$ $(j_i = 1, \ldots, 10)$ where the α_{j_i} are chosen equidistantly in $(0, 1]$ from $\alpha_1 = 0.01$ to $\alpha_{10} = 1$ with weights $p_{i,1} = 0.01$ and $p_{i,j_i} = \alpha_{j_i} - \alpha_{j_i - 1}$ for $j_i = 2, \ldots, 10$. Each of the ten intervals of $A^{(1)}$ is combined with each of the intervals of $A^{(2)}$. This leads to a discrete version of the joint random set A consisting of 100 rectangles A_{j_1, j_2} as focal sets with weights $p_{j_1, j_2} = p_{1,j_1} \cdot p_{2,j_2}$. The base rectangle $A_{1,1}$ is taken as the largest focal set as defined in Subsection 5.3 and is discretized by a grid of 141^2 points.

For each of these grid points approximations of the processes x_a are computed using the Order 1.5 Strong Taylor Scheme (see [97]). Each simulation involves 500 sample functions of the Wiener process on the time interval from $t_0 = 0$ up to time $t = 20$. Afterwards the system is no longer excited and the following free vibrations are observed during $5T_s$ seconds. The (constant) step size Δt of the time discretization is chosen as $1/20$. For each of the four components of the system response the approximations of the (interval) boundary processes are obtained by choosing in each time step, for each Wiener path and for each focal set of A among all $x_a(t)$ for which the grid point a is contained in the focal set the smallest and the largest value of the corresponding component. For reasons of comparison, the process $x_{\overline{a}}$ obtained by using the nominal parameter values $\overline{a} = (\overline{\zeta_d}, \overline{\omega_d})$ is considered as well.

Figure 44 (left) shows the bounds of a sample function of the non-dimensional displacement Y (bold lines, according to Equation (28)) obtained by choosing the largest focal set of A (which has probability assign-

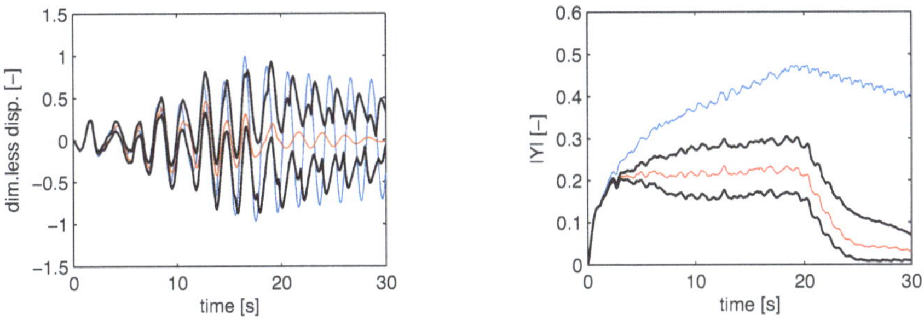

Figure 44. Left picture: Bounds of the sample function of the non-dimensional structural displacement Y (bold lines) corresponding to the largest focal set and sample functions of $y_{\overline{a}}$ (central thin line) and $\tilde{y}$ (outer thin lines, no TMD attached). Right picture: Mean value functions of the absolute values of the non-dimensional structural displacement $|Y|$ (bounding mean values of $|L|$ and $|U|$ in bold lines), $|y_{\overline{a}}|$ (central thin line) and $|\tilde{y}|$ (upper thin line, no TMD attached).

ment 0.0001) and $T_s = 2$ [s]. The thin lines represent the corresponding sample functions of $y_{\overline{a}}$ and $\tilde{y}$. One can see that although the TMD parameters differ from their nominal values by up to ±40% the TMD still has a positive impact on the displacement of the structure. In Figure 44 (right) the mean-value functions of $|Y|$, $|y_{\overline{a}}|$ and $|\tilde{y}|$ are plotted. Note that after a short period where the displacements of both systems are rapidly increasing the displacements $|Y|$ and $|y_{\overline{a}}|$ of the system damped by the TMD (bold lines and enclosed thin line) fluctuate around a constant value up to $t = 20$, whereas the displacement $|\tilde{y}|$ of the TMD-free SDOF system keeps growing. In the phase of free vibrations $|\tilde{y}|$ decays slowly whereas $|Y|$ and $|y_{\overline{a}}|$ fall off much more quickly.

The cumulative distribution functions of the first entrance times $\underline{\tau}^B$, $\tau_{\overline{a}}^B$ and the first inclusion time $\overline{\tau}^B$ of the processes $|Y|$ and $|y_{\overline{a}}|$ in the set $B = (0.5, \infty)$ are depicted in Figure 45 (left). Their expectation values are computed as

$$\mathrm{E}(\underline{\tau}^B) = 10.04, \quad \mathrm{E}(\tau_{\overline{a}}^B) = 13.67, \quad \mathrm{E}(\overline{\tau}^B) = 12.20.$$

In comparison to the first entrance and inclusion times one can consider the lower and the upper probability (see Equations (31) and (32)) of the set $B = (0.5, \infty)$ for the set-valued process $|Y|$ at each time $t \in [t_0, \overline{t}]$. These quantities are depicted in Figure 45 (right) by solid lines. The central line shows, for each $t \in [t_0, \overline{t}]$, the probability that $|y_{\overline{a}}(t)|$ is greater than 0.5.

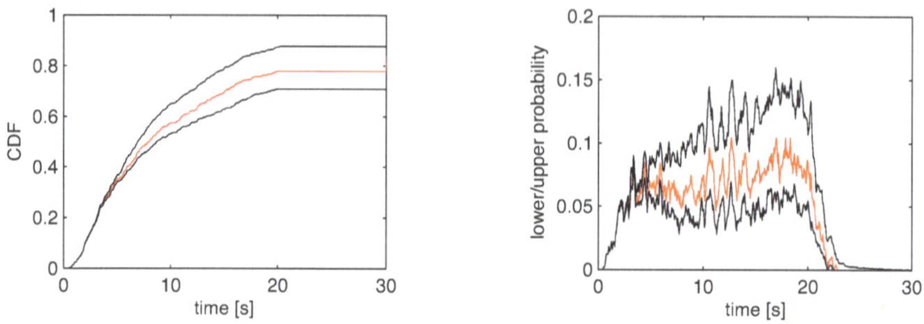

Figure 45. Left picture: Cumulative distribution functions of first entrance and inclusion times $\underline{\tau}^B$, $\overline{\tau}^B$ (outer lines) and $\tau^B_{\overline{a}}$ (central line) for $B = (0.5, \infty)$. Right picture: Plots of the lower and upper probabilities $\underline{P}(\cdot, B)$ and $\overline{P}(\cdot, B)$ for the set-valued process $|Y|$ (outer lines) and graph of $P(\cdot, B)$ for $|y_{\overline{a}}|$ (central line) ($B = (0.5, \infty)$.

6 Summary and conclusions

The number of proposed approaches to uncertainty modeling is enormous and ever increasing; some of them have been presented and explained in this lecture. The reader may well ask if there are some arguments to choose one or the other in a concrete problem at hand. Unquestionably, there is the normative aspect laid out in the Eurocodes, which would currently give preference to probabilistically based frameworks (which, not incidentally, also have been developed mathematically to the largest degree so far). Hopefully, the lectures have shown that interval based methods come in very naturally in engineering; a combination of both leads to various streams in imprecise probability.

One of the essential goals of uncertainty analysis is assessing the safety and reliability of structures. In as much as it gives predictions about singular structures (built just once with the same design), it cannot be put to test experimentally. It is the responsibility of the designing engineers and the decision makers to take the highest level of care and to properly apply the structure and the semantics of the chosen framework. There are no experimental tests that could prove or disprove that one or the other uncertainty theory is correct or not, appropriate or not. Put shortly, uncertainty analysis is not a scientific theory in the sense of Popper [158]. Rather, it should be seen as belonging to *decision theory* – however, rational decisions based on sound mathematical, scientific and engineering arguments.

6.1 Reliability Analysis and Decision Making in Engineering

The dream of a decision maker in engineering companies is to get a single number stating that the design of the structure to be built is safe. Classical probability appears to be able to meet this desire. This is reflected in the Eurocodes that purport and require a reliability $R = 1 - p_f$ of 0.999999. At best, such a number can be viewed as an *operational probability* (as explicitly stated in Eurocode EN 1990:2002 [48]). In a concrete application, such a number says nothing about the frequency of failure.

Thus the designing engineer is left with the question how big the distance to failure *actually* is. There is no answer to this question – there are many examples of structural failures due to reasons *not* taken into account in the model and the reliability analysis, or due to false reactions to unexpected behavior during construction, cf. [106]. The key to a well-founded decision about the safety of a structure is an interplay between model building, numerical calculations, laboratory experiments, in-situ investigations, and uncertainty analysis in a feed-back loop. (A showcase of such a procedure is described in [104].) This loop extends to monitoring during construction and possible re-design.

As mentioned, the available knowledge should be properly captured by the uncertainty model – thus the need for a framework going beyond probability and using various methods from imprecise probability. For example, if knowledge about a parameter consists in a lower and an upper bound, intervals should be used. If different plausibilities can be assigned to different intervals, random sets may be helpful. Fuzzy sets are especially useful in smaller models for getting a sketchy, but rapid assessment of the output uncertainties. To achieve this, simple three-point-estimates can be translated into triangular fuzzy numbers for a first overview of the behavior of the model under uncertainty.

Otherwise the choice of method is a question of the desired semantics. Generally, the basis for reliability analysis will be a sensitivity analysis of the model, identifying the most influential parameters. This may already result in a re-design in order to eliminate one or the other influence. A complete probabilistic or non-probabilistic reliability analysis is usually beyond available computing power and time. Instead, the analysis will usually be done with the aid of a reduced model, possibly given by interpolating or approximating the input-output function at a finite number of points (meta-model or response surface).

When proceeding with an imprecise probability model, the output parameters measuring the distance to failure come with upper and lower probabilities. In view of the lack of a total ordering, the designing engineer will have to base the decision about acceptance or rejection of the design on considerations as in multiple criteria decision making. In the end, a decision based on interval-valued, but accurate estimates will be more reliable than a decision based on sharp, but false numbers. As the methods become more widespread, this is more and more appreciated by practicing engineers who are aware of the need of a rational analysis of uncertainty.

6.2 Further Reading

Here is a selection of additional references on the aspects touched upon in the previous sections. It would be impossible to give a complete account of the hundreds of papers that have appeared so far on engineering applications.

The general concepts of probabilistic structural design are laid out in the Eurocode EN1990:2002 [48]; for a concise survey of computational methods, see [174]. As introductory literature to the probabilistic safety concept, see [10, 26]; for risk analysis concepts, see [116]. For models with random fields and stochastic processes, especially finite elements, see [67, 168]. A probabilistic approach to the uncertainty of the structural model has been developed by [114, 184, 185]; for an imprecise probability approach, see [55, 57, 74].

The topic of sensitivity analysis has been treated with sampling based methods [82] as well as with methods from imprecise probability theory [52, 72, 81, 144].

Various authors have critically appraised the probabilistic approach, notably the concept of failure probabilities [46, 140, 141, 142].

A monograph on random set models in engineering is [20]. These authors have used random set models in rock engineering, the assessment of seismic damage, and in reliability [18, 189, 191, 192]. See also the work [75, 164] on hydrological problems and on random sets for assessing the reliability: [59, 62, 143]. Random sets in finite element models have been developed in [156, 176, 177]; finite element analysis with probability boxes has been undertaken in [208] (for fuzzy methods and interval methods with finite elements, see [58, 124, 125, 154]; a recent survey is given in [117]).

A hybrid approach combining random sets and stochastic processes has been developed by [170] and applied to structures under earthquake excitation [171].

Evidence theory and p-boxes have been introduced into engineering models by [51, 148]. For upper and lower previsions and other imprecise probability models in reliability, see [31, 196, 197, 195]; a comparative survey has been recently given by [65], also covering the concept of clouds [130].

Interval probabilities for aiding engineering decisions have been proposed in [22, 73].

We should not fail to refer also to the literature using hybrid stochastic/fuzzy approaches [120]; fuzzy reliability [119, 121]; hybrid fuzzy set – random set modeling of seismic damage [19]. We also mention the monograph [163] comparing the fuzzy and the probabilistic approach in engineering applications. A representative survey of the rather extensive engineering literature using fuzzy sets only is beyond the scope of this summary. We recommend the reader to start from various references already given in the text or the collective works mentioned below. The same remark applies to the even vaster engineering literature using probabilistic methods.

Here are some more references on various approaches to fuzzy differential equations: [24, 27, 36, 39, 90, 91, 92, 136, 145, 153, 155, 160, 178].

Various alternative approaches to uncertainty quantification in engineering have been proposed: info-gap theory [16, 187], anti-optimization [47], robust statistics [6].

Finally, we mention four collections with various aspects of uncertainty modeling in engineering. The volume [9] emphasis fuzzy approaches, the volume [45] compares and discusses worst case scenarios, probability and random sets, while [134] contains a wealth of information on imprecise probability methods in engineering, including generalized information theory, evidence theory, interval methods, but also classical probabilistic ideas. The volume [49] critically appraises diverse aspects of uncertainty in civil engineering.

Acknowledgements. These lectures have been compiled from [139] and the published papers [135, 137, 138, 144, 146, 171]. I wish to thank my coauthors Christoph Adam (Unit for Applied Mechanics, University of Innsbruck), Wolfgang Fellin (Unit for Geotechnical and Tunnel Engineering, University of Innsbruck), Thomas Fetz, Bernhard Schmelzer (Unit for Engineering Mathematics, University of Innsbruck), Julian King (Breath Research Institute, Austrian Academy of Sciences) and Francesco Russo (Unité de Mathémathiques appliqués, Ecole Nationale Supérieure des Techniques Avancés, Paris) for their contributions and the kind permission to use the jointly obtained results. I also thank Isaac Elishakoff for the kind invitation to lecturing at the Advanced School on Nondeterministic Mechanics.

Bibliography

[1] H. Adeli and B. Möller, editors, *Special Issue: Fuzzy Modeling. Computer Aided Civil and Infrastructure Engineering* 14(2), 1999.

[2] D. A. Alvarez. On the calculation of the bounds of probability of events

using infinite random sets. *International Journal of Approximate Reasoning*, 43:241-267, 2006.

[3] D. A. Alvarez. Nonspecificity for infinite random sets of indexable type. *Fuzzy Sets and Systems*, 159:289–306, 2008.

[4] D. A. Alvarez. Reduction of uncertainty using sensitivity analysis methods for infinite random sets of indexable type. *International Journal of Approximate Reasoning*, 50:750–762, 2009.

[5] L. Arnold. *Stochastic Differential Equations: Theory and Applications.* Wiley, New York, 1974.

[6] T. Augustin and R. Hable. On the impact of robust statistics on imprecise probability models: a review. *Structural Safety*, 32:358–365, 2010.

[7] G. Auvinet and J. L. González. Three-dimensional reliability analysis of earth slopes. *Computers and Geotechnics* 26:247–261, 2000.

[8] E. O. Ayorinide and G. B. Warburton. Minimizing structural vibrations with absorbers. *Earthquake Engineering and Structural Dynamics* 8:219–236, 1980.

[9] B. M. Ayyub, editor, *Uncertainty Modelling and Analysis in Civil Engineering.* CRC Press, Boca Raton, 1998.

[10] G. W. Baecher and J. T. Christian. *Reliability and Statistics in Geotechnical Engineering.* John Wiley & Sons Ltd, Chichester, 2003.

[11] H. Bandemer. *Mathematics of Uncertainty. Ideas, methods, application problems.* Studies in Fuzziness and Soft Computing, vol. 189. Springer-Verlag, Berlin, 2006.

[12] H. Bandemer and S. Gottwald. *Einführung in Fuzzy-Methoden.* 4. Auflage, Akademie Verlag, Berlin, 1993.

[13] H. Bandemer and W. Näther. *Fuzzy Data Analysis.* Kluwer Academic Publishers, Dordrecht, 1992.

[14] A. Bárdoss and, L. Duckstein. *Fuzzy Rule-Based Modeling with Applications to Geophysical, Biological and Engineering Systems.* CRC Press, Boca Raton, 1995.

[15] T. Bedford and R. Cooke. *Probabilistic Risk Analysis.* Cambridge University Press, Cambridge, 2001.

[16] Y. Ben-Haim. *Information-gap decision theory. Decisions under severe uncertainty.* Series on Decision and Risk. Academic Press Inc., San Diego, CA, 2001.

[17] Y. Ben-Haim and I. Elishakoff. *Convex models of uncertainty in applied mechanics.* Studies in applied mechanics vol. 25, Elsevier, Amsterdam,

1990.

[18] A. Bernardini. Qualitative and quantitative measures in seismic damage assessment and forecasting of masonry buildings. In: A. Bernardini, editor, *Seismic Damage to Masonry Buildings*. A. A. Balkema, Rotterdam, 1999, pages 169–177.

[19] A. Bernardini, S. Lagomarsino, A. Mannella, A. Martinelli, L. Milano, and S. Parodi. Forecasting seismic damage scenarios of residential buildings from rough inventories: a case-study in the Abruzzo Region (Italy). *Journal of Risk and Reliability*, 224(4):279–296, 2010.

[20] A. Bernardini and F. Tonon. *Bounding Uncertainty in Civil Engineering. Theoretical Background*. Springer-Verlag, Berlin, 2010.

[21] C. Bertoluzza, M. Á. Gil and D. A. Ralescu, editors, *Statistical Modeling, Analysis and Management of Fuzzy Data*. Physica-Verlag, Heidelberg, 2002.

[22] D. Blockley and I. Ferraris. Managing the quality of a project. In Academia Nacional de Ingeniería, editor, *Seguridad en Ingeniería*. Editorial Dunken, Buenos Aires, 2000, pages 69–89.

[23] V. V. Bolotin. *Statistical Methods in Structural Mechanics*. Holden-Day Inc., San Francisco, 1969.

[24] A. Bonarini and G. Bontempi. A qualitative simulation approach for fuzzy dynamical models. *ACM Transactions on Modeling and Computer Simulation* 4:285–313, 1994.

[25] E. Bösinger. Serienuntersuchungen zum Vergleich verschiedenener Rahmenschergeräte. Institut für Bodenmechanik und Felsmechanik, Universität Karlsruhe, 1996.

[26] C. Bucher. *Computational Analysis of Randomness in Structural Mechanics*, volume 3 of *Structures and Infrastructures*. CRC Press/Balkema, Leiden, 2009.

[27] J. J. Buckley. Solving fuzzy equations. *Fuzzy Sets and Systems*, 50:1–14, 1992.

[28] J. J. Buckley, E. Eslami, T. Feuring. *Fuzzy Mathematics in Economics and Engineering*. Physica-Verlag, Heidelberg, 2002.

[29] G. Choquet. Theory of capacities. *Annales de l'Institut Fourier* 5:131–295, 1953/54.

[30] C. Christopoulos and A. Filitrault. *Principles of Passive Supplemental Damping and Seismic Isolation*. IUSS Press, Pavia, 2006.

[31] G. de Cooman. On modeling possibilistic uncertainty in two-state reliability theory. *Fuzzy Sets and Systems*, 83(2):215–238, 1996.

[32] G. de Cooman, T. Fine, T. Seidenfeld, editors, *ISIPTA'01, Proceedings of the Second International Symposium on Imprecise Probabilities and Their Applications.* Shaker Publishing BV, Maastricht, 2001.

[33] B. de Finetti. Teoria delle probabilità: sintesi introduttiva con appendice critica. Volumi primo e secondo. Nuova Biblioteca Scientifica Einaudi, 25* et 25**. Giulio Einaudi Editore, Torino, 1970. English translation: Theory of probability: a critical introductory treatment. Vol. 1 and Vol. 2. John Wiley & Sons, London, 1974/1975.

[34] A. P. Dempster. Upper and lower probabilities induced by a multivalued mapping. *Annals of Mathematical Statistics* 38:325–339, 1967.

[35] D. Denneberg. *Non-additive Measure and Integral.* Kluwer Acadademic Publishers, Dordrecht, 1994.

[36] P. Diamond and P. Kloeden. *Metric Spaces of Fuzzy Sets.* World Scientific, Singapore, 1994.

[37] R. Diedrich. *Entscheidungen bei Ungewißheit.* Physica-Verlag, Heidelberg, 1999.

[38] O. Ditlevsen, N. J. Tarp-Johansen and H. Denver. Bayesian soil assessments combining prior with posterior censored samples. *Computers and Geotechnics* 26:187–198, 2000.

[39] D. Dubois and H. Prade. Towards fuzzy differential calculus, part 3: differentiation. *Fuzzy Sets and Systems* 8:223–233, 1982.

[40] D. Dubois and H. Prade. *Possibility theory. An approach to computerized processing of uncertainty.* Plenum Press, New York, 1988.

[41] D. Dubois, H. Prade and R. Sabbadin. Decision-theoretic foundations of qualitative possibility theory. *European Jornal of Operational Res.* 128:459–478, 2001.

[42] D. Dubois, H. Prade and S. Sandri. On possibility/probability transformations. In: R. Lowen, M. Roubens, editors, *Fuzzy Logic: State of the Art.* Kluwer Academic Publishers, Dordrecht, 1993, pages 103–112.

[43] H. H. Einstein. Quantifying uncertain engineering geologic information. *Felsbau,* 19(5):72–84, 2001.

[44] F. Eisenführ and M. Weber. *Rationales Entscheiden.* 3. Auflage, Springer-Verlag, Berlin, 1999.

[45] I. Elishakoff, editor, *Whys and Hows in Uncertainty Modelling: Probability, Fuzziness and Anti-Optimization,* volume 388 of *CISM Courses and Lectures.* Springer-Verlag, Wien/New York, 1999.

[46] I. Elishakoff. Are probabilistic and anti-optimization approaches compatible? In I. Elishakoff, editor, *Whys and Hows in Uncertainty Modelling.* Springer-Verlag, Wien/New York, 1999, pages 263–355.

[47] I. Elishakoff and M. Ohsaki. *Optimization and Anti-Optimization of Structures Under Uncertainty*. Imperial College Press, London, 2010.

[48] European Committee for Standardization. *EN 1990:2002. Eurocode – Basis of Structural Design*. CEN, Brussels, 2002.

[49] W. Fellin, H. Lessmann, M. Oberguggenberger and R. Vieider. *Analyzing Uncertainty in Civil Engineering*. Springer-Verlag, Berlin, 2005.

[50] W. Fellin and A. Ostermann. Parameter sensitivity in finite element analysis with constitutive models of the rate type. *International Journal for Numerical and Analytical Methods in Geomechanics* 30:91–112, 2006.

[51] S. Ferson, V. Kreinovich, L. Ginzburg, D. S. Myers, and K. Sentz. Constructing probability boxes and Dempster-Shafer structures. Technical Report SAND2002–4015, Sandia National Laboratories, 2003.

[52] S. Ferson and W. T. Tucker. Sensitivity analysis using probability bounding. *Reliability Engineering and System Safety*, 91:1435–1442, 2006.

[53] T. Fetz. Sets of joint probability measures generated by weighted marginal focal sets. In: G. de Cooman, T. Fine, T. Seidenfeld, editors, *ISIPTA'01, Proceedings of the Second International Symposium on Imprecise Probabilities and Their Applications*. Shaker Publishing BV, Maastricht, 2001, pages 171–178.

[54] T. Fetz. Multi-parameter models: rules and computational methods for combining uncertainties. In: W. Fellin, H. Lessmann, M. Oberguggenberger and R. Vieider, editors, *Analyzing Uncertainty in Civil Engineering*. Springer-Verlag, Berlin, 2005, pages 73–99.

[55] T. Fetz. Modelling uncertainties in limit state functions. In F. Coolen, G. de Cooman, T. Fetz, and M. Oberguggenberger, editors, *ISIPTA'11: Proceedings of the Seventh International Symposium on Imprecise Probability: Theory and Applications*. Studia Verlag, Innsbruck, 2011, pages 179–188.

[56] T. Fetz. Multivariate Models and Variability Intervals: A Local Random Set Approach. *International Journal of Uncertainty, Fuzziness and Knowledge-based Systems* 19(5):799–823, 2011.

[57] T. Fetz. Modelling uncertainties in limit state functions. *International Journal of Approximate Reasoning* 53(1):1–23, 2012.

[58] T. Fetz, J. Jäger, D. Köll, G. Krenn, M. Oberguggenberger and H. Lessmann. Fuzzy models in geotechnical engineering and construction management. *Computer-Aided Civil & Infrastructure Engineering*, 14(2):93–106, 1999.

[59] T. Fetz and M. Oberguggenberger. Propagation of uncertainty through multivariate functions in the framework of sets of probability measures. *Reliability Engineering and System Safety*, 85:73–87, 2004.

[60] T. Fetz and M. Oberguggenberger. Multivariate models of uncertainty: A local random set approach. *Structural Safety*, 32:417–424, 2010.

[61] T. Fetz, M. Oberguggenberger and S. Pittschmann. Applications of possibility and evidence theory in civil engineereing. *International Journal of Uncertainty, Fuzziness and Knowledge-Based Systems* 8:295–309, 2000.

[62] T. Fetz and F. Tonon. Probability bounds for series systems with variables constrained by sets of probability measures. *International Journal of Reliability and Safety*, 2:309–339, 2008.

[63] T. L. Fine. *Theories of Probability. An Examination of Foundations.* Academic Press, New York, 1973.

[64] A. N. Freudenthal. Safety and the probability of structural failure. *Transactions ASCE*, 121:1337–1397, 1956.

[65] M. Fuchs. Clouds, p-boxes, fuzzy sets, and other uncertainty representations in higher dimensions. *Acta Cybernetica*, 19:61–92, 2009.

[66] A. B. Gelman, J. S. Carlin, H. S. Stern and D. B. Rubin. *Bayesian Data Analysis.* Chapman & Hall/CRC, Boca Raton, 1997.

[67] R. G. Ghanem and P. D. Spanos. *Stochastic finite elements: a spectral approach.* Springer-Verlag, New York, 1991.

[68] I. I. Gikhman and A. V. Skorokhod. *Stochastische Differentialgleichungen.* Akademie-Verlag, Berlin, 1971.

[69] I. R. Goodman and H. T. Nguyen. Fuzziness and randomness. In C. Bertoluzza, M. Á. Gil, and D. A. Ralescu, editors, *Statistical Modeling Analysis and Management of Fuzzy Data.* Physica-Verlag, Heidelberg, 2002, pages 3–21.

[70] J. Goutsias, R. P. S. Mahler and H. T. Nguyen, editors, *Random Sets. Theory and Applications.* Springer-Verlag, New York, 1997.

[71] K. Haldar and R. K. Reddy. A random-fuzzy analysis of existing structures. *Fuzzy Sets and Systems*, 48:201–210, 1992.

[72] J. Hall. Uncertainty-based sensitivity indices for imprecise probability distributions. *Reliability Engineering & System Safety*, 91:1443–1451, 2006.

[73] J. Hall, D. I. Blockley, and J. P. Davis. Uncertain inference using interval probability theory. *International Journal of Approximate Reasoning*, 19:247–264, 1998.

[74] J. Hall and J. Lawry. Imprecise probabilities of engineering system failure from random and fuzzy set reliability analysis. In G. de Cooman, T. L. Fine, and T. Seidenfeld, editors, *ISIPTA'01: Proceedings of the Second International Symposium on Imprecise Probability: Theories and Applications*. Shaker Publishing BV, Maastricht, 2001, pages 195–204.

[75] J. Hall, E. Rubio, and M. Anderson. Random sets of probability measures in slope hydrology and stability analysis. *Journal of Applied Mathematics and Mechanics*, 84(10–11):710–720, 2004.

[76] P. R. Halmos. *Measure Theory*. Van Nostrand, New York, 1950.

[77] J. Hanisch and W. Struck. Charakteristischer Wert einer Boden- oder Materialeigenschaft aus Stichprobenergebnissen und zusätzlicher Information. *Bautechnik* 10:338–348, 1985.

[78] M. Hanss. *Applied Fuzzy Arithmetic. An Introduction with Engineering Applications*. Springer-Verlag, Berlin, 2005.

[79] J. C. Helton and F. J. Davis. Sampling-based methods for uncertainty and sensitivity analysis. Technical Report SAND99-2240, Sandia National Laboratories, Albuquerque, 2000.

[80] J. C. Helton and F. J. Davis. Latin hypercube sampling and the propagation of uncertainty in analyses of complex systems. Technical Report SAND2001-0417, Sandia National Laboratories, Albuquerque, 2002.

[81] J. C. Helton, J. D. Johnson, W. L. Oberkampf, and C. J. Sallaberry. Sensitivity analysis in conjunction with evidence theory representations of epistemic uncertainty. *Reliability Engineering and System Safety*, 91:1414–1434, 2006.

[82] J. C. Helton, J. D. Johnson, C. J. Sallaberry, and C. B. Storlie. Survey of sampling-based methods for uncertainty and sensitivity analysis. *Reliability Engineering & System Safety*, 91:1175–1209, 2006.

[83] R. Heuer and C. Adam. Passive control of structural elements considering nonlinear response. In: R. Flesch, H. Irschik and M. Krommer, editors, *Proceedings of the Third European Conference on Structural Control (3ECSC), July 12 - 15, 2004, Vienna University of Technology*. Schriftenreihe der TU-Wien, Wien, 2005, pages 191–194.

[84] P. J. Huber. *Robust Statistics*. John Wiley & Sons, New York, 1981.

[85] R. L. Iman and W. J. Conover. The use of the rank transformation in regression. *Technometrics*, 21:499–509, 1979.

[86] R. L. Iman and W. J. Conover. A distribution-free approach to inducing rank correlation among input variables. *Communications in Statistics - Simulation and Computation*, 11:311–334, 1982.

[87] R. L. Iman, M. J. Shortencarier and J. D. Johnson. A FORTRAN 77 program and user's guide for the calculation of partial correlation and standardized regression coefficients. Technical Report SAND85-0044, Sandia National Laboratories, Albuquerque, 1985.

[88] J.-Y. Jaffray. Linear utility theory for belief functions. *Operations Research Letters* 8:107–112, 1989.

[89] E. J. Jung and J. H. Kim. On set-valued stochastic integrals. *Stochastic Analysis and Applications.* 21:401–418, 2003.

[90] O. Kaleva. Fuzzy differential equations. *Fuzzy Sets and Systems* 24:301–307, 1987.

[91] O. Kaleva. The Cauchy problem for fuzzy differential equations. *Fuzzy Sets and Systems* 35:389–396, 1990.

[92] A. Kandel. *Fuzzy Mathematical Techniques with Applications.* Addison-Wesley, Reading, 1986.

[93] I. Karatzas and S. E. Shreve. *Brownian Motion and Stochastic Calculus.* Springer-Verlag, New York, 2005.

[94] L. Kleinrock. *Queueing Systems, vol. I: Theory.* John Wiley, New York, 1975.

[95] O. Klingmüller and U. Bourgund. *Sicherheit und Risiko im Konstruktiven Ingenieurbau.* Verlag Vieweg, Braunschweig, 1992.

[96] G. Klir and M. Wierman. *Uncertainty-Based Information: Elements of Generalized Information Theory.* Physica-Verlag, Heidelberg, 1998.

[97] P. E. Kloeden and E. Platen. *Numerical Solution of Stochastic Differential Equations.* Springer-Verlag, Berlin, 1992.

[98] A. Kolmogoroff. *Grundbegriffe der Wahrscheinlichkeitsrechnung.* Springer-Verlag, Berlin, 1933.

[99] V. Kreinovich. Random sets unify, explain, and aid known uncertainty methods in expert systems. In: J. Goutsias, R. P. S. Mahler and H. T. Nguyen, editors, *Random Sets. Theory and Applications.* Springer-Verlag, New York, 1997, pages 321–345.

[100] E. Kriegler, H. Held. Global mean temperature projections for the 21st century using random sets. In: J.-M. Bernard, T. Seidenfeld and M. Zaffalon, editors, *ISIPTA'03. Proceedings of the Third International Symposium on Imprecise Probabilities and Their Applications.* Proceedings in Informatics 18, Carleton Scientific, Kitchener, Ontario, 2003, pages 345–360.

[101] W. Kruskal. Relative importance by averaging over orderings. *The American Statistician* 41:6–10, 1987.

[102] G. Kühn. *Der maschinelle Erdbau.* Teubner-Verlag, Stuttgart, 1984.

[103] R. Langley. Unified approach to probabilistic and possibilistic analysis of uncertain systems. *Journal of Engineering Mechanics* 126(11):1163–1172, 2000.

[104] H. Lehar, G. Niederwanger, and G. Hofstetter. FE ultimate load analysis of pile-supported pipelines – tackling uncertainty in a real design problem. In W. Fellin, H. Lessmann, M. Oberguggenberger, and R. Vieider, editors, *Analyzing Uncertainty in Civil Engineering.* Springer-Verlag, Berlin, 2005, pages 129–163.

[105] H. Lessmann, J. Mühlögger and M. Oberguggenberger. Netzplantechnik mit unscharfen Methoden. *Bauingenieur* 69:469–478, 1994.

[106] H. Lessmann and R. Vieider. The implications of geotechnical model uncertainty for construction management. In: W. Fellin, H. Lessmann, M. Oberguggenberger and R. Vieider, editors, *Analyzing Uncertainty in Civil Engineering.* Springer-Verlag, Berlin, 2005, pages 165–182.

[107] J. Li and S. Li. Itô type set-valued stochastic differential equations. *Journal of Uncertain Systems* 3:52–63, 2009.

[108] R.-J. Li and E. S. Lee. Analysis of fuzzy queues. *Computers & Mathematics with Applications* 17:1143–1147, 1989.

[109] S. Li and A. Ren. Representation theorems, set-valued and fuzzy set-valued Itô integral. *Fuzzy Sets and Systems* 158:949–962, 2007.

[110] B. S. L. P. de Lima and N. F. F. Ebecken. A comparison of models for uncertainty analysis by the finite element method. *Finite Elements in Analysis and Design* 34:211–232, 2000.

[111] M. Loève. *Probability Theory, vol. II.* 4th edition, Springer-Verlag, New York, 1978.

[112] H. F. Martz and R. A. Waller. *Bayesian Reliability Analysis.* John Wiley & Sons Ltd., Chichester, 1982.

[113] M. A. Meyer and J. M. Booker. *Eliciting and Analyzing Expert Judgement.* Society for Industrial and Applied Mathematics, Philadelphia PA, 2001.

[114] M. P. Mignolet and C. Soize. Stochastic reduced order models for uncertain geometrically nonlinear dynamical systems. *Computer Methods in Applied Mechanics and Engineering,* 197(45-48):3951–3963, 2008.

[115] E. Miranda, I. Couso and P. Gil. Extreme points of credal sets generated by 2-alternating capacities. *International Journal of Approximate Reasoning* 33:95–115, 2003.

[116] M. Modarres. *Risk Analysis in Engineering. Techniques, Tools, and Trends.* Taylor & Francis, Boca Raton, 2006.

[117] D. Moens and D. Vandepitte. A survey of non-probabilistic uncertainty treatment in finite element analysis. *Computer Methods in Applied Mechanics and Engineering*, 194:1527–1555, 2005.

[118] I. Molchanov. *Theory of Random Sets.* Springer-Verlag, Berlin, 2005.

[119] B. Möller. Fuzzy-Modellierung in der Baustatik. *Bauingenieur*, 72(5):75–84, 1997.

[120] B. Möller and M. Beer. *Fuzzy Randomness. Uncertainty in Civil Engineering and Computational Mechanics.* Springer-Verlag, Berlin, 2004.

[121] B. Möller, M. Beer, W. Graf, and A. Hoffmann. Possibility theory based safety assessment. *Computer-Aided Civil & Infrastructure Engineering*, 14(2):81–91, 1999.

[122] P.-A. Monney. *A Mathematical Theory for Statistical Evidence.* Physica-Verlag, Heidelberg, 2003.

[123] D. C. Montgomery, E. A. Peck and G. G. Vining: *Introduction to Linear Regression Analysis.* 3rd edition, John Wiley & Sons, New York, 2001.

[124] R. L. Muhanna and R. L. Mullen. Formulation of fuzzy finite-element methods for solid mechanics problems. *Computer-Aided Civil & Infrastructure Engineering*, 14(2):107–117, 1999.

[125] R. L. Muhanna, H. Zhang and R. L. Mullen. Interval finite elements as a basis for generalized models of uncertainty in engineering mechanics. *Reliable Computing*, 13(2):173–194, 2007.

[126] R. Nau. Extensions of the Subjective Expected Utility Model. In: W. Edwards, R. Miles Jr. and D. von Winterfeldt, editors, *Advances in Decision Analysis: From Foundations to Applications*, Cambridge University Press, Cambridge, 2007, pages 253–278.

[127] R. B. Nelsen. *An introduction to copulas.* 2nd edition. Springer Series in Statistics. Springer, New York, 2006.

[128] A. Neumaier. *Interval Methods for Systems of Equations.* Cambridge University Press, Cambridge, 1990.

[129] A. Neumaier. Fuzzy modeling in terms of surprise. *Fuzzy Sets and Systems* 135:21–38, 2003.

[130] A. Neumaier. Clouds, fuzzy sets and probability intervals. *Reliable Computing*, 10:249–272, 2004.

[131] H. T. Nguyen, T. Wang. Belief functions and random sets. In: J. Goutsias, R. P. S. Mahler, H. T. Nguyen, editors, *Random Sets. Theory and Applications.* Springer-Verlag, New York, 1997, pages 243–255.

[132] H. T. Nguyen. *An Introduction to Random Sets.* Chapman & Hall,

Boca Raton, 2006.

[133] G. Niederwanger. Über die Horizontalkraft schwingender Glocken mit horizontal verschiebbarem Auflager. Projektbericht, Universität Innsbruck, Innsbruck, 1991.

[134] E. Nikolaidis, D. M. Ghiocel and S. Singhal, editors, *Engineering Design Reliability Handbook*, CRC Press, Boca Raton, 2005.

[135] M. Oberguggenberger. Fuzzy data in geotechnical models. In: D. Kolymbas, editor, *Advanced Mathematical and Computational Geomechanics*. Springer, Berlin, 2003, pages 301–315.

[136] M. Oberguggenberger. Fuzzy and weak solutions to differential equations. Proceedings of the Tenth International Conference IPMU 2004, Perugia. Editrice Università La Sapienza, Rome, 2004, pages 517–524.

[137] M. Oberguggenberger. The mathematics of uncertainty: models, methods and interpretations. In: W. Fellin, H. Lessmann, M. Oberguggenberger and R. Vieider, editors, *Analyzing Uncertainty in Civil Engineering*. Springer-Verlag, Berlin, 2005, pages 51–72.

[138] M. Oberguggenberger. Queueing models with fuzzy data in construction management. In: W. Fellin, H. Lessmann, M. Oberguggenberger and R. Vieider, editors, *Analyzing Uncertainty in Civil Engineering*. Springer-Verlag, Berlin, 2005, pages 197–210.

[139] M. Oberguggenberger. Engineering. In: T. Augustin, F. Coolen, G. de Cooman and M. Troffaes, editors, *Introduction to Imprecise Probabilities*. John Wiley & Sons Ltd., Chichester, to appear.

[140] M. Oberguggenberger and W. Fellin. From probability to fuzzy sets: the struggle for meaning in geotechnical risk assessment. In R. Pöttler, H. Klapperich, and H. F. Schweiger, editors, *Probabilistics in Geotechnics: Technical and Economic Risk Estimation*, pages 29–38. Verlag Glückauf, Essen, 2002.

[141] M. Oberguggenberger and W. Fellin. Assessing the sensitivity of failure probabilities: a random set approach. In: G. Augusti, G. I. Schuëller and M. Ciampoli, editors, *Safety and Reliability of Engineering Systems and Structures, ICOSSAR 2005 – Rome*. Millpress, Rotterdam, 2005, pages 1755–1760.

[142] M. Oberguggenberger and W. Fellin. The fuzziness of failure probabilities. In W. Fellin, H. Lessmann, M. Oberguggenberger, and R. Vieider, editors, *Analyzing Uncertainty in Civil Engineering*, pages 33–49. Springer-Verlag, Berlin, 2005.

[143] M. Oberguggenberger and W. Fellin. Reliability bounds through random sets: nonparametric methods and geotechnical applications. *Com-*

puters & Structures, 86:1093–1101, 2008.

[144] M. Oberguggenberger, J. King and B. Schmelzer. Classical and imprecise probability methods for sensitivity analysis in engineering: A case study. *International Journal of Approximate Reasoning*, 50(4):680–693, 2009.

[145] M. Oberguggenberger and S. Pittschmann. Differential equations with fuzzy parameters. *Mathematical and Computer Modelling of Dynamical Systems*, 5:181-202, 1999.

[146] M. Oberguggenberger and F. Russo. Fuzzy, probabilistic and stochastic modelling of an elastically bedded beam. In: G. de Cooman, T. Fine and T. Seidenfeld, editors, *ISIPTA '01, Proceedings of the Second International Symposium on Imprecise Probabilities and Their Applications.* Shaker Publishing BV, Maastricht, 2001, pages 293–300.

[147] M. Oberguggenberger, G. I. Schuëller and K. Marti, editors, *Special Issue: Fuzzy Sets, Imprecise Probability and Stochastics in Engineering. Journal of Applied Mathematics and Mechanics* 84(10-11), 2004.

[148] W. L. Oberkampf and J. C. Helton. Investigation of Evidence Theory for engineering applications. Technical Report AIAA 2001–1569, American Institute of Aeronautics and Astronautics, Reston VA, 2002.

[149] W. L. Oberkampf, J. C. Helton and K. Sentz. Mathematical Representation of Uncertainty. Technical Report AIAA 2001–1645, American Institute of Aeronautics and Astronautics, Reston VA, Non-Deterministic Approaches Forum, Seattle WA, 2001.

[150] A. O'Hagan. Eliciting expert beliefs in substantial practical applications. *The Statistician* 47:21–35, 1998.

[151] ÖNORM B4435-2. Erd- und Grundbau, Flächengründungen, EUROCODE-nahe Berechnung der Tragfähigkeit. Österreichisches Normungsinstitut, Wien, 1999.

[152] A. Ostermann. Sensitivity analysis. In: W. Fellin, H. Lessmann, M. Oberguggenberger and R. Vieider, editors, *Analyzing Uncertainty in Civil Engineering.* Springer-Verlag, Berlin, 2005, pages 101–114.

[153] K. N. Otto, A. D. Lewis, and E. K. Antonsson. Membership induced on manifolds by vector fields and flows. *Fuzzy Sets and Systems* 70:15–29, 1995.

[154] K. Ovaere, G. Deschrijver and E. E. Kerre. Application of fuzzy decision making to the fuzzy finite element method. *Fuzzy Information and Engineering*, 11:27–36, 2009.

[155] R. Palm and D. Driankov. Fuzzy inputs. *Fuzzy Sets and Systems* 70:315–335, 1995.

[156] G. M. Peschl. *Reliability Analyses in Geotechnics with the Random Set Finite Element Method*, volume 25 of *Mitteilungshefte Gruppe Geotechnik Graz*. Technische Universität Graz, 2005.

[157] K. Popper. The propensity interpretation of probability. *British Journal for the Philosophy of Science* 10:25–42, 1959.

[158] K. Popper. *Logik der Forschung*. 10. Auflage, J. C. B. Mohr (Paul Siebeck), Tübingen, 1994.

[159] F. Pukelsheim. The Three Sigma Rule. *The American Statistician* 48:88–91, 1994.

[160] M. L. Puri and D. Ralescu. Differentials of fuzzy functions. *Journal of Mathematical Analysis and Applications* 91:552–558, 1983.

[161] R. Rackwitz. Reviewing probabilistic soils modelling. *Computers and Geotechnics*, 26:199–223, 2000.

[162] T. J. Ross. *Fuzzy Logic with engineering applications*. 2nd edition. Wiley India, New Delhi, 2008.

[163] T. J. Ross, J. M. Booker. and W. J. Parkinson, editors, *Fuzzy Logic and Probability Applications. Bridging the Gap*, Philadelphia, PA/Alexandria, VA, 2002. Society for Industrial and Applied Mathematics/Americal Statistical Association.

[164] E. Rubio, J. Hall and M. G. Anderson. Uncertainty analysis in a slope hydrology and stability model using probabilistic and imprecise information. *Computers and Geotechnics*, 31:529–536, 2004.

[165] A. Saltelli and I. M. Sobol. About the use of rank transformation in sensitivity analysis of model output. *Reliability Engineering & System Safety*, 50:225-239, 1995.

[166] A. Saltelli, M. Ratto, T. Andres, F. Campolongo, J. Cariboni, D. Gatelli, M. Saisana and S. Tarantola. *Global sensitivity analysis. The primer*. John Wiley & Sons Ltd., Chichester, 2008.

[167] L. J. Savage. *The Foundations of Statistics*. Wiley, New York, 1954.

[168] C. A. Schenk and G. I. Schuëller. *Uncertainty Assessment of large Finite Element Systems*, volume 24 of *Lecture Notes in Applied and Computational Mechanics*. Springer, Berlin, 2005.

[169] B. Schmelzer. On solutions of stochastic differential equations with parameters modelled by random sets. In: T. Augustin, F. P. A. Coolen, S. Moral and M. C. M. Troffaes, editors, *ISIPTA'09, Proceedings of the Sixth International Symposium on Imprecise Probability: Theories and Applications*. SIPTA, Durham, 2009, pages 367–376.

[170] B. Schmelzer. On solutions to stochastic differential equations with parameters modeled by random sets. *International Journal of Approx-*

imate Reasoning, 51:1159–1171, 2010.

[171] B. Schmelzer, M. Oberguggenberger and C. Adam. Efficiency of tuned mass dampers with uncertain parameters on the performance of structures under stochastic excitation. *Journal of Risk and Reliability*, 224(4):297–308, 2010.

[172] H. R. Schneider. Definition and determination of characteristic soil properties. In: *Proceedings of the Fourteenth International Conference on Soil Mechanics and Foundation Engineering, ICSMFE 14, Hamburg 1997.* A.A. Balkema, Rotterdam, 1999, Vol. 4, pages 2271–2274.

[173] G. I. Schuëller. On the treatment of uncertainties in structural mechanics and analysis. *Computers & Structures*, 85:235–243, 2007.

[174] G. I. Schuëller and H. J. Pradlwarter. Uncertainty analysis of complex structural systems. *International Journal for Numerical Methods in Engineering*, 80(6-7):881–913, 2009.

[175] G. I. Schuëller, H. J. Pradlwarter and P. S. Koutsourelakis. A critical appraisal of reliability estimation procedures for high dimensions. *Probabilistic Engineering Mechanics,* 19:463–474, 2004.

[176] H. F. Schweiger and G. M. Peschl. Reliability analysis in geotechnics with the random set finite element method. *Computers and Geotechnics*, 32:422–435, 2005.

[177] H. F. Schweiger, G. M. Peschl, and R. Pöttler. Application of the random set finite element method for analysing tunnel excavation. *Georisk*, 1:43–56, 2007.

[178] S. Seikkala. On the fuzzy initial value problem.*Fuzzy Sets and Systems* 24:319–330, 1987.

[179] R. Seising, editor, *Fuzzy Theorie und Stochastik.* Vieweg, Braunschweig, 1999.

[180] G. L. S. Shackle. *Decision – order and time in human affairs.* Cambridge University Press, Cambridge, 1969.

[181] G. Shafer. *A Mathematical Theory of Evidence.* Princeton University Press, Princeton NJ, 1976.

[182] J. Shao and D.-S. Tu. *The Jackknife and Bootstrap.* Springer Verlag, New York, 1995.

[183] I. M. Sobol. Sensitivity analysis for nonlinear mathematical models. *Mathematical Modeling & Computational Experiment*, 1:407–414, 1993.

[184] C. Soize. A comprehensive overview of a non-parametric probabilistic approach of model uncertainties for predictive models in structural dynamics. *Journal of Sound and Vibration*, 288(3):623–652, 2005.

[185] C. Soize. Random matrix theory for modeling uncertainties in computational mechanics. *Computer Methods in Applied Mechanics and Engineering*, 194(12-16):1333–1366, 2005.

[186] T. T. Soong. *Active Structural Control: The Theory and Practice*. Longman Scientific and Technical Series, New York, 1990.

[187] I. Takewakia and Y. Ben-Haim. Info-gap robust design with load and model uncertainties. *Journal of Sound and Vibration* 288:551–570, 2005.

[188] F. Tonon. On the use of random set theory to bracket the results of Monte Carlo simulations. *Reliable Computing* 10:107–137, 2004.

[189] F. Tonon. Using random set theory to propagate epistemic uncertainty through a mechanical system. *Reliability Engineering and System Safety*, 85:169–181, 2004.

[190] F. Tonon, A. Bernardini and I. Elishakoff. Hybrid Analysis of uncertainty: probability, fuzziness and anti-optimization *Chaos, Solitons and Fractals* 12:1403–1414, 2001.

[191] F. Tonon, A. Bernardini, and A. Mammino. Determination of parameters range in rock engineering by means of random set theory. *Reliability Engineering and System Safety*, 70:241–261, 2000.

[192] F. Tonon, A. Bernardini, and A. Mammino. Reliability analysis of rock mass response by means of random set theory. *Reliability Engineering and System Safety*, 70:263–282, 2000.

[193] F. Tonon and C. L. Pettit. Toward a definition and understanding of correlation for variables constrained by random relations. In G. Augusti, G. I. Schuëller, and M. Ciampoli, editors, *Safety and Reliability of Engineering Systems and Structures, ICOSSAR 2005 - Rome*. Millpress, Rotterdam, 2005, pages 1741–1745.

[194] L. V. Utkin. An uncertainty model of structural reliability with imprecise parameters of probability distributions. Journal of Applied Mathematics and Mechanics 84(10-11):688–699, 2004.

[195] L. V. Utkin and F. P. A. Coolen. Imprecise reliability: an introductory overview. In G. Levitin, editor, *Computational Intelligence in Reliability Engineering, Volume 2: New Metaheuristics, Neural and Fuzzy Techniques in Reliability*, Springer-Verlag, Berlin, 2007, pages 261–306.

[196] L. V. Utkin and I. O. Kozine. Computing the reliability of complex systems. In G. de Cooman, T. L. Fine, and T. Seidenfeld, editors, *ISIPTA'01: Proceedings of the Second International Symposium on Imprecise Probability: Theories and Applications*. Shaker Publishing BV, Maastricht, 2001, pages 324–331.

[197] L. V. Utkin and I. O. Kozine. Stress-strength reliability models under incomplete information. *International Journal of General Systems*, 31(6):549–568, 2002.

[198] R. Viertl. *Einführung in die Stochastik.* Mit Elementen der Bayes-Statistik und der Analyse unscharfer Information. 3. Auflage, Springer-Verlag, Wien, 2003.

[199] P. Walley. *Statistical Reasoning with Imprecise Probabilities.* Chapman & Hall, London, 1991.

[200] P. Walley. Towards a unified theory of imprecise probability. *International Journal of Approximate Reasoning* 24:125–148, 2000.

[201] Z. Wang and G. J. Klir. *Fuzzy Measure Theory.* Plenum Press, New York, 1992.

[202] Z. Wang and G. J. Klir. *Generalized Measure Theory.* IFSR International Series on Systems Science and Engineering, vol. 25. Springer, New York, 2009.

[203] K. Weichselberger. *Elementare Grundbegriffe einer allgemeinen Wahrscheinlichkeitsrechnung I.* Physica-Verlag, Heidelberg, 2001.

[204] O. Wolkenhauer. *Data Engineering. Fuzzy Mathematics in Systems Theory and Data Analysis.* John Wiley & Sons, New York, 2001.

[205] L. A. Zadeh. Fuzzy sets. *Information and Control* 8:338–353, 1965.

[206] L. A. Zadeh. The concept of a linguistic variable and its applications to approximate reasoning, Part 1 - 3. *Information Sci.* 8:199–250, 8:301–357; 9:43–80, 1975.

[207] L. A. Zadeh. Fuzzy sets as a basis for a theory of possibility. *Fuzzy Sets and Systems* 1:3–28, 1978.

[208] H. Zhang, R. L. Mullen and R. L. Muhanna. Finite element structural analysis using imprecise probabilities based on p-box representation. In M. Beer, R. L. Muhanna and R. L. Mullen, editors, *4th International Workshop on Reliable Engineering Computing (REC 2010). Robust Design – Coping with Hazards, Risk and Uncertainty.* Research Publishing Services, Singapore, 2010, pages 211–225.

[209] H.-J. Zimmermann. *Fuzzy Sets, Decision Making, and Expert Systems.* Kluwer Academic Publishers, Boston, 1987.

[210] H.-J. Zimmermann. *Fuzzy Set Theory – and Its Applications.* 2nd edition, Kluwer Academic Publishers, Boston, 1991.

Kolibri 360 GmbH
Ferdinand-Jühlke-Straße 7
99095 Erfurt, Deutschland
produktsicherheit@kolibri360.de